INSECTS AND OTHER TERRESTRIAL ARTHROPODS: BIOLOGY, CHEMISTRY AND BEHAVIOR

BEETLES

BIODIVERSITY, ECOLOGY AND ROLE IN THE ENVIRONMENT

INSECTS AND OTHER TERRESTRIAL ARTHROPODS: BIOLOGY, CHEMISTRY AND BEHAVIOR

Additional books in this series can be found on Nova's website under the Series tab.

Additional e-books in this series can be found on Nova's website under the e-book tab.

INSECTS AND OTHER TERRESTRIAL ARTHROPODS: BIOLOGY, CHEMISTRY AND BEHAVIOR

BEETLES

BIODIVERSITY, ECOLOGY AND ROLE IN THE ENVIRONMENT

CAMILLA STACK
EDITOR

nova publishers
New York

Additional color graphics may be available in the e-book version of this book.

Library of Congress Cataloging-in-Publication Data

Beetles : biodiversity, ecology and role in the environment / editor: Camilla Stack.
pages cm. -- (Insects and other terrestrial arthropods: biology, chemistry and behavior)
Includes index.
ISBN 978-1-63463-380-2 (hardcover)
1. Beetles--Ecology. I. Stack, Camilla, editor. II. Series: Insects and other terrestrial arthropods--biology, chemistry and behavior series.
QL573.B438 2015
595.76--dc23
2014040968

Published by Nova Science Publishers, Inc. † New York

CONTENTS

PREFACE

Insects have been shown to be appropriate ecological indicators for evaluating changes in the environment, due to several characteristics that they possess. An excellent group are beetles (Coleoptera), the largest group of insects, with about four hundred thousand known species, distributed in more than a hundred families occupying practically all ecosystems and feeding on a variety food sources. This book examines several species of beetles. It examines the biodiversity, ecology and role in the environment of beetles.

Chapter 1 - Before modern forestry started, forest fire was the dominating large-scale disturbance in boreal forest. Prescribed burning of clear-cuts is increasingly used as a conservation measure but is potentially less beneficial for fire-associated species than burning of intact forest. The authors investigated the temporal effect of prescribed burning of clear-cuts in two regions in central Sweden: one region with a relatively unbroken forest-fire continuity (Orsa) and one region with a broken fire continuity (Långshyttan) more typical for Fennoscandia. The dead wood associated (saproxylic) beetle fauna was sampled with ten trunk-window traps in each of 16 clear-cuts. The authors used a paired design with two clear-cuts that had been logged at the same time and where one had been subjected to prescribed burning one or two years after clear-felling. The authors sampled clear-cuts either the same year as burning (0-year old) or 1-, 2- or 4-years post-burning. In Orsa the authors sampled five pairs (0-4 years old) and in Långshyttan three pairs (0-2 years old).

The authors caught 147 species (5908 individuals) of saproxylic beetles and 33 of those species were fire-associated. There was a pronounced temporal effect of prescribed burning on beetle abundance and species richness. The year of the fire there was a higher abundance and species richness in the burnt clear-cuts than in the unburnt clear-cuts. This positive effect disappeared already after one year. This pattern remained when the authors restricted their analysis to only fire-associated species.

There were several regional differences consistent with a better fire continuity in Orsa: (1) In four out of five clear-cut pairs in Orsa there was a positive effect of burning on abundance and species richness of fire-associated species. In Långshyttan there was no such effect. (2) In Orsa the species composition of the fire-associated species differed between burnt and unburnt clear-cuts in four out of five pairs. In Långshyttan there was a difference only in one out of three pairs and that difference was smaller than the differences in Orsa. (3) When the authors restricted their analysis to 0- to 2-year old clear-cuts in order to have identical sample sizes in both regions, the authors found 725 individuals of 25 species of fire-associated species in Orsa and 218 individuals of 24 species in Långshyttan. (4) The four fire-dependent species that the authors found were the most specialised species and three of these, *Melanophila acuminata*, *Sphaeriestes stockmanni*, and *Acmaeops septentrionis* were only found in Orsa whereas *Henoticus serratus* was found in both regions.

To conclude, prescribed burning of clear-cuts only had a short-lived positive effect on saproxylic beetle diversity. This result casts doubt on the value of using prescribed burning of clear-cuts as a conservation measure. The better effect of prescribed burning in Orsa than in Långshyttan points to a differential effect of the value of prescribed burning. It also indicates that a long period of restoration burning might be needed in many parts of Sweden in order to increase the populations of fire-associated species.

Chapter 2 - Community assembly is the trajectory exhibited over time regarding species composition and community structure as species colonize an empty patch. Chase (2007) distinguish two basic types of assembly where a niche-assembled community is constrained by species interactions and competitive hierarchies whereas a dispersal-assembled community is unaffected by competition. The first wood-living (saproxylic) beetles to colonize a dead tree are sub-cortical, cambial-living species with a larval development-time of one year or less. Important community assembly events could therefore take place the first summer after tree death.

Community assembly was studied during the eight weeks immediately after tree death. By exposing 0.5 m long stem sections (bolts) of Norway spruce, *Picea abies*, to colonization for increasingly longer times (0.5, 1, 2, 4, and 8 weeks) a temporal sequence was created. A total of 140 bolts at four forest sites in central Sweden were studied with six to eight replicates per site. Two of the sites were old-growth forest and the other two were forest sites subjected to burning right before the experiment started. The author used community ecology measures to describe assembly patterns and emphasize findings that would have gone unnoticed in an analysis confined to the 8-week final state.

In total 8802 beetles of 56 species were reared from the bolts. Two fire-dependent species were found in burnt forest, this is noteworthy since unburnt bolts were used in both forest types. The total number of emerged beetles per bolt was similar in the two forest types in early assembly but later this density was higher in the old-growth forest. During assembly the community-level carrying capacity was reached in old-growth forest but not in burnt forest. During assembly, community evenness and taxonomic richness was higher in the burnt forest than in old-growth forest but at eight weeks these community measures had converged. The two forest types became more dissimilar regarding species composition as the community assembly progressed. Old-growth forest also had a lower bolt-to-bolt variation in species composition than burnt forest. Cambivores was the functional group that made up the largest proportion of individuals in both forest types but the proportion was highest in old-growth forest. The proportion fungivores was highest in burnt forest. The proportion predators peaked early in both forest types and then decreased during assembly. In old-growth forest, 74 % of the bolts were numerically dominated by the cambivorous scolytine *Dryocoetes autographus*. In burnt forest only 33 % of the bolts were dominated by that species. Furthermore, burnt forest also had fungivores as bolt dominants, for example the fire-favoured corticarid *Corticara rubripes*, whereas old-growth forest bolts were always dominated by cambivores.

In conclusion, the beetle assemblages in the two forest types followed different assembly trajectories. The beetle assemblage in burnt forest conforms to a dispersal-assembled community whereas that in old-growth forest conforms to a niche-assembled community (*sensu* Chase 2007). The biology of fire-adapted beetles is discussed in light of these findings.

Chapter 3 - Emerging concerns with environmental issues has led researchers to look for indicators capable of providing information on the condition of ecosystems. Several animal species have been used for this role. Among these organisms, ground-beetles were found to be appropriate indicators, since they are adapted to all terrestrial habitats, being associated with ecosystem function and sensitive to ongoing changes. Several families of ground beetles, e.g. Carabidae, Cicindelidae, Scarabaeidae, Cerambycidae, Histeridae, Staphylinidae and Curculionidae were used in diverse ecological assessment studies due to its main characteristics, such as habitat fidelity and taxonomic diversity, and because they provide cheap and easy measurements from functionally diverse trophic chains.

Although the growing number of studies on ecological indicators little is known about the ecological patterns in the communities of ground beetles of tropical ecosystems. The present study provides a review on the use of ground beetles as ecological indicators in tropical ecosystems, with focus on the Brazilian Amazon region. The authors discuss its use for monitoring the effects of anthropogenic induced changes, namely the conversion of forests to agriculture and farming practices, among others on the functioning of tropical systems. The authors also highlight the major advances and applications employed, as well as proposals for future studies with this group.

Chapter 4 - Cerambycids constitute one of the most diverse groups of insects in the world, with approximately 35,000 described species. They are commonly cited as models in several areas of Entomology, especially regarding their interactions with host plants and their economic impacts on agriculture. Females of the tribe Onciderini (Cerambycidae: Lamiinae) are twig-girdlers. They cut the branches of healthy plants, perforate the bark and insert their eggs. Usually, the branch falls to the ground and larvae develop inside the branch on the forest floor. Although some information is available for non tropical species, the biology of twig-girdler beetles and their influence on the floral communities in the neotropics remains largely unknown. This chapter will cover the current available data on the diversity of twig-girdler beetles (Onciderini) of the Atlantic rainforest of Brazil, with special emphasis on community composition, population dynamics, host plants range and ecological interactions. The first two sections are an introduction of this chapter and description of the study site. The third section will focus on diversity and host-plant preference. The community structure and the phenology will be also presented in this sections. In the fourth section the authors will discourse on the effects of twig-girdler beetles on host-plants population dynamics and community structure in their natural habitats. In the fifth and final section the authors will discuss the host utilization and microhabitat specialization of these beetles.

Chapter 5 - Beetle species belonging to the coleopteran families Bruchidae, Curculionidae, Laemophoeidae, Silvanidae and Tenebrionidae, as well as beetle-like insects from the psocopteran family Liposcelidae, are responsible for serious damages to agricultural products and resources. These beetles can be primary and/or secondary pests, feeding on integral and healthy grains or attacking those already damaged. The affected grains lose weight and germination power, are decreased in nutritive value and vigor, and are impaired by hygiene and sanitary conditions. This chapter summarizes information on biological and ecological aspects of stored product pests such as life cycle, fecundity, longevity, growth rate, voracity, natural habitats and hosts, and infestation focuses of beetles. These aspects are important for the development and choice of control measures. Classic strategies include mechanical methods, biological control, and the use of insecticide formulations such as powders, emulsions, aerosols, and microcapsules. Fumigation with phosphine has been the main strategy for control but this insecticide is highly volatile and toxic and increasingly there are reports of insect resistance to it. In this sense, natural insecticides, such as plant extracts, secondary metabolites, essential oils and lectins, have been investigated for insecticidal activity on destructive beetles. In summary, this chapter provides a brief and updated view of the state of the art on beetles that act as stored grain pests.

Chapter 6 - In terrestrial ecosystems, a rich assemblage of Coleoptera species contributes to the breakdown of plant and animal organic materials such as dead wood, fruits, leaf litter, feces, and carcasses, including human cadavers. This taxon represents a key component of energy flow in many ecosystems both because heterotrophic Coleoptera species contribute to nutrient cycling and because they are important trophic links in food webs. However, the taxon as a whole has not received much consideration in the literature and similarities between terrestrial heterotrophic systems have yet to be investigated. This chapter examines the diversity and ecological roles of Coleoptera fauna involved in terrestrial heterotrophic communities and reviews the main mechanisms and models of heterotrophic succession. The impacts on heterotrophic Coleoptera of resource patch characteristics (e.g., abundance, distribution, and resource quality) and of anthropogenic alterations of habitats supporting resource patches (e.g., agriculture, forest management, urbanization, global changes) are described and discussed.

Chapter 7 - Dung beetles of the subfamily Scarabaeinae (Coleoptera: Scarabaeidae) are one of the most important insect groups in tropical regions because their relevant ecosystem role as organic recycling agents, biological pest controllers and secondary seed dispersers. However, the author's knowledge of the different mechanisms that maintain assemblage structure and allow species coexistence is poor. The authors lack such information because dung beetle assemblages are often sampled over the short term, there are few studies that cover extended periods of time and broad spatial areas, and it is common to use just one type of trap and bait. To better explore the mechanisms underlying dung beetle coexistence, it was studied a beetle assemblage found in a tropical rain forest located in Tinigua National Park in Colombia. Sampling was conducted in August 1992, May-July and December 1994, January and December 1995, January and July 1996, January-July 1997, and January and July 1998. Dung beetles were sampled in the three main habitats found in the study area: mature mainland forest, flooded lowland forest, and river beaches. Specimens were collected using baited pitfall traps, flight interception traps, and baited elevated traps placed at different heights using different type of baits. In addition, feces from different vertebrates, vegetation, logs, fungi, litter, and fallen fruit were checked for the presence of dung beetles. A total of 70 species belonging to 19 genera and 7 tribes were collected. Compared to other areas, previously sampled in Colombia, this area demonstrates a remarkable diversity and has a very high level of species richness (one of the highest in the country). Nevertheless, it cannot be considered a hotspot for endemism because there is a strong influence and contribution of species of nearby biogeographic provinces to its own fauna. Some of the mechanisms that may affect the richness and structure of the assemblage, favoring the coexistence of species are: spatial partitioning (habitats and microhabitats), temporal partitioning (seasonality and dial activity), resource partitioning, and guild-specific resource relocation behavior. The values of the Jaccard similarity index reveal low levels of species overlap. Therefore, these coexistence mechanisms appear to act in tandem to partition the ecosystem in a way that decreases overlap among similar species, thus allowing a large number of dung beetle species to coexist in the same area. These results demonstrate that it is necessary to carry out broad-ranging and exhaustive sampling to fully characterize the structure of tropical dung beetle assemblages. Finally, to completely describe the assemblage composition at this locality, the following additional improvements should be made to the sampling protocol: carrying out exhaustive sampling over one full year; using more specialized baits; studying the forest canopy; searching in specialized microhabitats, such as bromeliads and ant and termite nests; and characterizing potential phoretic interactions with mammals.

Chapter 8 - The species of the genus *Adalia* are highly variable and demonstrate a clearly notable polymorphism in color and pattern of elytra and pronotum. This makes a problem to identify taxonomic status of some forms. Black-colored individuals (melanic morphs) are always present in populations of the most widely distributed Eurasian species, *A. bipunctata*. The authors studied frequency distribution of the melanics in population from France, Germany, Italy, Bulgaria, the Baltic States, Ukraine, and Russia in comparison with the data reported earlier for the population of the E Europe.

Several centers of melanism with high frequency of black individuals were identified. The highest percentage of the melanics (about 80%) was found in Saint Petersburg and in Rome, confirming the assumption of Y. Lusis that the melanic morphs are better adapted to the conditions of sea climate and polluted atmosphere in big cities.

The authors studied intraspecies and interspecies DNA polymorphism: the sequence of the mitochondrial gene *COI* and the nuclear sequence ITS2 were analyzed. It was shown that the forms often considered as distinct species, *A. revelierei, A. fasciatopunctata* and *A. turanica*, are in fact the forms of geographically polymorphic populations of the species *A. bipunctata*. As indicated by DNA markers, *A. tetraspilota, A.decempunctata, A.frigida*, showing sympatric distribution with *A. bipunctata* should be considered as distinct species, *A. frigida* being the closest species to *A. bipunctata*. In geographically remote populations of *A. bipunctata* symbiotic male killing bacteria are found: *Spiroplasma* and *Rickettsia*. These bacteria infect the individuals of specific mitotypes (H1 and related mitotypes in case of *Spiroplasma*, H10 in case of *Rickettsia*).

Chapter 9 - A regular prospection was carried out from 1993 to 2014 to identify the ladybeetles predators of fruits trees and vegetables crops aphids in Tunisia. The study allows detecting five Coccinellids with a frequency of 6% were the most abundant predators: *Scymnus aptezi* Mulsant, *Oenopia doublieri* Mulsant, *Hippodamia variegata* Goeze, *Coccinella undecimpunctata* L. *currently Hippodamia (Semiadalia) undecimnotata* Schneider, *Hippodamia tredecimpunctata* L., *Psyllobora 22-punctata* L. and *Coccinella algerica* Kovar. This latest was the most abundant coccinellids and was observed as larva and adult respectively on population of *Aphis gossypii* Glover, *Aphis fabae* Scopoli, *Aphis punicae* Shinji, *Capitophorus elaeagni* Del Guercio, *Myzus persicae* Sulzer, *Macrosiphum euphorbiae Thomas, Macosiphum rosae L., Hyalopterus pruni Geoffrey and Pterochloroides persicae* Cholodkovsky *on several host plant: eggplant, pepper,* pomegranate, tomato, lettuce, melon, peach, almond, plum...

The association's plants, aphids and preys of *C. algerica* which has an important action on the control of aphids were analyzed. *C. algerica* is found on several host plants belonging to 22 botanical families who may be vegetable crops, fruit trees or spontaneous plants. *C. algerica* consumed alternative or essential food present on hosts plants. These plants supported eggs, larva and nymph of *C. algerica*. This predator consumes twenty species of aphids. The consumption varies depending on the state of physiological activity: It feeds on *Aphis*, *Rhopalosiphum* and *Uroleucom* genera at quiescence. *C. algerica* shows a preference for *C. elaeagni* in autumn and *H. pruni* in spring. Furthermore, this ladybird on diapauses was collected in the summer nears the root of *Stipa tenacissima* (Gramineae).

Also, it's interesting to mention that *C. algerica* was attacked by several parasites like *Dinocampus coccinellae* Schrank (Braconidae, Hymenoptera), *Homolotylus flaminius* Dalm (Encyrtidae, Hymenoptera) and *Podapolipus* Rovelli et Grassi (Acarina).

In: Beetles: Biodiversity, Ecology and Role in the Environment ISBN: 978-1-63463-380-2
Editor: Camilla Stack

Chapter 1

The Positive Effects of Prescribed Burning of Clear-Cuts on Saproxylic Beetle Diversity Are Short-Lived and Depend on Forest-Fire Continuity

Jonas Victorsson[1,2,*], Lars-Ove Wikars[1,3] and Stefan Ås[1]
[1]Animal Ecology, Evolutionary Biology Centre, Uppsala University, Norbyvägen 18 D, SE-752 36 Uppsala, Sweden
[2]Department of Ecology, Swedish University of Agricultural Sciences, Box 7044, SE-750 07 Uppsala, Sweden
[3]Hovgården 66, SE -781 91 Borlänge, Sweden

Abstract

Before modern forestry started, forest fire was the dominating large-scale disturbance in boreal forest. Prescribed burning of clear-cuts is increasingly used as a conservation measure but is potentially less beneficial for fire-associated species than burning of intact forest. We investigated the temporal effect of prescribed burning of clear-cuts in two regions in central Sweden: one region with a relatively unbroken forest-fire continuity (Orsa) and one region with a broken fire continuity (Långshyttan) more typical for Fennoscandia. The dead wood associated (saproxylic) beetle fauna was sampled with ten trunk-window traps in each of 16 clear-cuts. We used a paired design with two clear-cuts that had been logged at the same time and where one had been subjected to prescribed burning one or two years after clear-felling. We sampled clear-cuts either the same year as burning (0-year old) or 1-, 2- or 4-years post-burning. In Orsa we sampled five pairs (0-4 years old) and in Långshyttan three pairs (0-2 years old).

We caught 147 species (5908 individuals) of saproxylic beetles and 33 of those species were fire-associated. There was a pronounced temporal effect of prescribed burning on beetle abundance and species richness. The year of the fire there was a higher abundance and species richness in the burnt clear-cuts than in the unburnt clear-cuts. This positive effect disappeared already after one year. This pattern remained when we restricted our analysis to only fire-associated species.

[*] Corresponding author: E-mail: jonas.victorsson@slu.se Fax: (+46)18-672890 City/Country: Uppsala/Sweden.

There were several regional differences consistent with a better fire continuity in Orsa: (1) In four out of five clear-cut pairs in Orsa there was a positive effect of burning on abundance and species richness of fire-associated species. In Långshyttan there was no such effect. (2) In Orsa the species composition of the fire-associated species differed between burnt and unburnt clear-cuts in four out of five pairs. In Långshyttan there was a difference only in one out of three pairs and that difference was smaller than the differences in Orsa. (3) When we restricted our analysis to 0- to 2-year old clear-cuts in order to have identical sample sizes in both regions, we found 725 individuals of 25 species of fire-associated species in Orsa and 218 individuals of 24 species in Långshyttan. (4) The four fire-dependent species that we found were the most specialised species and three of these, *Melanophila acuminata*, *Sphaeriestes stockmanni*, and *Acmaeops septentrionis* were only found in Orsa whereas *Henoticus serratus* was found in both regions.

To conclude, prescribed burning of clear-cuts only had a short-lived positive effect on saproxylic beetle diversity. This result casts doubt on the value of using prescribed burning of clear-cuts as a conservation measure. The better effect of prescribed burning in Orsa than in Långshyttan points to a differential effect of the value of prescribed burning. It also indicates that a long period of restoration burning might be needed in many parts of Sweden in order to increase the populations of fire-associated species.

INTRODUCTION

Before the start of modern forestry, forest fire was the dominating large-scale disturbance in boreal forest ecosystems (Zackrisson 1977; Bonan & Shuggart 1989; Niklasson & Granström 2000) and 0.5-2 % of the forest burned annually (Bonan & Shuggart 1989; Zackrisson & Östlund 1991). Fire intervals varied from 30-50 years for dry sites to an average of 160 years for wet sites (Angelstam 1996). Fire maintains structural complexity within forest stands and this complexity promotes plant and animal diversity (Hansen et al. 1991). Especially among insects there are many species that depend on natural disturbances to create a habitat patch mosaic that ensures species survival (Schowalter 1985; McCullough et al. 1998; Schowalter 2012) and among beetle species that depend on dead wood (saproxylic) there is a positive effect of increased fire frequency on diversity (Moretti et al. 2004). Saproxylic insects make up a large proportion of the boreal forest insect fauna (Stokland et al. 2012). In Sweden, where the beetle fauna is comparatively well known, at least 1257 species are saproxylic (Stokland et al. 2012). This is 28 % of the Swedish total of 4456 beetle species (Gärdenfors et al. 2003).

Among these saproxylic species some are fire dependent and have a distribution restricted to burnt forest. These pyrophilous (fire-loving) species have physiological or behavioural adaptations to fire, such as the infrared sensory organs of *Melanophila acuminata* (Evans 1966; Vondran et al. 1995; Schmitz et al. 1997), attraction to wood smoke (Schutz et al. 1999), or swarming behaviour at burning anthills (Wikars 1992). Other species are fire favoured and have a higher abundance in burnt forest than in other forest types (Wikars 1992; Wikars 2006; Johansson et al. 2011; Victorsson this volume). Again in Sweden, at least 116 beetle species are considered fire associated (Wikars 2006).

Active fire suppression in Fennoscandia generally started in the 19th century (Esseen et al. 1997) and has drastically decreased fire frequencies (Zackrisson 1977). Important factors in fire suppression are an increased density of human habitations, modern forestry, and road construction – all these factors have increased during the 20th century. Forest roads, apart from making forestry possible, also facilitate fire suppression, salvage logging, and snag removal. Snag removal is important since lightning-struck snags are often the start of forest fires (Esseen et al. 1997). Due to this fire suppression many fire-associated species are now threatened or locally extinct (Ehnström & Waldén 1986) and out of 507 threatened forest-living beetle species in Sweden 20 % are favoured by forest fires and/or extensive storm fellings (Berg et al. 1994). Many fire-dependent species have disappeared from southern Sweden and western Finland, the two areas in those countries with the longest history of habitation and forestry (Wikars 2006; Kouki et al. 2012).

Primarily during the 1950s and 60s, burning over of clear-cuts was used in Fennoscandia as a means of soil preparation before regeneration planting (Hörnsten et al. 1995). This method was used primarily in places with a thick humus layer to improve soil quality and prepare the ground for seedlings. Today prescribed burning of clear-cuts is performed more for conservation purposes than for soil preparation. There are no requirements concerning prescribed burning in the Swedish Forestry Act, but current FSC standards require that large forest owners burn 5% of their annual clear-cut area in order to benefit biodiversity (Anonymous 2013). As a result of this, forest companies now burn 2000-3000 ha annually in Sweden, mostly clear-cuts (Wikars 2006).

The conservation value of prescribed burning of clear-cuts is questionable. The effect on ground living (epigeal) beetles range from generally beneficial for the diversity of the group (Gongalsky et al. 2006; Hjältén et al. 2010a) to negative for individual fire-associated species (Wikars 1995). There is a clear conservation value of burning more or less intact forest (Wikars 2002; Toivanen & Kotiaho 2007b; Johansson et al. 2011; Victorsson this volume) but the value of burning logged sites seem to disappear over a period of 1-16 years after burning and depends strongly on the level of pre-fire tree retention (Toivanen & Kotiaho 2007a).

Here we wanted to test the short-term temporal effect of prescribed burning of clear-cuts on saproxylic beetles. Burnt forest and burnt clear-cuts share features, such as heat and smoke at the time of the fire, which could attract fire-associated species. This could lead to a different species composition in a newly burnt clear-cut compared to an unburnt clear-cut that will mainly harbour common species that can utilize stumps and small diameter logging residue (Jonsell 2008; Hjältén et al. 2010b; Victorsson & Jonsell 2013). We also expected a positive effect of burning on total abundance and species richness of saproxylic beetles in general. However, since clear-cuts contain only a small amount of dead wood we expect that this positive effect will fade, and that burnt clear-cuts soon will be indistinguishable from ordinary unburnt clear-cuts, especially regarding fire-favoured species.

Some areas in Sweden have experienced less intense fire suppression and are well suited as reference areas when analysing the effects of broken fire continuity. This study compares one such region with another region with a fire suppression normal for most parts of Fennoscandia. In this regard we expected a larger difference in species composition between burnt and unburnt clear-cuts in the region with an unbroken fire continuity than in the region with the broken continuity, since in the "good" region a more complete fauna of fire-associated species could be expected (Wikars 1997). These species could be attracted to burnt clear-cuts which would lead to an increased difference between burnt and unburnt clear-cuts.

MATERIAL AND METHODS

The Study Sites

The study was performed in two regions in the province of Dalecarlia in south-central Sweden. The regions are separated by only 200 km but differ in their forestry history and fire continuity. The region near the village of Långshyttan (60° 32'N, 16° 06'E) has a broken fire continuity. Intense mining activities, not least from the copper mine in Falun, led to intense forestry between the years 1600 and 1900, after that modern forestry started with forest road construction and efficient fire suppression. In this region, virtually all signs of forest fires such as fire scars in trees are absent (Lundqvist 1994). The other region, Orsa Finnmark (61° 30'N, 14° 50'E) has a good fire continuity. Orsa escaped the intensive logging during the mining period, and the region was indeed considered virgin forest as late as in 1881 (Linder & Östlund 1992). Slash-and-burn agriculture, a practice that would increase the fire frequency, was common in this region during the 18th and 19th century. Road construction and modern forestry started much later here due to local economic conditions. Rather unique for Sweden, the principal forest owner in this region has continued with prescribed burning of clear-cuts to enhance forest regeneration and today they burn 25-100 ha annually (Wikars 1997). Accordingly, signs of fires are very common in the Orsa region and fire scars from up to five consecutive fires can be found in individual trees (Anonymous 1995; Wikars 1997).

Today both regions are dominated by typical Swedish forestry including clear-felling and thinning operations. The two dominating tree species are the conifers Norway spruce *Picea abies* and Scots pine *Pinus sylvestris*. Some deciduous tree species, such as birch *Betula* spp. alder *Alnus* spp. and aspen *Populus tremula* occur as admixture.

We used a paired design with two clear-cuts that had been logged at the same time and where one of the clear-cuts had been subjected to prescribed burning one or two years after clear-felling. We collected data on clear-felling and burning date from the principal forest owners in the two regions (Table 1). The intra-pair distance was 600 m or shorter in all but one pair. The proportion of *P. abies* and *P. sylvestris* before clear-felling was similar within the pairs according to the persons responsible for harvesting at the forest companies. We identified five pairs in Orsa and three pairs in Långshyttan. The sampled sites included all burnt clear-cuts in Långshyttan where prescribed burning had just started. In Orsa we used a selection of the clear-cuts available. Age refers to the number of years since the prescribed burning: in a 0-year pair the burnt clear-cut was burnt the year of investigation, in a 1-year pair the year before and so on (Table 1).

Sampling

Each clear-cut was sampled with ten trunk window traps mounted on dead wood substrates along a 90 m transect. The traps were operated during 60 days from the beginning of July to the beginning of September in 1995. The traps consisted of a 15 x 25 cm transparent plastic plate fastened perpendicularly onto dead wood substrates and a small aluminium container fastened underneath to collect the insects (as described in Wikars et al. 2005). The container was filled with 50 % ethylene glycol and a few drops of dish wash detergent and Bitrex added. To avoid subjectively choosing "good" substrates the traps were put up every 10 m along the transect on the first encountered dead wood substrate that fulfilled the following standards: (1) either *P. abies* or *P. sylvestris*; (2) snags, logs, or logging waste; (3) a minimum diameter of 8 cm. On standing substrates the trap was placed 40 cm above ground, on logs at "breast height" and on logging waste at least 50 cm from any of the end parts. The proportion of trap substrates of the two conifer species was similar in all clear-cut pairs except Hirvilampi (Table 1). When possible, the trap was put up facing south on the dead wood substrate, if necessary west or east. Transect routes were chosen such that the ground slope of the route as well as the distance to the nearest forest edge was the same in both clear-cuts in each pair. Since many beetles are more active at higher temperatures, these steps were taken to ensure that the traps would be exposed to the sun in a similar way.

Analysis

All beetles were determined to species or in a few case genus, according to Lundberg and Gustafsson (1995) and Ehnström and Holmer (2007) by JV and Stig Lundberg. Species were classified as saproxylic or ground living (epigeal) according to: Hansen (1964), Koch (1989-1992), and Palm (1951, 1959) from a database made available by Mats Jonsell. Species were classified as associated with fire based on the most recent classification of Swedish insects (Wikars 2006). These fire-associated species can be divided into two groups: (1) *fire-dependent species* have strong behavioural specializations for using burnt forest and almost exclusively occur in burnt forest; and (2) *fire-favoured species* are adapted to fire and/or are more abundant in burnt forest compared to other forest habitats. In all analyses described below we performed separate analyses for (1) all saproxylic beetles and for (2) fire-associated species.

We analysed the temporal effect of prescribed burning on beetle abundance and species richness with two-way ANOVAs with the fixed factors clear-cut type (burnt or unburnt) and clear-cut age (0, 1, 2, or 4 years after buring) and a random factor consisting of individual clear-cuts nested within the interaction clear-cut type x clear-cut age. This means that we had individual traps as observations but eight degrees of freedom in the denominator in the ANOVAs. In the analysis of beetle abundance we used Poisson error and log link as well as quasi-likelihood models to compensate for overdispersion (Quinn & Keough 2002). These analyses were done in procedure GLIMMIX in software SAS ver. 9.3 (SAS, Cary, NC). The temporal effect on species composition was tested with a PERmutational MANOVA (PERMANOVA) (Andersson 2001) using Sørensen similarity which is based on presence/absence data. We used the same fixed and random effects as above and did the analyses in software PRIMER using 9999 randomizations (Clarke & Gorley 2006).

To look for effects of fire continuity on the effectiveness of prescribed burning we analysed the effect of burning on abundance and species richness in each clear-cut pair separately. This was one-way ANOVAs with 18 degrees of freedom in the denominator. These analyses were done in software SAS and abundance data was treated as mentioned above. We performed ANalysis Of SIMilarity (ANOSIM) to get a measure of the magnitude of the difference in species composition (again using Sørensen similarity) between burnt and unburnt clear-cuts in each pair. This analysis gives a test statistic R which goes from 0 to 1 where larger values indicate a greater difference in species composition. This was done in software PRIMER using 9999 randomizations. Furthermore, we performed a group average linkage cluster analysis of Sørensen similarity of all clear-cuts with an attached SIMilarity PROFile test (SIMPROF) analysis in software PRIMER. For this analysis data from the ten traps was pooled for each clear-cut. From this a dendrogram was prepared where solid lines indicate groups with a significant species composition difference at the $P < 0.05$ level and dashed lines indicate non-significant differences (based on 9999 randomizations). Finally, we performed sample-based rarefaction comparing burnt and unburnt clear-cuts in each pair using 9999 randomizations in the software EstimateS (Colwell 2006) to construct 95 % confidence limits. We conclude that there is a significant treatment effect in a rarefaction when the mean of one of the treatments lies outside the 95 % confidence limits of the other treatment.

RESULTS

Overview

We caught 147 species (5908 individuals) of saproxylic beetles. At burnt clear-cuts we recorded 118 species (3648 individuals) and at unburnt clear-cuts 114 species (2260 individuals) (Appendix 1). Of the saproxylic species, 85 species (58 %) were recorded from both clear-cuts types whereas 33 species (22 %) were unique to burnt clear-cuts and 29 species (20 %) were unique to unburnt clear-cuts. We also caught 16 epigeal species (167 individuals) and 11 species (26 individuals) with an unknown habitat association (Appendix 1). We caught five fire-dependent species (four of those were saproxylic), and 30 fire-favoured species (29 saproxylic species) (Appendix 1). Seven species were on the current Swedish red list, from the year 2010 (Appendix 1). Three species, *Tragosoma depsarium*, *Acmaeops septentrionis*, and *Stagetus borealis*, were both saproxylic, red listed and fire-associated. These species were only found the region with a better fire continuity (Orsa).

Temporal Effects of Prescribed Burning

There was a pronounced temporal effect of prescribed burning on saproxylic beetle abundance and species richness (Figure 1). The year the prescribed burning took place (year 0) there was a higher abundance and species richness of all saproxylic species in the burnt clear-cuts than in the unburnt clear-cuts (Figure 1, Table 2). This positive effect of burning disappeared already after one year. The statistical differences were the same for the fire-associated species but for them there was also a (non-significant) tendency toward a positive effect one year after burning (Figure 1, Table 2). The species composition for all saproxylic species differed between burnt and unburnt clear-cuts only in the 2-year old clear-cuts (Table 2). The species composition for the fire-associated species differed between burnt and unburnt clear-cuts only the year of burning (Table 2).

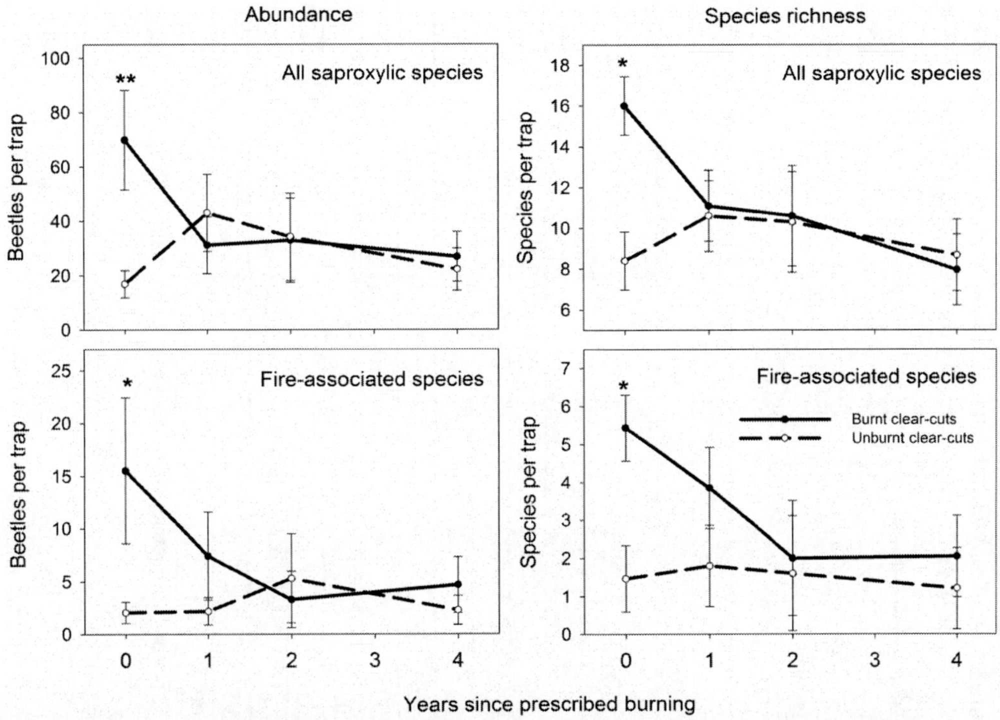

Figure 1. The temporal effect of prescribed burning of clear-cuts on beetle abundance and species richness. The upper panels show results for all 147 saproxylic species. The lower panels show results for 33 species of fire-associated species. Means and SE from ANOVAs. Asterisks denote a significant effect of prescribed burning in that age class (simple effects from the ANOVAs): ** $P < 0.01$, * $P < 0.05$.

Effects of Fire Continuity

There was a positive effect of burning on the abundance of all saproxylic species in the 0-year old clear-cuts both in the region with a better fire continuity (Orsa) and in the region with the broken fire continuity (Långshyttan) (Figure 2). The number of saproxylic species was positively affected by burning in the 0- to 1-year old clear-cuts in Orsa but not in Långshyttan. In four out of five clear-cut pairs in Orsa there was a positive effect of burning on abundance and species richness of fire-associated species (Figure 2). The magnitude of the difference decreased over time and was quite small or non-existent in the 4-year old clear-cuts. In Långshyttan there was no effect of burning on the fire associated species abundance or richness (Figure 2).

The species composition of all saproxylic species differed between burnt and unburnt clear-cuts in all pairs except Lillhamra (Figure 2). The magnitude of the difference was largest in the 0- to 1-year old clear-cuts in Orsa and in the 0-year old pair in Långshyttan. In the 4-year old clear-cuts and in the 1- to 2-year clear-cuts in Långshyttan the difference was smaller. In Orsa the species composition of the fire-associated species differed between burnt and unburnt clear-cuts in all pairs except Lillhamra (Figure 2). In Långshyttan there was a difference only in Dammsjön and that difference was smaller than the differences in Orsa.

Table 1. Clear-cuts sampled in a study of the effects of prescribed burning on saproxylic beetles in central Sweden. The study had a paired design with one burnt and one unburnt clear-cut in each pair. Data on clear-felling and burning date are from forest companies

Region	Clear-cut name	Clear-felling	Burning	Altitude (meters above sea level)	Age	Distance between B and U (meters)	Number of traps with Scots pine as substrate**
Orsa	Avenlam burnt	1992-11-01	1995-06-01	280	0	3200	8
	unburnt	1992-01-01		290			7
	Sälgtjärn burnt	1993	1995	520	0	500	3
	unburnt	1993		550			1
	Pilka burnt	1992	1994	545	1	600	3
	unburnt	1992		550			4
	Hirvilampi burnt	*	1991	280	4	300	6
	unburnt	*		260			10
	Lillhamra burnt	*	1991	350	4	250	9
	unburnt	*		350			7
Långshyttan	Dammsjön burnt	1993-06-01	1995-06-01	200	0	250	5
	unburnt	1993-06-01		205			4
	Nordvik burnt	1993-11-01	1994-05-01	265	1	600	6
	unburnt	1993-11-01		270			6
	Rudtjärn burnt	1993-05-01	1993-05-01	195	2	300	7
	unburnt	1992-12-01		200			6

Notes: * Exact dates could not be obtained from the forestry company. In this region it is normal practise to wait two years after clear-felling before prescribed buring. The probable cutting date is the winter 1988/1989.

** Number of traps with Scots pine *Pinus sylvestris* as substrate out of 10 traps per clear-cut. The other trap substrate was Norway spruce *Picea abies*.

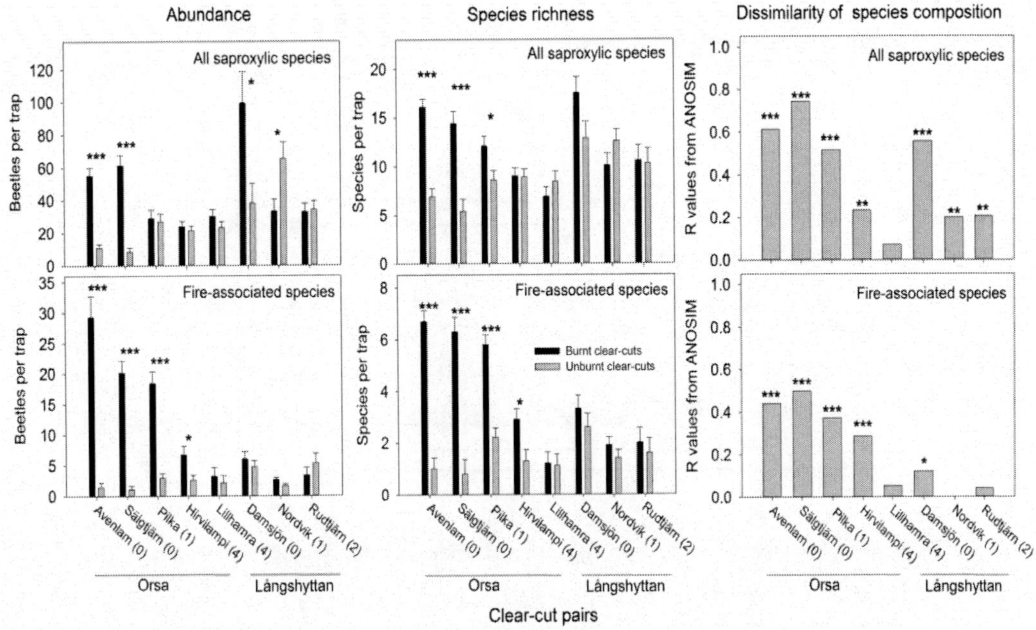

Figure 2. The effect of prescribed burning on beetle abundance, species richness and dissimilarity of species composition in each sampled pair of clear-cuts. The upper panels show results for all 147 saproxylic species. The lower panels show results for 33 species of fire-associated species. The figure in parenthesis after each clear-cut pair name is the number of years since the prescribed burning. The two sampled regions, Orsa with a better fire continuity and Långshyttan with broken fire continuity are indicated on the x-axis. For abundance and species richness the means and SEs are from ANOVAs. For dissimilarity of species composition the values are the test statistic R from an ANOSIM analysis for each pair. R varies between zero and one with larger values indicating an increased dissimilarity. Asterisks denote a significant effect of prescribed burning in that pair: *** $P < 0.001$, ** $P < 0.01$, * $P < 0.05$.

The cluster analysis of all saproxylic species identified five groups based on species composition (Figure 3). Five of the six clear-cuts in Långshyttan separated into one group that was different from the Orsa sites. Within that group there was a (non-significant) grouping of the burnt clear-cuts. Within the Orsa region the burnt 0- to 1-year old clear-cuts and one unburnt clear-cut were separated into one group that was different from a group comprising of all the 4-year old clear-cuts and two younger unburnt clear-cuts (Figure 3). In the cluster analysis of fire-associated species a group with all the 4-year old clear-cuts in Orsa separated from the rest (Figure 3). The next group consisted of all four Sälgtjärn and Pilka clear-cuts in Orsa. Within that group there was a (non-significant) grouping of the burnt clear-cuts. The next group consisted of the burnt Avenlam clear-cut in Orsa and all six clear-cuts in Långshyttan. Within that group there were three sub-groups and the two 0-year old burnt sites, Avenlam (Orsa) and Dammsjön (Långshyttan) clustered together.

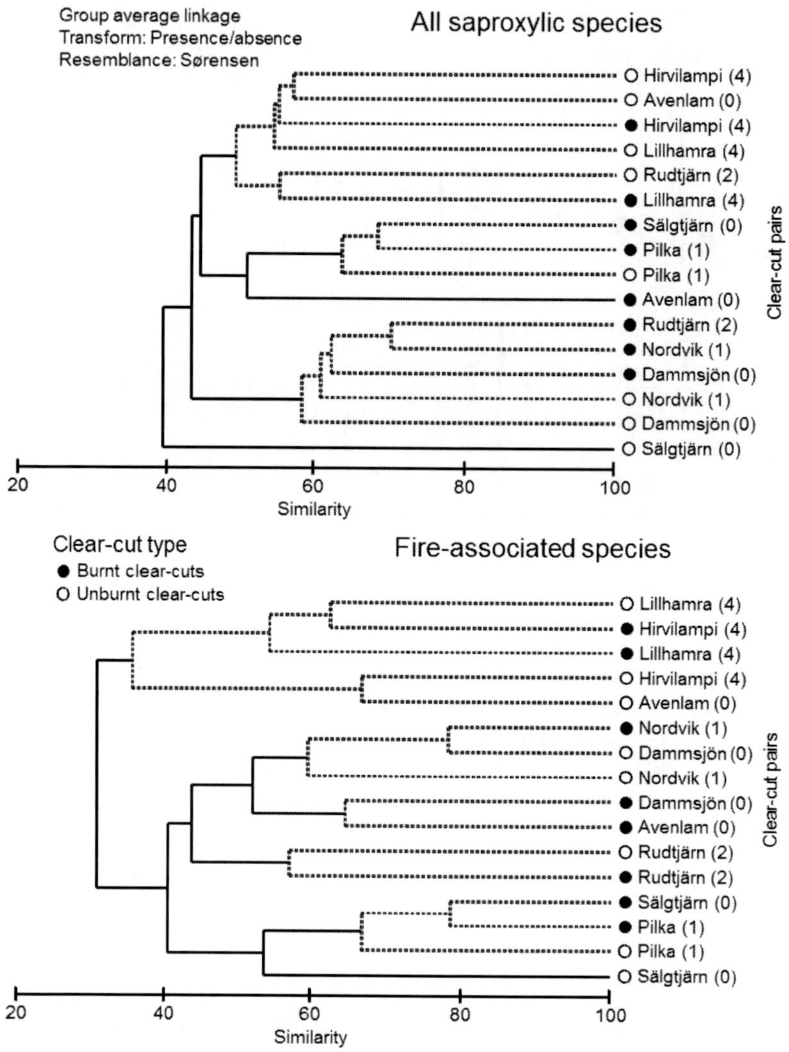

Figure 3. Dendrograms showing similarities in species composition for the 16 clear-cuts. The upper panel shows the result for all 147 saproxylic species. The lower panel shows the result for 33 species of fire-associated species. The tree was produced using group average linkage clustering of Sørensen similarities (presence/absence data). Solid lines indicate groups supported (at $P < 0.05$) by a SIMPROF analysis. The figure in parenthesis after each clear-cut pair name is the number of years since the prescribed burning.

In the rarefaction analysis of all saproxylic species there was higher species richness in the burnt clear-cuts in the two 0-year old clear-cut pairs in Orsa (Figure 4) and there was higher species richness in the unburnt clear-cut in one of the 4-year old clear-cut pairs. In Långshyttan there was no effect of burning on the assemblage of all saproxylic beetles (Figure 4). In the rarefaction analysis of the fire-associated species there was higher species richness in the burnt clear-cuts in four out of five clear-cut pairs in Orsa (Figure 4).

The magnitude of this difference was highest in 0- to 1- year old clear-cuts. In Långshyttan there was a positive effect of burning in one of the three clear-cut pairs (Figure 4).

As an indication of the value of the prescribed burning for fire-associated and threatened species we here compare the species found in the 0- to 2- year old clear-cuts in the two regions since the number of sampled clear-cuts in this age class was the same in both regions. In total we found 725 individuals of 25 species of fire-associated species in Orsa and 218 individuals of 24 species in Långshyttan (Appendix 1). The four fire-dependent species that we found are the most specialised species and they were only found in 0- and 1-year old burnt clear-cuts (Appendix 1). Three of the species, *Melanophila acuminata* (3 individuals in total), *Sphaeriestes stockmanni* (2), and *A. septentrionis* (1) were only found in the Orsa region whereas *Henoticus serratus* (2) was found in both regions. Of the four red-listed saproxylic species we found in these clear-cuts all were found in Orsa and one in Långshyttan (Appendix 1).

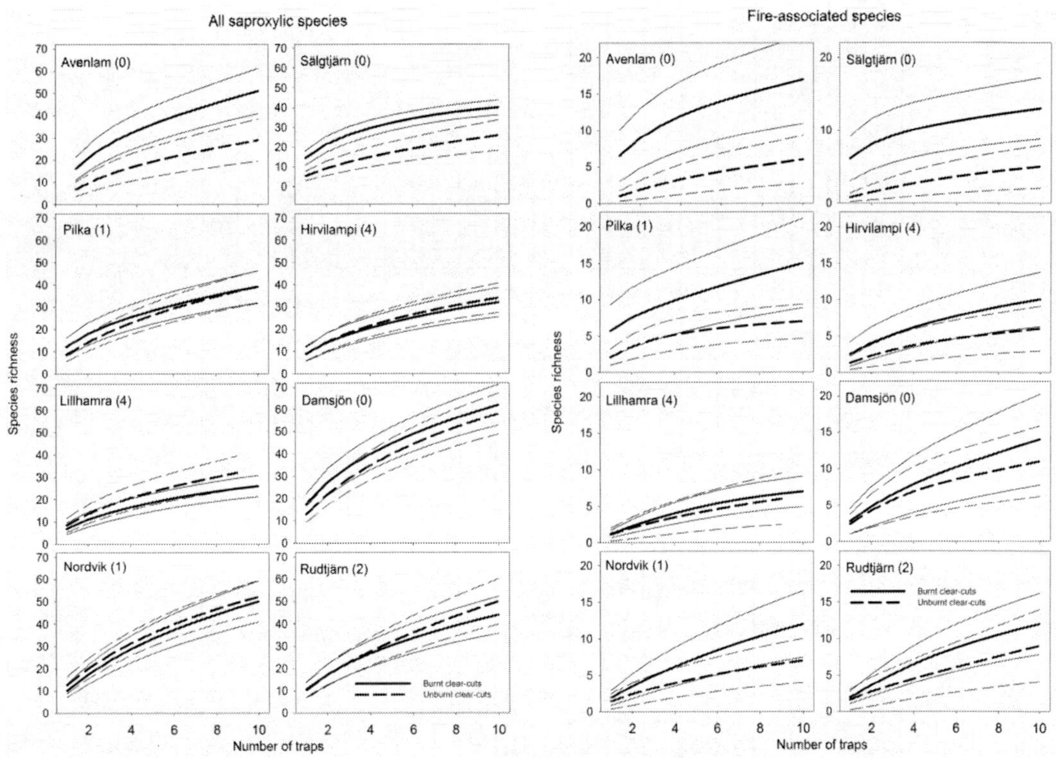

Figure 4. Sample based rarefaction for each inventoried pair of clear-cuts. Thick lines are mean rarefied species richness and thin lines indicate 95 % confidence limits from 9999 randomizations in software EstimateS. Solid lines are burnt clear-cuts and dashed lines are unburnt clear-cuts. The figure in parenthesis after each clear-cut pair name is the number of years since the prescribed burning.

Table 2. Results from tests of the temporal effect of prescribed burning on beetle abundance, species richness, and species composition at 16 clear-cuts in central Sweden

Source of variation	**df**	***F***	***P***	***F***	***P***	***F/t***	***P***
All saproxylic species		***Abundance***		***Species richness***		***Species composition***	
Clear-cut type	1,8	1.48	0.26	2.07	0.19	1.06	0.37
Clear-cut age	3,8	0.57	0.65	1.99	0.19	1.28	0.21
type x age	3,8	3.19	0.08	3.03	0.09	0.69	0.81
Simple effect: burnt clear-cuts	3,8(4)	2.19	0.17	4.62	**0.037**	1.24	0.28
Simple effect: unburnt clear-cuts	3,8(4)	1.68	0.25	0.41	0.75	0.72	0.83
Simple effect: 0-year clear-cuts	1,8	12.98	**0.0069**	14.23	**0.0054**	*1.29*	0.10
Simple effect: 1-year clear-cuts	1,8	0.48	0.51	0.04	0.84	*0.80*	0.67
Simple effect: 2-year clear-cuts	1,8	0	0.96	0.01	0.93	*1.59*	**0.0041**
Simple effect: 4-year clear-cuts	1,8	0.16	0.7	0.09	0.77	*0.88*	0.50
Fire species		*Abundance*		*Species richness*		*Species composition*	
Clear-cut type	1,8	3.92	0.08	4.95	0.06	1.64	0.15
Clear-cut age	3,8	0.38	0.77	1.4	0.31	2.15	**0.014**
type x age	3,8	1.34	0.33	1.19	0.37	0.93	0.53
Simple effect: burnt clear-cuts	3,8(4)	1.44	0.3	2.54	0.13	2.72	**0.013**
Simple effect: unburnt clear-cuts	3,8(4)	0.37	0.78	0.05	0.98	0.84	0.64
Simple effect: 0-year clear-cuts	1,8	9.19	**0.0163**	10.34	**0.012**	*1.55*	**0.022**
Simple effect: 1-year clear-cuts	1,8	2.23	0.17	1.84	0.21	*0.64*	1.00
Simple effect: 2-year clear-cuts	1,8	0.17	0.69	0.04	0.85	*1.10*	0.28
Simple effect: 4-year clear-cuts	1,8	0.75	0.41	0.31	0.59	*1.68*	0.17

The upper part shows results for all 147 saproxylic species and the lower part show results for 33 species of fire-associated species. Abundance and species richness tested with ANOVAs. Species composition tested with PERMANOVAs on Sørensen similarities. The test of simple effects on burnt and unburnt clear-cuts has eight df for abundance and species richness but four df for species composition. Test statistics in italics are t-values and all other are F-values. Statistically significant differences ($P < 0.05$) are shown in bold

DISCUSSION

Temporal Effects of Prescribed Burning

Prescribed burning of clear-cuts had a positive but very short lived effect on the saproxylic beetles in this study. The positive effect on the abundance and species richness disappeared already one year after the fire when we tested both regions together. The reason for the positive effect on abundance and species richness in the year of the burning is probably because a prescribed burning in some ways mimics a natural forest fire. Heat and smoke at the time of fire is perhaps the feature that most closely resembles a natural forest fire.

Both fire-dependent species (Schmitz et al. 1997; Schmitz & Trenner 2003) and other saproxylic beetles (Schutz et al. 1999; Allison et al. 2004) are attracted by smoke which may explain the higher number of individuals and species the year of burning. The scorched and blackened ground after a prescribed burning also mimics the situation in a burnt forest and this creates a warmer microclimate on the burnt clear-cut (Holliday 1991). There is a positive correlation between species richness and sun exposure (and thus temperature) in boreal saproxylic beetles (Kaila et al. 1997; Jonsell et al. 1998; Lindhe et al. 2005). Therefore the increased temperature in burnt clear-cuts could also help explain the higher number of individuals and species the year of burning.

On the other hand, the reason why the positive effect of burning disappeared so quickly is probably because a prescribed burning differs in several ways from a natural forest fire. For example, the blackening of the ground, evident in newly burnt clear-cuts, disappeared rapidly. In all 2- and 4-year old localities the ground was instead to a large degree covered with fireweed *Epilobium angustifolium* that effectively hides the dark ground (personal observation).

This rapid colonization of *E. angustifolium* would result in a lowered ground temperature already two years post fire and could decrease saproxylic beetle abundance at burnt clear-cuts. This is in contrast to a burnt forest where the input of frass from the fire-associated longicorn *Monochamus scutellatus* to the soil decreases germination of *E. angustifolium* in North America (Cobb et al. 2010). This inhibitory effect on *E. angustifolium* would be less important on our clear-cuts because the Swedish congeneric *Monochamus sutor*, that has a similar conspicuous frass production as the North American species, prefers snags over stumps (Ehnström & Axelsson 2002). Due to the low number of snags on our clear-cuts the *Monochamus*-induced inhibition of *E. angustifolium* would not be effective.

The largest difference between prescribed burning and natural forest fires is probably the much larger amount of dead wood in burnt forest. Burning a clear-cut kill trees left after cutting (Toivanen & Kotiaho 2007a) and this increases the local amount of dead wood compared to the situation in an unburnt clear-cut. Some retention trees on the clear-cuts in this investigation had indeed been killed by fire (personal observation). However, burnt clear-cuts contain much less dead wood than burnt forest simply because there are not that many retention trees on a clear-cut that can suffer fire-induced mortality.

Therefore, the burnt clear-cut can only act as a provider of dead wood to small populations of saproxylic species, and as this substrate is consumed the numbers of individuals and species will decline.

The low level of dead wood after a prescribed burning could also lead to high levels of larval resource competition. Primarily because the high level of attraction of beetles the year of the fire that will lead to a high colonization pressure of the limited amount of dead wood available and also because most prescribed burnings tend to be low-intensity fires (Hjältén et al. 2010a) that do not kill that many of the beetles who have already established on the clear-cut during the one to two years that normally elapses between cutting and burning (Wikars 1995).

Saproxylic species arriving at a burnt clear-cut may therefore face intense resource competition. Priority-effects, when the establishment order of individual species affects the species composition of a community, is demonstrated in saproxylic beetles (Weslien et al. 2011; Victorsson 2012) and might exclude fire-associated species under such circumstances.

Our results indicate that prescribed burning of clear-cuts is a conservation measure with only a short-term benefit. Other studies point in the same direction. Hyvärinen and co-authors studied the effects of prescribed burning of logged sites for two years after burning (Hyvärinen et al. 2005; Hyvärinen et al. 2006; Hyvärinen et al. 2009). They find a positive effect of prescribed burning on saproxylic beetle richness the same year as the burning and this positive effect remains also the first year after burning for a subset of rare and threatened species but only if at least 10 retention trees are left per hectare on the clear-cuts. Toivanen and Kotiaho (2007a) studied the effect of prescribed burning from one to 16 years after burning. They also find that species richness and abundance of saproxylic beetles decrease over time.

Our results therefore add to the pattern found in previous studies by documenting a rapid decrease in species richness already one year after buring. The conservation value of prescribed burning increases with increasing tree retention on the clear-cut (Hyvärinen et al. 2005; Hyvärinen et al. 2006; Toivanen & Kotiaho 2007a). This makes sense since studies of fires in intact forest point to a real long-term value for saproxylic beetles (Boulanger & Sirois 2007; Nappi et al. 2010; Boucher et al. 2012). For example, in a retrospective study of burnt sites there was an effect of forest fire up to 28 years post fire (Boulanger & Sirois 2007). The value of burning intact forest can therefore not be overstated.

Effects of Fire Continuity

The species composition of all saproxylic species differed between the two regions and most clear-cuts in Långshyttan separated from the Orsa clear-cuts in the cluster analysis. In the cluster analysis of the fire-associated species the strongest effect was that the 4-year old sites were different from all others. Among the 0- to 2-year old clear-cuts there was then also some regional differentiation. These results point to both regional differences and effects of age on the species composition.

There were several regional differences consistent with a better fire continuity in Orsa. There was a clear benefit of prescribed burning to the saproxylic beetle assemblage in Orsa whereas there was less of a benefit in Långshyttan.

This was seen for example in the rarefaction analysis that showed that in equal-sized samples there were more fire-associated

species in burnt clear-cuts in four out of five clear-cut pairs in Orsa but there was such a pattern in only one out of three pairs in Långshyttan. It was seen also in the larger difference between burnt and unburnt clear-cuts in species composition of fire-associated species in Orsa than in Långshyttan.

The more positive effect in Orsa could be an effect of a more intact fauna of fire-associated species in that region due to a better fire continuity and this is supported by the fact that we found four fire-dependent saproxylic species in Orsa but only one in Långshyttan. Furthermore, more red-listed saproxylic species were found in Orsa than in Långshyttan. The pattern for these rare species is based on relatively few individuals and should be interpreted with caution. However, the pattern for more common species also indicate a higher quality of Orsa since many common fire-associated species, such as *Acmaeops pratensis*, *Asemum striatum*, *Corticaria rubripes* and *Atomaria pulchra*, showed a pattern with a higher abundance in Orsa than in Långshyttan.

The total number of fire-associated species was similar in both regions but it seems that the populations of these species were larger in Orsa as indicated by the higher abundance of these species. In total we caught three times as many individuals of fire-associated species in Orsa compared to Långshyttan when we compared only the 0- to 2- year old clear-cuts to achieve similar sample sizes in the two regions. Taken together, this indicates that in Orsa there was a saproxylic beetle fauna that could respond positively to prescribed burning, whereas in Långshyttan the fauna in burnt clear-cuts was more similar to that in ordinary clear-cuts.

Furthermore, the positive effect of prescribed burning was possibly more long lived in the north, but the number of replicates in our investigation does not permit a firm conclusion in this respect. We just note that significant differences in rarefied species richness and species composition of fire-associated species was found in the 1-year old clear-cut and in one of the 4-year old clear-cuts in Orsa. A larger sample is needed to clarify this.

Our results are similar to those of a Finish study where restoration burning of young forest is more beneficial in a region where modern forestry started later (Kouki et al. 2012). Other studies also show that regional differences in forestry history can affect the species composition of saproxylic beetles in other habitats than burnt forest (Siitonen & Martikainen 1994; Martikainen et al. 1996; Victorsson 2009) also the geographical proximity to high quality habitat in Southern Sweden affect saproxylic beetles, at least in birch (Lindbladh et al. 2007; Abrahamsson et al. 2009). Our study therefore adds to the emerging picture that the regional species pool is important for local diversity in saproxylic beetles.

The better effect of prescribed burning in Orsa than in Långshyttan points to a differential effect of the value of prescribed burning. It also indicate that a long period of restoration burning might be needed in many parts of Sweden in order to augment the fire-associated fauna. It seems likely that restoration burning of intact forest or of clear-cuts with a high level of tree retention has the best chance of success (Toivanen & Kotiaho 2007b; Kouki et al. 2012). Such attempts at reintroducing forest fire as an important disturbance will probably succeed earlier if they are performed in regions with better fire continuity (Kouki et al. 2012).

The interpretation of the results should be tempered by two facts. Firstly, the southern region is situated at lower latitude *and* altitude and therefore should have more days with a favourable weather for insects. However, it is not evident that this should affect the comparison within pairs of burnt and unburnt clear-cuts that forms the basis for this chapter. Secondly, the difference in fire continuity is linked to forest history. There could be other differences between the regions more directly related to forest history than fire continuity affecting these species. Only further landscape level studies could clarify that.

CONCLUSION

Prescribed burning of clear-cuts mimics natural forest fires in some ways and this was probably the explanation for the strong positive effect on species richness and abundance of saproxylic beetles the same year as the fire. However, clear-cuts also differ from burnt forest in important ways and this probably explains the short-lived response of saproxylic beetles to prescribed burning.

Already one year after burning species richness and abundance was similar in burnt and unburnt clear-cuts. Our study complement Toivanen and Kotiaho (2007a) who find a decreasing species richness of saproxylic beetles over a period of 1-16 years after prescribed burning. Here we demonstrated a pronounced decrease of species richness already the first year after prescribed burning.

However, the positive effect of prescribed burning for fire-associated species was larger in the region with a better fire continuity than in the region with a poor fire continuity more representative for Fennoscandia. This fact points to a differential effect of the value of prescribed burning and to the need for restoration of the fire-associated beetle fauna in many parts of Sweden.

In the "new forestry" in Sweden there is a growing awareness that dead wood must be left or created at clear-felling (Ekbom et al. 2006). In order to maximise the value of conservation measures it is important to understand how they affect forest organisms.

This chapter shows that the beneficial effects of burning of clear-cuts will disappear in a very short period of time. Considering the long-term effect on this fauna after natural fires it seems reasonable to draw the conclusion that natural forest-fire dynamics is not mimicked by this procedure. If the goal is to create conditions similar to those in natural forest, then it is essential to burn more or less intact forest on a regular basis. Nevertheless it should be stressed that burning of clear-cuts can be an important complement, especially in regions with good fire continuity.

ACKNOWLEDGMENT

We thank Staffan Ulfstrand and Anna Sandberg for comments on earlier versions of the manuscript and Stig Lundberg for identification of beetles. Thanks to Anna Sandberg for field assistance. Thanks to Mats Jonsell for making his saproxylic database available.

APPENDIX 1. SPECIES LIST

Abundance of species caught in trunk window traps in eight pairs of clear-cuts in two regions in central Sweden: Orsa with a better fire continuity and Långshyttan with a broken fire continuity. One clear-cut in each pair was subjected to prescribed burning (B) and the other was unburnt (U). The figure in parenthesis after each clear-cut pair name is the number of years since the prescribed burning. Species relation to fire from Wikars (2006) and its history of red listing is indicated after the species names followed by the publishing year of the Swedish red list (2000, 2005, or 2010).

		Region: Orsa										Region : Långshyttan						
family	Species	Avenlam (0)		Sälgtjärn (0)		Pilka (1)		Hirvilampi (4)		Lillhamra (4)		Dammsjön (0)		Nordvik (1)		Rudtjärn (2)		Total
	Saproxylic species	B	U	B	U	B	U	B	U	B	U	B	U	B	U	B	U	
Histeridae	*Plegaderus vulneratus*												3					**3**
	Platysoma lineare (NT2010)			2									1					**3**
Leiodidae	*Anisotoma humeralis*															2	4	**6**
	Anisotoma axillaris	8	5		1			43	32	17	11	11	2	3	24	20	41	**218**
	Anisotoma castanea											2	2	2	12	4	4	**26**
	Anisotoma glabra	3	11			3	4	32	19	34	18	11	12	6	13	13	11	**190**
	Agathidium rotundatum									1								**1**
	Agathidium atrum	1																**1**
	Agathidium seminulum														1			**1**
	Agathidium laevigatum										1							**1**
Staphylinidae	*Gabrius splendidulus*											2	2				1	**5**
	Quedius xanthopus					1							1					**2**
	Nudobius lentus														1		1	**2**
	Tyrus mucronatus											1						**1**
	Acrulia inflata								1									**1**
	Hapalarea clavigera (NT2005) FF	4	2					12	5	5	1							**29**
	Phloeonomus planus	2		1		3			1			1		3	2	1		**14**
	Phloeonomus lapponicus FF	6		26	1	11	5					6	3	2	1			**61**
	Phloeonomus pussilus FF	1											2	1	1	2		**7**
	Phloeonomus sjoebergi FF	1		3								1	5	1				**11**
	Coryphium angusticolle	1	1											1				**3**
	Scaphisoma boleti		1			1		7	16		6	8	5	1	1	6		**52**

(Continued)

		Region: Orsa										Region : Långshyttan						
family	Species	Avenlam (0)		Sälgtjärn (0)		Pilka (1)		Hirvilampi (4)		Lillhamra (4)		Dammsjön (0)		Nordvik (1)		Rudtjärn (2)		Total
	Lordithon thoracicus								2					1			1	**4**
	Lordithon lunulatus								2									**2**
	Phloeopara corticalis							1										**1**
	Atheta sodalis	2	3		18	1	3		1			6	1	1	6		3	**45**
	Dinaraea aequata										1	2						**3**
	Anomagnathus cuspidatus FF					1		1		4	1						1	**8**
	Homalota plana FF	1										1	1	1	2	2		**8**
	Placusa depressa						1					1	1					**3**
Scarabeidae	*Trichius fasciatus*	74	22	11	10	27		29	12	10	19	18		8	1	14	9	**264**
Lycidae	*Lygistopterus sanguineus*	10	1				5	3		1		33		7		3	11	**74**
Cantharidae	*Malthodes fuscus*		4	6	1	2	2		6		1		4	1	4		1	**32**
	Malthodes marginatus	1	1	1	4	1	2	2	2		4	2	3	1		1	1	**26**
	Malthodes spathifer								1						1			**2**
	Malthodes brevicollis	1		1			1	1										**4**
Elateridae	*Ampedus balteatus* FF									1						1		**2**
	Ampedus tristis FF				1	1		1		1	2	1			1		1	**9**
	Ampedus nigrinus		1		1	3	2			3		1				1	3	**15**
	Melanotus castanipes	9					1	2	1	1	1	1	2	2	2	1	1	**24**
	Cardiophorus ruficollis											1				1		**2**
Buprestidae	*Buprestis rustica* FF	4	1					4	3							1	2	**15**
	Melanophila acuminata (NT2000) FD	3																**3**
	Phaenops cyanea FF		1									1	1	1		3		**7**
	Anthaxia quadripunctata FF												1	1		5	1	**8**
	Chrysobothris chrysostigma FF												1					**1**
	Agrilus viridis												1					**1**
Anobiidae	*Ptinus subpilosus*										1							**1**
	Ernobius nigrinus						2											**2**
	Hadrobregmus pertinax							1	7					1	1	1		**11**
	Stagetus borealis (NT2010) FF							1	1		1							**3**
	Dorcatoma punctulata																1	**1**
Cleridae	*Thanasimus formicarius*											2	8	1	1			**12**
Melyridae	*Dasytes niger*	26	17	3	14	2	1	5	5	160	89	657	46	205	314	145	130	**1819**
Nitidulidae	*Epuraea laeviuscula*											2						**2**

family	Species	Region: Orsa										Region : Långshyttan						Total
		Avenlam (0)		Sälgtjärn (0)		Pilka (1)		Hirvilampi (4)		Lillhamra (4)		Dammsjön (0)		Nordvik (1)		Rudtjärn (2)		
	Epuraea angustula												3					**3**
	Epuraea boreella			2			1						1		1			**5**
	Epuraea marseuli						1					1	1		2			**5**
	Epuraea biguttata	1													1			**2**
	Epuraea muehli			3														**3**
	Thalycra fervida								1									**1**
	Glischrochilus quadripunctatus FF			1														**1**
	Pityophagus ferrugineus	1		4		1	1								1			**8**
Aspidiphoridae	*Sphindus dubius*				1	1	1		2	2	1	1	2	1			1	**13**
Monotomidae	*Rhizophagus ferrugineus*	4		10		5	1		1		1	1	3	1	3	3		**33**
	Rhizophagus dispar							1									1	**2**
	Rhizophagus bipustulatus		1															**1**
Cryptophagidae	*Henoticus serratus* FD	1										1						**2**
	Cryptophagus fallax (NT2010)	1																**1**
	Cryptophagus badius	2	1								1							**4**
	Cryptophagus dentatus									1								**1**
	Caenoscellis ferruginea FF	1				1						2						**4**
	Atomaria subangulata (NT2000)				2										1			**3**
	Atomaria bella	1									1	3	1	1	5	23	6	**41**
	Atomaria pulchra FF	28		18		1						1					1	**49**
Erotylidae	*Triplax aena*	1																**1**
	Dacne bipustulata FF									14								**14**
Cerylonidae	*Cerylon histeroides*														1			**1**
Latridiidae	*Latridius consimilis*																1	**1**
	Enicmus rugosus	2	5	3	1	3	4	19	5	11	6	15	22	3	22	7	6	**134**
	Aridius nodifer											2	1		1			**4**
	Corticaria rubripes FF	25	3	9		67	1	2				15	6	11	8			**147**
	Corticaria longicollis																2	**2**
	Corticaria ferruginea FF	90		58		2						8						**158**
Cisidae	*Cis jacquemartii*											1						**1**
	Cis glabratus												1					**1**
	Hadreule elongatula FF															2		**2**
	Sulcacis affinis													1				**1**
Salpingidae	*Sphaeriestes stockmanni* (NT2000) FD			1		1												**2**

(Continued)

family	Species	Region: Orsa										Region : Långshyttan						Total
		Avenlam (0)		Sälgtjärn (0)		Pilka (1)		Hirvilampi (4)		Lillhamra (4)		Dammsjön (0)		Nordvik (1)		Rudtjärn (2)		
Aderidae	*Euglenes pygmaeus*	15	1					2	5			24	14	4	1			**66**
	Anidorus nigrinus											10	4	5	3	12	2	**36**
Tenebrionidae	*Corticeus linearis*			6			2						2		2			**12**
Scraptiidae	*Anaspis bohemica*	1	1	2	4	3	1	1	2		1	2	2	1			1	**22**
	Anaspis marginicollis										1			1	2			**4**
	Anaspis thoracica							1				37		16	1	1	1	**57**
	Anaspis flava											10	1	6	4	1		**22**
Mordelidae	*Curtimorda maculosa*												1		1			**2**
Melandryidae	*Hallomenus axillaris* (NT2005)								4			1	1					**6**
	Orchesia micans	1																**1**
	Abdera triguttata FF							1				4		1				**6**
	Xylita laevigata					2	2			1							1	**6**
Cerambycidae	*Tragosoma depsarium* (VU2010) FF	3																**3**
	Arhopalus rusticus FF									2		1	4	2	1	1	2	**13**
	Asemum striatum FF	1		41		20	7		1									**70**
	Tetropium castaneum			2			2											**4**
	Rhagium mordax			7	1	2	5			3	6						1	**25**
	Rhagium inquisitor FF			3	4	21	3	2								1		**34**
	Oxymirus cursor		1				1				1							**3**
	Pachyta lamed	1																**1**
	Acmaeops septentrionis (NT2010) FD					1												**1**
	Acmaeops pratensis FF	11		7	4	35	2	38		5	1	4		2		1	1	**111**
	Stictoleptura maculicornis							1										**1**
	Stictoleptura rubra											11	1	3	3	1	4	**23**
	Anastrangalia sanguinolenta	5							1	14	3	30	1	1	5	9	2	**71**
	Anastrangalia reyi													1				**1**
	Lepturobosca virens	36	2	2	2	7			4	1	2						1	**57**
	Judolia sexmaculata		1		1	2						2			1			**7**
	Leptura quadrifasciata	1	6				1	1		1	3	3		2			1	**19**
	Stenurella melanura				1			3	1	1		6		5	2	12	1	**32**
	Molorchus minor				1													**1**
	Monochamus sutor FF	3	1	1		1	1	1	1				1		2	2	1	**15**
	Pogonocherus fasciculatus	1		16			2	1			1	2	1	1	2	2	4	**33**

family	Species	Region: Orsa										Region : Långshyttan						
		Avenlam (0)		Sälgtjärn (0)		Pilka (1)		Hirvilampi (4)		Lillhamra (4)		Dammsjön (0)		Nordvik (1)		Rudtjärn (2)		Total
Curculionidae	*Rhyncolus sculpturatus*															1		**1**
	Magdalis duplicata	11	4	2		16	35	1	7	5	6	1	4		2		1	**95**
	Hylobius abietis FF	108	6	29	1	18	9		14		13	1	19	2		11	43	**274**
	Pissodes castaneus											1						**1**
	Pissodes harcyinae											1						**1**
	Pissodes piniphilus	1					1		1			1	4			1		**9**
	Hylurgops palliatus												13					**13**
	Hylastes brunneus	3	1	1			3				3	1			4	3	1	**20**
	Hylastes cunicularius			2		1								1		3	2	**9**
	Tomicus piniperda												14	2	3	1		**20**
	Polygraphus poligraphus	2		104														**106**
	Polygraphus punctifrons				1													**1**
	Pityogenes chalcographus			121		1	4					2	1				1	**130**
	Pityogenes quadridens			3	1		1								1			**6**
	Pityogenes bidentatus	30	4	77		14	139	18	45	2		12	12	4	174	1	22	**554**
	Orthotomicus proximus											1			1			**2**
	Orthotomicus suturalis FF			2		1								1				**4**
	Orthotomicus laricis											1	4					**5**
	Dryocoetes autographus			21	5	6	10					1		1	1	2	1	**48**
	Dryocoetes hectographus			1	2								2				1	**6**
	Crypturgus cinereus												1					**1**
	Crypturgus pusillus	1											10	2	2	1	1	**17**
	Trypodendron proximum												6					**6**
	Trypodendron lineatum												110					**110**
	Pityophthorus micrographus				1													**1**
	Saproxylic species total:	**551**	**109**	**613**	**84**	**290**	**270**	**238**	**212**	**301**	**208**	**993**	**382**	**333**	**653**	**329**	**342**	**5908**
	Epigeal species																	
Carabidae	*Dromius schneideri*												1					**1**
	Dromius fenestratus	1																**1**
Staphylinidae	*Quedius tenellus*			1	1								3	1	6		2	**14**
	Proteinus brachypterus														1			**1**
	Acidota crenata	16															1	**17**
	Trichophya pilicornis											1						**1**
	Atheta myrmecobia				1							1		1	2			**5**
Scirtidae	*Cyphon punctipennis*	1																**1**
Cantharidae	*Absidia sp*	11	2	10	26	5	8		1	3	8	12	12	2	5			**105**

(Continued)

		Region: Orsa										Region : Långshyttan						
family	Species	Avenlam (0)		Sälgtjärn (0)		Pilka (1)		Hirvilampi (4)		Lillhamra (4)		Dammsjön (0)		Nordvik (1)		Rudtjärn (2)		Total
	Malthodes minimus													1				**1**
Nitidulidae	*Epuraea binotata*											1			1			**2**
Silvanidae	*Silvanoprus fagi*			1														**1**
Cryptophagidae	*Cryptophagus abietis*								1									**1**
Latridiidae	*Cartodere constricta* FF											13	1					**14**
	Corticaria pubescens										1							**1**
	Corticarina similata																1	**1**
	Epigeal species total:	**29**	**2**	**12**	**28**	**5**	**8**	**0**	**2**	**3**	**9**	**28**	**17**	**5**	**15**	**0**	**4**	**167**
	Species with unknown habitat affiliation																	
Scydmaenidae	*Microscydmus minimus*														1			**1**
Staphylinidae	*Oxypoda hansseni*		1															**1**
	Thyasophila wockii	1					1											**2**
	Pentanota meuseli (NT2010)														1			**1**
	Aloconota subgrandis				1													**1**
	Paranopleta inhabilis (NT2010) FD			1									2			1		**4**
Nitidulidae	*Epuraea placida*								1									**1**
Cucujidae	*Pediacus fuscus* FF	2		1		2	1	4										**10**
Cryptophagidae	*Cryptophagus angustus*					1												**1**
	Cryptophagus saginatus					1												**1**
Latridiidae	*Corticaria elongata*			2													1	**3**
	Unknown habitat species total:	**3**	**1**	**4**	**1**	**4**	**2**	**4**	**1**	**0**	**0**	**0**	**2**	**0**	**2**	**1**	**1**	**26**

Abbreviations: B, burnt clear-cut; U, unburnt clear-cut; FD, fire-dependent; FF, fire favoured; NT, near threatened; VU, vulnerable.

Red listing from: Gärdenfors, U., editor. 2000. The 2000 red list of Swedish species. ArtDatabanken, SLU, Uppsala, Sweden.

Gärdenfors, U., editor. 2005. The 2005 red list of Swedish species. ArtDatabanken, SLU, Uppsala, Sweden.

Gärdenfors, U., editor. 2010. The 2010 red list of Swedish species. ArtDatabanken, SLU, Uppsala, Sweden.

Species relation to fire from: Wikars, L.-O. 2006. Åtgärdsprogram för bevarande av brandinsekter i boreal skog. Naturvårdsverket, Stockholm, Sweden.

REFERENCES

Abrahamsson, M., Jonsell, M., Niklasson, M. & Lindbladh, M. (2009). Saproxylic beetle assemblages in artificially created high-stumps of spruce (*Picea abies*) and birch (*Betula pendula/pubescens*) - does the surrounding landscape matter? *Insect Conservation and Diversity*, *2*, 284-294.

Allison, J. D., Borden, J. H. & Seybold, S. J. (2004). A review of the chemical ecology of the Cerambycidae (Coleoptera). *Chemoecology*, *14*, 123-150.

Andersson, M. J. (2001). A new method for non-parametric multivariate analysis of variance. *Austral Ecology*, *26*, 32-46.

Angelstam, P. (1996). The ghost of forest past-natural disturbance regimes as a basis for reconstruction of biologically diverse forests in Europe. In: *Conservation of faunal diversity in forested landscapes* (eds. DeGraaf RM & Miller RI). Chapman & Hall New York, 287-337.

Anonymous, (1995). Skyddsvärd natur i Orsa Länstyrelsen Dalarna.

Anonymous, (2013). The contribution of FSC-certification to biodiversity i Swedish forests, Report 2. Forest Stewardship Council FSC Sweden.

Berg, A., Ehnström, B., Gustafsson, L., Hallingback, T., Jonsell, M. & Weslien, J. (1994). Threatened plant, animal, and fungus species in Swedish forests - distribution and habitat associations. *Conservation Biology*, *8*, 718-731.

Bonan, G. B. & Shuggart, H. H. (1989). Environmental factors and ecological processes in boreal forests. *Annual Review of Ecology and Systematics*, *20*, 1-28.

Boucher, J., Azeria, E. T., Ibarzabal, J. & Hebert, C. (2012). Saproxylic beetles in disturbed boreal forests: Temporal dynamics, habitat associations, and community structure. *Ecoscience*, *19*, 328-343.

Boulanger, Y. & Sirois, L. (2007). Postfire succession of saproxylic arthropods, with emphasis on coleoptera, in the north boreal forest of Quebec. *Environmental Entomology*, *36*, 128-141.

Clarke, K. R. & Gorley, R. N. (2006). PRIMER v6: user manual/tutorial. PRIMER-E Ltd., Plymoth.

Cobb, T. P., Hannam, K. D., Kishchuk, B. E., Langor, D. W., Quideau, S. A. & Spence, J. R. (2010). Wood-feeding beetles and soil nutrient cycling in burned forests: implications of post-fire salvage logging. *Agricultural and Forest Entomology*, *12*, 9-18.

Colwell, R. K. (2006). EstimateS: Statistical estimation of species richness and shared species from samples. Version 8.0. Users guide and application published at: http://purl.oclc.org/estimates.

Ehnström, B. & Axelsson, R. (2002). Insektsgnag i bark och ved. ArtDatabanken, SLU, Uppsala.

Ehnström, B. & Holmer, M. (2007). Nationalnyckeln till Sveriges flora och fauna. Skalbaggar: Långhorningar. Coleoptera: Cerambycidae. ArtDatabanken, SLU, Uppsala.

Ehnström, B. & Waldén, H. W. (1986). Faunavård i skogsbruket, Del 2 - Den lägre faunan. Skogsstyrelsen, Jönköping.

Ekbom, B., Schroeder, L. M. & Larsson, S. (2006). Stand specific occurrence of coarse woody debris in a managed boreal forest landscape in central Sweden. *Forest Ecology and Management*, *221*, 2-12.

Esseen, P. A., Ehnström, B., Ericson, L. & Sjöberg, K. (1997). Boreal forests. *Ecological Bulletins*, *46*, 16-47.

Evans, W. G. (1966). Perception of infrared radiation from forest fires by *Melanophila acuminata* de geer (Buprestidae Coleoptera). *Ecology*, *47*, 1061-1067.

Gongalsky, K. B., Midtgaard, F. & Overgaard, H. J. (2006). Effects of prescribed forest burning on carabid beetles (Coleoptera : Carabidae): a case study in south-eastern Norway. *Entomologica Fennica*, *17*, 325-333.

Gärdenfors, U., Hall, R., Tomas, H., G. H. H. & Lars, H. (2003). Djur, svampar och växter i Sverige 2003. Förteckning över antalet arter per familj. ArtDatabanken, SLU, Uppsala.

Hansen, A. J., Spies, T. A., Swanson, F. J. & Ohmann, J. L. (1991). Conserving biodiversity in managed forests - lessons from natural forests. *Bioscience*, *41*, 382-392.

Hansen, V. (1964). Fortegnelse over Danmarks biller 1. og 2. del. (Catalogue of the Coleoptera of Denmark 1st and 2nd part). *Entomologiske Meddelelser*, *33*, 1-507.

Hjältén, J., Gibb, H. & Ball, J. P. (2010a). How will low-intensity burning after clear-felling affect mid-boreal insect assemblages? *Basic and Applied Ecology*, *11*, 363-372.

Hjältén, J., Stenbacka, F. & Andersson, J. (2010b). Saproxylic beetle assemblages on low stumps, high stumps and logs: Implications for environmental effects of stump harvesting. *Forest Ecology and Management*, *260*, 1149-1155.

Holliday, N. J. (1991). Species responses of carabid beetles (Coleoptera, Carabidae) during postfire regeneration of boreal forest. *Canadian Entomologist*, *123*, 1369-1389.

Hyvärinen, E., Kouki, J. & Martikainen, P. (2006). Fire and green-tree retention in conservation of red-listed and rare dead-wood-dependent beetles in Finnish boreal forests. *Conservation Biology*, *20*, 1711-1719.

Hyvärinen, E., Kouki, J. & Martikainen, P. (2009). Prescribed fires and retention trees help to conserve beetle diversity in managed boreal forests despite their transient negative effects on some beetle groups. *Insect Conservation and Diversity*, *2*, 93-105.

Hyvärinen, E., Kouki, J., Martikainen, P. & Lappalainen, H. (2005). Short-term effects of controlled burning and green-tree retention on beetle (Coleoptera) assemblages in managed boreal forests. *Forest Ecology and Management*, *212*, 315-332.

Hörnsten, L., Nohlgren, E. & Aldentun, Y. (1995). Natural forest fires and controlled burning - a study of the literature. SkogForsk, Uppsala.

Johansson, T., Andersson, J., Hjältén, J., Dynesius, M. & Ecke, F. (2011). Short-term responses of beetle assemblages to wildfire in a region with more than 100 years of fire suppression. *Insect Conservation and Diversity*, *4*, 142-151.

Jonsell, M. (2008). Saproxylic beetle species in logging residues: which are they and which residues do they use? . *Norwegian Journal of Entomology*, *55*, 109-122.

Jonsell, M., Weslien, J. & Ehnström, B. (1998). Substrate requirements of red-listed saproxylic invertebrates in Sweden. *Biodiversity and Conservation*, *7*, 749-764.

Kaila, L., Martikainen, P. & Punttila, P. (1997). Dead trees left in clear-cuts benefit saproxylic Coleoptera adapted to natural disturbances in boreal forest. *Biodiversity and Conservation*, *6*, 1-18.

Koch, K. (1989-1992). Die Käfer Mitteleuropas. Ökologie. Band 1-3. Goecke & Evers, Krefeld.

Kouki, J., Hyvärinen, E., Lappalainen, H., Martikainen, P. & Similä, M. (2012). Landscape context affects the success of habitat restoration: large-scale colonization patterns of saproxylic and fire-associated species in boreal forests. *Diversity and Distributions*, *18*, 348-355.

Lindbladh, M., Abrahamsson, M., Seedre, M. & Jonsell, M. (2007). Saproxylic beetles in artificially created high-stumps of spruce and birch within and outside hotspot areas. *Biodiversity and Conservation*, *16*, 3213-3226.

Linder, P. & Östlund, L. (1992). Förändringar i norra Sveriges skogar 1870-1991. *Svensk Botanisk Tidskrift*, *86*, 199-215.

Lindhe, A., Lindelöw, Å. & Åsenblad, N. (2005). Saproxylic beetles in standing dead wood density in relation to substrate sun-exposure and diameter. *Biodiversity and Conservation*, *14*, 3033-3053.

Lundberg, S. & Gustafsson, B. (1995). Catalogus Coleoperorum Sueciae. Naturhistoriska riksmuseet & Entomologiska föreningen Stockholm, Stockholm.

Lundqvist, R. (1994). Inventering av områden viktiga för den biologiska mångfalden i skogarna inom Hedemora kommun. Hedemora kommun, Hedemora.

Martikainen, P., Siitonen, J., Kaila, L. & Punttila, P. (1996). Intensity of forest management and bark beetles in non-epidemic conditions: A comparison between Finnish and Russian Karelia. *Journal of Applied Entomology-Zeitschrift Fur Angewandte Entomologie*, *120*, 257-264.

McCullough, D. G., Werner, R. A. & Neumann, D. (1998). Fire and insects in northern and boreal forest ecosystems of North America. *Annual Review of Entomology*, *43*, 107-127.

Moretti, M., Obrist, M. K. & Duelli, P. (2004). Arthropod biodiversity after forest fires: winners and losers in the winter fire regime of the southern Alps. *Ecography*, *27*, 173-186.

Nappi, A., Drapeau, P., Saint-Germain, M. & Angers, V. A. (2010). Effect of fire severity on long-term occupancy of burned boreal conifer forests by saproxylic insects and wood-foraging birds. *International Journal of Wildland Fire*, *19*, 500-511.

Niklasson, M. & Granström, A. (2000). Numbers and sizes of fires: Long-term spatially explicit fire history in a Swedish boreal landscape. *Ecology*, *81*, 1484-1499.

Palm, T. (1951). Die Holz- und Rinden-Käfer der nordschwedischen Laubbäume. Esselte, Stockholm.

Palm, T. (1959). Die Holz- und Rinden-Käfer der Sud- und Mittelschwedischen Laubbäume. *Opuscula Entomologica*, Supplementum, 1-374.

Quinn, G. P. & Keough, M. J. (2002). Experimental design and data analysis for biologists. Cambridge University Press, Cambridge.

Schmitz, H., Bleckmann, H. & Murtz, M. (1997). Infrared detection in a beetle. *Nature*, *386*, 773-774.

Schmitz, H. & Trenner, S. (2003). Electrophysiological characterization of the multipolar thermoreceptors in the "fire-beetle" *Merimna atrata* and comparison with the infrared sensilla of *Melanophila acuminata* (both Coleoptera, Buprestidae). *Journal of Comparative Physiology a-Neuroethology Sensory Neural and Behavioral Physiology*, *189*, 715-722.

Schowalter, T. D. (1985). Adaptations of insects to disturbance. In: *The ecology of natural disturbance and patch dynamics* (eds. Pickett STA & White EP). Academic press London, 235-252.

Schowalter, T. D. (2012). Insect Responses to Major Landscape-Level Disturbance. *Annual Review of Entomology*, *57*, 1-20.

Schutz, S., Weissbecker, B., Hummel, H. E., Apel, K. H., Schmitz, H. & Bleckmann, H. (1999). Insect antenna as a smoke detector. *Nature*, *398*, 298-299.

Siitonen, J. & Martikainen, P. (1994). Occurrence of rare and threatened insects living on decaying *Populus tremula* - a comparison between Finnish and Russian Karelia. *Scandinavian Journal of Forest Research*, *9*, 185-191.

Stokland, J. N., Siitonen, J. & Jonsson, B. G. (2012). Biodiversity in dead wood. Cambridge University Press, Cambridge.

Toivanen, T. & Kotiaho, J. S. (2007a). Burning of logged sites to protect beetles in managed boreal forests. *Conservation Biology*, *21*, 1562-1572.

Toivanen, T. & Kotiaho, J. S. (2007b). Mimicking natural disturbances of boreal forests: the effects of controlled burning and creating dead wood on beetle diversity. *Biodiversity and Conservation*, *16*, 3193-3211.

Weslien, J., Djupström, L. B., Schroeder, M. & Widenfalk, O. (2011). Long-term priority effects among insects and fungi colonizing decaying wood. *Journal of Animal Ecology*, *80*, 1155-1162. .

Victorsson, J. (2009). Community assembly and spatial ecology of saproxylic Coleoptera. PhD theis from the Department of Animal Ecology, Uppsala University, Uppsala.

Victorsson, J. (2012). Semi-field experiments investigating facilitation: arrival order decides the interrelationship between two saproxylic beetle species. *Ecological Entomology*, *37*, 395-401.

Victorsson, J. (this volume). Community assembly of saproxylic beetles in old-growth forest and recently burnt forest. In C. Stacks (ed) : *Beetles: Biodiversity, Ecology and Role in the Environment*. Nova.

Victorsson, J. & Jonsell, M. (2013). Effects of stump extraction on saproxylic beetle diversity in Swedish clear-cuts. *Insect Conservation and Diversity*, *6*, 483-493.

Wikars, L. O. (1992). Skogsbränder och insekter. *Entomologisk Tidskrift*, *113*, 1-11.

Wikars, L. O. (1995). Clear-cutting before burning prevents establishment of the fire-adapted *Agonum quadripunctatum* (Coleoptera:Carabidae). *Annales Zoologici Fennici*, *32*, 375-384.

Wikars, L. O. (1997). Pyrophilous insects in Orsa Finnmark, Central Sweden: biology, distribution, and conservation. *Entomologisk Tidskrift*, *118*, 155-169.

Wikars, L. O. (2002). Dependence on fire in wood-living insects: an experiment with burned and unburned spruce and birch logs. *Journal of Insect Conservation*, *6*, 1-12.

Wikars, L. O. (2006). Åtgärdsprogram för bevarande av brandinsekter i boreal skog. Naturvårdsverket, Stockholm.

Wikars, L. O., Sahlin E. & Ranius T. (2005). A comparison of three methods to estimate species richness of saproxylic beetles (Coleoptera) in logs and high stumps of Norway spruce. *Canadian Entomologist*, *137*, 304-324.

Vondran, T., Apel, K. H. & Schmitz, H. (1995). The infrared receptor of *Melanophila acuminata* De Geer (Coleoptera: Buprestidae): Ultrastructural study of a unique insect thermoreceptor and its possible descent from a hair mechanoreceptor. *Tissue & Cell*, *27*, 645-658.

Zackrisson, O. (1977). Influence of forest fires on the North Swedish boreal forest. *Oikos*, *29*, 22-32.

Zackrisson, O. & Östlund, L. (1991). Branden formade skogslandskapets mosaik. *Skog & Forskning*, *4*, 13-17.

In: Beetles: Biodiversity, Ecology and Role in the Environment ISBN: 978-1-63463-380-2
Editor: Camilla Stack

Chapter 2

COMMUNITY ASSEMBLY OF SAPROXYLIC BEETLES IN OLD-GROWTH FOREST AND RECENTLY BURNT FOREST

Jonas Victorsson*

Animal Ecology, Evolutionary Biology Centre, Uppsala University, Norbyvägen 18 D, SE-752 36 Uppsala, Sweden, Department of Ecology, Swedish University of Agricultural Sciences, Box 7044, SE-750 07 Uppsala, Sweden

ABSTRACT

Community assembly is the trajectory exhibited over time regarding species composition and community structure as species colonize an empty patch. Chase (2007) distinguish two basic types of assembly where a niche-assembled community is constrained by species interactions and competitive hierarchies whereas a dispersal-assembled community is unaffected by competition. The first wood-living (saproxylic) beetles to colonize a dead tree are sub-cortical, cambial-living species with a larval development-time of one year or less. Important community assembly events could therefore take place the first summer after tree death.

Community assembly was studied during the eight weeks immediately after tree death. By exposing 0.5 m long stem sections (bolts) of Norway spruce, *Picea abies*, to colonization for increasingly longer times (0.5, 1, 2, 4, and 8 weeks) a temporal sequence was created. A total of 140 bolts at four forest sites in central Sweden were studied with six to eight replicates per site. Two of the sites were old-growth forest and the other two were forest sites subjected to burning right before the experiment started. I used community ecology measures to describe assembly patterns and emphasize findings that would have gone unnoticed in an analysis confined to the 8-week final state.

In total 8802 beetles of 56 species were reared from the bolts. Two fire-dependent species were found in burnt forest, this is noteworthy since unburnt bolts were used in both forest types. The total number of emerged beetles per bolt was similar in the two forest types in early assembly but later this density was higher in the old-growth forest. During assembly the community-level carrying capacity was reached in old-growth forest but not in burnt forest. During assembly, community evenness and taxonomic richness

* Corresponding author: E-mail: jonas.victorsson@slu.se Fax: (+46)18-672890.

was higher in the burnt forest than in old-growth forest but at eight weeks these community measures had converged. The two forest types became more dissimilar regarding species composition as the community assembly progressed. Old-growth forest also had a lower bolt-to-bolt variation in species composition than burnt forest. Cambivores was the functional group that made up the largest proportion of individuals in both forest types but the proportion was highest in old-growth forest. The proportion fungivores was highest in burnt forest. The proportion predators peaked early in both forest types and then decreased during assembly. In old-growth forest, 74 % of the bolts were numerically dominated by the cambivorous scolytine *Dryocoetes autographus*. In burnt forest only 33 % of the bolts were dominated by that species. Furthermore, burnt forest also had fungivores as bolt dominants, for example the fire-favoured corticarid *Corticara rubripes*, whereas old-growth forest bolts were always dominated by cambivores.

In conclusion, the beetle assemblages in the two forest types followed different assembly trajectories. The beetle assemblage in burnt forest conforms to a dispersal-assembled community whereas that in old-growth forest conforms to a niche-assembled community (*sensu* Chase 2007). The biology of fire-adapted beetles is discussed in light of these findings.

INTRODUCTION

Community assembly (Diamond 1975) is the process deciding species inclusion/exclusion as well as determining the relative species abundance in a local community (Weiher and Keddy 1999). Community assembly implies change over time and can be viewed as the trajectory, regarding species composition and community structure, a community exhibits over time (Drake et al. 1999). Studying the final state of communities is important but by ignoring community assembly, the final states might seem idiosyncratic, especially if the species acting during assembly, such as a key-stone predator (Warren et al. 2003), is no longer present in the final state (Drake 1991). Community patterns can either be created by environmental differences between patches or by biotic interactions within patches. In this respect community assembly can be viewed as sequential species filtering (Weiher and Keddy 1999) where abiotic filtering removes species for which the patch does not fulfil environmental niche criteria and biotic filtering, i.e. community ecology interactions such as resource competition and predation, exclude further species.

Saproxylic beetles depend on dead wood and are an important group in the heterotrophic succession in coarse woody debris (CWD) (Stokland et al. 2012). After tree-death, insects colonize the thin, nutrient-rich sub-cortical layer (phloem, cambium, and outer sapwood) (Dajoz 2000). Beetle larvae are the main primary consumers (cambivores) (Esseen et al. 1992) with scolytines and cerambycids being two important groups. The sub-cortical layer is consumed in less than two years and is the most ephemeral resource in dead trees. Most cambial-living saproxylic beetles have a larval development-time of one year or less (Ehnström and Axelsson 2002), so the time for community assembly is short and significant assembly events could be expected to be separated by weeks only.

Forest fire is the major large-scale disturbance in natural boreal forest (Zackrisson 1977, Niklasson and Granström 2000). The newly burnt forest is rich in sun-exposed coarse woody debris (CWD) for many years (Boulanger and Sirois 2007, Stokland et al. 2012). The forest then develop through a young-forest stage with very little CWD (Stokland et al. 2012). Later, when the forest becomes an old-growth stand, more CWD is again produced during small-scale tree-mortality events (self-thinning, drought, wind throw, or fungus infestations) (Siitonen 2001). For saproxylic insects, a natural boreal landscape is therefore a forest-patch mosaic where most available dead wood is found in patches of old-growth forest and burnt forest. Therefore, community assembly in saproxylic beetles is probably a result of interactions developed in these two forest stages.

If community structure and composition can be recorded non-invasively then community assembly can be studied directly. For example Jenkins and Buikema (1998) did this when they used biweekly plankton sampling to study community assembly in ponds. Saproxylic insects, however, complete their larval development hidden within the dead wood and inspecting the community by bark removal would destroy the process under study. In this experiment, I therefore exposed stem sections, bolts (0.5 m long), of Norway spruce *Picea abies* to natural colonization for increasingly longer times (0.5 to 8 weeks). Beetles reared from the bolts offered community assembly "snapshots", providing quantitative community data along the assembly trajectory. This methodology has been used to study insects in dung and carrion (Koskela and Hanski 1977, Kuusela and Hanski 1982).

Chase (2007) distinguish two community types based on the species-interaction intensity. (1) A *niche-assembled community* is constrained by species interactions in the patch. Competitive hierarchies and species traits determine community assembly. An increased species-composition similarity during assembly is predicted. (2) In a *dispersal-assembled community*, colonization and extinction are unaffected by species competitive traits (as in the neutral theory of Hubbell 2001). These communities have a high patch-to-patch variation in species composition. I predicted that burnt forest would have a dispersal-assembled community whereas old-growth forest would have a niche-assembled community. The reasoning behind this was: (1) that the intensity of density-dependent effects (such as larval resource competition) is predicted to be lower in burnt forest than in old-growth forest (Esseen et al. 1992, Wikars 1992); and (2) that burnt forest attracts colonizers on a landscape scale (Boulanger et al. 2010) which could lead to a wider taxonomic range of colonizing species. I used increasingly fine-scaled community-measures, such as evenness, importance of nutritional functional groups and species composition in an attempt to describe important community-level patterns during community assembly. I emphasize patterns that would not have been found in an analysis of first-year final states only.

METHODS

Study Sites

The study was performed in central Sweden, an area dominated by managed, boreal forest. The dominating tree species in the area are Scots pine *Pinus sylvestris* and Norway spruce *Picea abies*. Some deciduous species, mainly birch *Betula* spp., aspen *Populus tremula*, and alder *Alnus* spp., also occur but in low densities. The managed forest has a very low level of dead wood, only 2-6 m^3 ha^{-1} (Fridman and Walheim 2000). The experiment was performed in two pairs of burnt and old-growth forest sites (inter-pair distance 135 km). The burnt sites were surrounded by managed forest and in both pairs, the only available old-growth site within a 60 km radius of the burnt site were used. The first burnt site, Eggåsen (B1) (61°05' N, 16°00' E), was burnt for conservation purposes on June 11 1997. Before the fire, the site was mature, managed forest that was thinned the winter before burning, leaving 25 mature Scots pine, trees per hectare. Approximately 20 % of the stand was left untouched before the fire and the study area was established in this part (see Wikars and Schimmel 2001, for details of Eggåsen). The old-growth site, Rödmyrberget (OG1) (61°04' N, 16°03' E), was situated 3 km from Eggåsen.

This was a set-aside, excluded from forest management due to high conservation values. Rödmyrberget had a mixed tree-species composition and an uneven tree-age structure (see Wikars 2002, for details of Rödmyrberget). The second burnt site, Galhån (B2) (62°00' N, 14°38' E), was burnt by a wildfire that started June 3 1997 and continued until June 11. Galhån was managed forest, with a stand age of approximately 60 years (the rotation period is 90 years). The corresponding old-growth site, Ryggen (OG2) (62°12' N, 14°20' E), was situated 15 km from the burnt site. Just like Rödmyrberget this was a set-aside with a mixed tree-species composition and an uneven tree-age structure.

Field Experiment

The experiment had a two-way factorial design with forest type and exposure time as factors. The experiment ran in the field from June 12 to August 7 and started one day after the fire at the burnt sites and one day later still at the old-growth sites. The experimental units were 140 bolts (0.5 m long, diameter 15-25 cm) from eight Norway spruce trees taken at site B1. This is a common host tree in both forest types. From each tree, 20 bolts were taken and divided so that each individual tree was represented as a 5-bolt group at each site. The bolts in each such group were randomly subjected to one of five exposure treatments: 0.5, 1, 2, 4, or 8 weeks. Each tree was therefore a blocking variable.

The bolt bark area was calculated after measuring the circumference and length of each bolt. The bark area was a measure of the amount of substrate available to saproxylic insects and it was used as a covariate in all analyses where that was possible. The bark area was 0.324 ± 0.058 m^2 (mean ± SD), but there was no treatment differences (ANOVA; B1/OG1: $F_{9,70} = 0.70$, $P = 0.70$; B2/OG2: $F_{9,50} = 0.59$, $P = 0.80$).

The bolt-groups were placed within a 250 x 250 m study area at the centre of each site with a minimum distance between groups of 10 m. Distance between bolts within group was 0.5 m. When the experiment started, all bolts were exposed to colonization. Terminating the exposure was done by enclosing the bolt in a cloth mesh cage (mesh size 1 mm). The number of exposure replicates was eight in B1/OG1 and six in B2/OG2. Please note that neither the bolts in the old-growth forest nor the bolts in the burnt forest were scorched by fire.

The 8-week bolts were the first-year final stage. This is not the end of the heterotrophic succession but it is the end of most of the season for the focal species that year (Vanderwel et al. 2006). If a longer exposure time than eight weeks had been used then the species with the shortest larval development time would have left the bolts before the experiment ended (e.g. *Pityogenes chalcographus*).

In natural logs, the colonization would undoubtedly commence anew the following year (Boulanger and Sirois 2007) but my study focused on the first-year community assembly.

The bolts were hanged in individual emergence traps at each original bolt-exposure site on August 8-10. Traps were emptied four times: (1) November 1997; (2) July 1998; (3) October 1998; and (4) October/November 1999. Regrettably, at emptying number four, most traps at study site B1 had been destroyed. Therefore, the analyses were made on beetles from emptying 1-3 in B1/OG1 and from emptying 1-4 in B2/OG2.

Species Ecology

The species reared with this method belong to successional stage one in the 4-stage successional sequence in Norway spruce (Esseen et al. 1992). All beetles except 423 *Dryocoetes spp.* were determined to species (these individuals belonged to either *Dryocoetes autographus* or *Dryocoetes hectographus*). Species were assigned to one of four functional groups based on the literature (Martikainen et al. 1999, Dajoz 2000, Ehnström and Axelsson 2002, Lieutier et al. 2004, Johansson et al. 2007, Hjältén et al. 2010): (1) *cambivores* feed directly on inner bark or outer sapwood; (2) *fungivores* feed on subcortical mycelium; (3) *predators* feed on other arthropods; and (4) *facultative predators* are predators but also feed as fungivores or as detritivores (feeding on the remains of other saproxylics). Out of all 56 species, 51 could be assigned (99.8 % of all individuals).

Species were classified as fire-dependent or fire-favoured based on the most recent classification of Swedish insects in relation to fire (Wikars 2006). Fire-favoured species are regularly abundant in burnt forest but also occur in other habitats whereas fire-dependent species are confined to burnt forest.

Community Variables and Analysis

A range of community variables were calculated. The *total abundance* (N_{tot}) and the *number of species* of all beetles emerging from each bolt were calculated. *Evenness*, in each bolt was measured with the index E_{var} (Smith and Wilson 1996). *Taxonomic diversity* was measured with average taxonomic distinctness (AvTD) (Warwick and Clarke 1995). This index measures the average path length, along a taxonomic tree, between species in the sample. AvTD was calculated using software PRIMER (version 6.1.10, PRIMER-E, Plymouth).

I used the taxonomic information in Lundberg and Gustafsson (1995). Both AvTD and E_{var} are relatively unaffected by total species number and abundance (Magurran 2004) which is advantageous since species number and abundance was positively correlated with exposure time. I tested the effect on the *relative abundance of each functional group* for the four functional groups. For example, for cambivores this was the effect on $N_{cambivores}/N_{tot}$. As exposure times increase so does total abundance in the bolts and the relative abundance control for this in order to get an unbiased measure of the importance of the different functional groups.

All tests on those community variables were performed using procedure GLIMMIX in SAS version 9.3. (SAS institute, Cary, NC). They were analysed with two-way factorial ANCOVAs with the treatment factors *forest type* and *exposure time*. The *bark area* for each bolt was used as a covariate. The variables *substrate tree* (the eight Norway spruce trees) and *study site pair* (B1/OG1 and B2/OG2 respectively) were used as random variables (blocking variables).

Further tests of simple effects (slices) were done as described in Littell et al. (2006). For N_{tot} a Poisson error distribution (suitable for counts) and log link was used. A quasi-likelihood model was used to compensate for overdispersion (Quinn and Keough 2002). Another preplanned test, a contrast between the 4- and 8-week exposures to detect a "levelling-off" in beetle density, was also done for this measure. The two indices, AvTD and E_{var}, were arcsine square root transformed, with additional multiplication by $2/\pi$ to rescale values to the original scale as described by McCune and Grace (2002). The functional groups were tested using the events/trials syntax in procedure GLIMMIX (Littell et al. 2006). For example, when testing the effect on cambivores, an event was $N_{cambivores}$ and a trial was N_{tot}. For tests on the two functional groups fungivores and facultative predators, there was too few observation in the 0.5-week exposure for the model to converge. Therefore a model without the 0.5-week exposure was fitted instead.

More detailed measures took species identity into account. Bolt-dominants, the most abundant species, in each bolt were visualized using a mosaic plot. The most detailed community measure, species composition, uses species identity and abundance of all species in a bolt. I tested the difference in species composition between forest types in the 8-week bolts with an analysis of similarity (ANOSIM) (Clarke and Warwick 2001) using Bray-Curtis similarity and 4th root transformed abundances (to dampen the influence of the most abundant species) in software PRIMER.

The test statistic, R, ranges from 0 to 1 and is a measure of difference in species composition. Further testing was done using similarity percentage analysis (SIMPER) (Clarke and Warwick 2001) to decompose the forest-type difference into the percentage contribution of each species to that dissimilarity. Species with a high SIMPER value for a particular habitat are said to characterize that habitat (Clarke and Gorley 2006). I also used within-treatment Bray-Curtis similarity to test for differences in bolt-to-bolt similarity along the assembly trajectory. This was modelled with a two-way ANOVA with study site pair as random factor.

To visualize the assembly trajectory, regarding species composition, I performed a non-metric multidimensional scaling (NMDS) ordination of R-values from an ANOSIM testing for differences between all treatments. This analysis involved four steps (all in software PRIMER): (1) ANOSIM testing for differences between treatments; (2) a similarity-matrix was constructed using these R-values; (3) this matrix was used in an NMDS ordination with Euclidian distances between treatments now showing R-values; (4) the matrix was also used in a group-average linking (UPGMA) cluster analysis; (5) circles indicating R-distances, obtained from the clustering, were added to the NMDS ordination to increase interpretability.

I used as many bolts as was meaningful in each analysis. This meant using all 140 bolts when testing N_{tot} and number of species. In the more derived measures fewer bolts could be used. For example, calculating E_{var} is not meaningful with too few individuals in a bolt. In the analyses of species composition I restricted the analysis to bolts with > 6 individuals reared. This criterion was met for 88 bolts, 46 in old-growth forest and 42 in burnt forest (for the exposure times 0.5-8 weeks the sample sizes in burnt forest were 5, 8, 6, 10, 13 and in the old-growth forest 4, 7, 8, 13, and 14).

Environmental Variables and Analysis

I kept bolt-group sun-exposure similar in the two forest types by setting out bolts in sun-exposed sites at the old-growth sites. I took a canopy-cover picture at each bolt-group and calculated sun-exposure as the proportion sky in the pictures using software ImageJ (version 1.38). Trees and CWD were inventoried in November 1997, the same year as the experiment. Diameter at breast height and species was recorded for all trees and CWD (with a diameter > 10 cm) within five circular plots (10 m radius) placed in a pre-determined grid at each forest site. CWD was classified into two classes: (1) fresh CWD (0-2 years old), and (2) old CWD (> 2 years old). I summed the basal area of each class for each of the five circular plots (as m^2 ha^{-1}). I tested for forest-type differences in with one-way ANOVAs for each factor and study-site pair.

RESULTS

More individuals were reared from the old-growth forest, 6787 individuals, than from the burnt forest, 2015 individuals. Among the 56 species reared, the dominating groups were Scolytinae (81 % of all individuals), other Curculionidae (7 %), Cerambycidae (6 %), and Staphylinidae (3 %) (Appendix 1). Two fire-dependent species, *Henoticus serratus* and *Cryptophagus corticinus* were reared in burnt forest. Nine fire-favoured species were found, five of those only in burnt forest (Appendix 1).

Burnt forest and old-growth forest showed forest-type specific community assembly regarding all nine univariate community measures (Figure 1) since there were significant (or in the case of evenness borderline significant, $P = 0.052$) exposure x forest type interactions for all measures (Table 1). The total abundance of beetles reared per bolt increased during assembly in both forest types (Table 1, Figure 1). Between the 2-week and 8-week exposures the abundance rose and reached a plateau in old-growth forest (difference between 4- and 8-week exposure, $t_{2,121} = 1.00$, $P = 0.32$) but not in burnt forest ($t_{2,121} = -2.99$, $P = 0.0034$). The number of species increased 2.6 times (from 1.6 to 4.2 species) during assembly in burnt forest and 5.2 times in old-growth forest (from 1.4 to 7.3 species) (Table 1, Figure 1). The taxonomic diversity was significantly higher in burnt forest than in old-growth forest in the 0.5-week exposure and marginally so in the 2-week exposure (Table 1, Figure 1). During assembly there was then a forest type convergence in taxonomic diversity. Evenness remained the same in burnt forest during assembly whereas it decreased during assembly in old-growth forest. Evenness was therefore significantly higher in burnt forest in the 4-week exposure but in the 8-week exposure there was forest-type convergence (Table 1, Figure 1). Within exposure treatment, the old-growth bolts were more similar regarding species composition than the burnt forest bolts since the Bray-Curtis similarity was higher in old-growth bolts in the 0.5-, 4-, and 8-week exposures (Table 1, Figure 1).

The forest-type differences in the relative abundance of each functional group were larger during assembly than in the first-year final states where there was a tendency towards forest-type convergence (Table 1, Figure 1). Cambivores and fungivores differed most between the forest types during assembly (Table 1, Figure 1). Cambivores made up the largest proportion in all treatments but the proportion was significantly lower in burnt forest than in old-growth forest during assembly. The proportion fungivores was significantly higher in burnt forest than in old-growth forest during assembly. The proportion predators peaked early during assembly and made up 5-20 % of total abundance in the 1- and 2-week exposures (Table 1, Figure 1). After that, the proportion decreased to 1-2 %. The peak came earlier on old-growth forest (1-week bolts) than in burnt forest (2-week bolts). The proportion facultative predators made up the smallest proportion overall (Table 1, Figure 1). Forest-type differences were not significant for this functional group, except for 8-week bolts, but the overall pattern was very similar to that of the predators.

The assemblages followed forest-type specific assembly trajectories regarding species composition (Figure 2). The forest types became more and more dissimilar as the community assembly progressed (Figure 2). There was a pronounced difference in species composition in the final state (8-week bolts, burnt forest versus old-growth forest: ANOSIM, $R = 0.42$, $P = 0.001$) and nine species contributed > 4 % each to this difference according to the SIMPER analysis. Old-growth forest was characterized by four cambivores *Dryocoetes autographus* (17.0 % contribution to the dissimilarity according to the SIMPER), *Tetropium castaneum* (9.8 %), *Hylastes cunicularius* (7.5 %), and *Dryocoetes hectographus* (5.2 %); one fire-favoured predator *Phloeonomus sjoebergi* (6.7 %); and one facultative predator *Rhizophagus ferrugineus* (5.2 %). Burnt forest was characterized by two cambivores *Pissodes gyllenhalii* (10.4 %) and *Pityogenes chalcographus* (7.1 %); and one fire-favoured fungivore, *Atomaria pulchra* (4.2 %). Of the 20 most abundant species, two were only reared from burnt forest, the fire-favoured fungivores *A. pulchra* and *Corticaria rubripes*, whereas four species were only reared from old-growth forest, *Dendroctonus micans*, *D. hectographus*, *Rhizophagus dispar*, and *Leptusa pulchella*, (Appendix 1). A total of 25 species were only reared from exposure times shorter than eight weeks (Appendix 1).

In old-growth forest only cambivores were bolt dominants (the most abundant species in a bolt) and *D. autographus* dominated 74 % of the bolts (Figure 3). There was a wider selection of bolt-dominant species in burnt forest where *D. autographus* dominated only 33 % of the bolts. In burnt forest another cambivore, *P. gyllenhalii*, dominated 31 % of the bolts and two fire-favoured fungivores, *A. pulchra* and *C. rubra,* and a fire-favoured predator, *P. sjoebergi*, dominated a total of six bolts (Figure 3).

Sun exposure at the bolt-groups was similar in the two forest types (% sky in the canopy-cover pictures) (B1/OG1: $F_{1,14} = 1.40$, $P = 0.26$; B2/OG2: $F_{1,10} = 0.36$, $P = 0.56$) (Table 2). In OG1 there was a larger living-tree basal area than in B1 ($F_{1,8} = 9.18$, $P = 0.016$) but there was no difference in B2/OG2 ($F_{1,8} = 1.46$, $P = 0.26$) (Table 2). The amount of old CWD was similar in the two forest types (B1/OG1: $F_{1,8} = 0.03$, $P = 0.86$; B2/OG2: $F_{1,8} = 0.05$, $P = 0.83$) (Table 2). The amount of fresh CWD was probably severely underestimated in the burnt sites and is not presented. There was more living Norway spruce in the old-growth sites than in the burnt sites (Table 2).

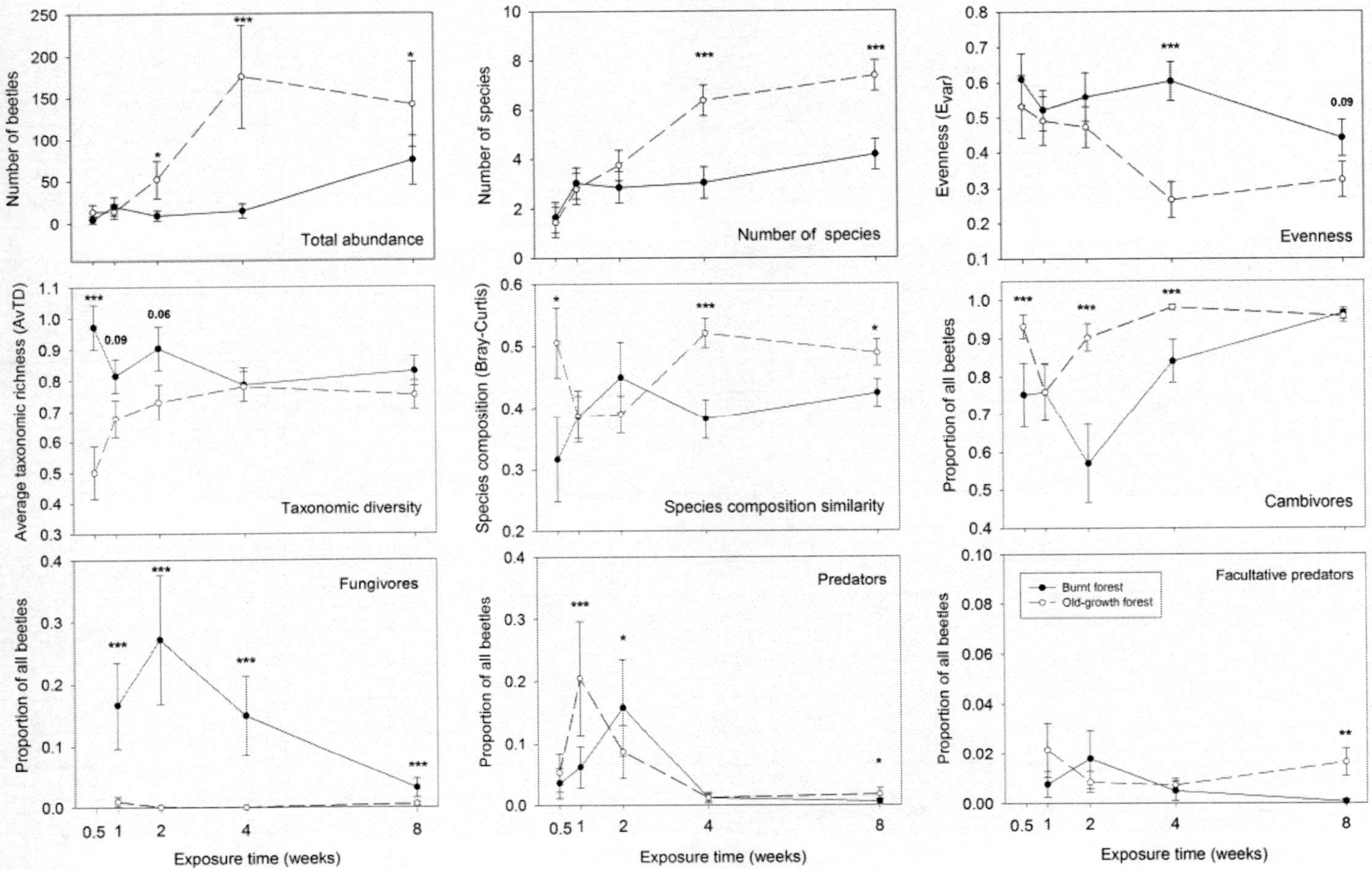

Figure 1. Community structure changes during community assembly. Means and SE from the ANCOVA/ANOVAs. Note different scale on Y-axes for the functional groups. Asterisks denote significant forest-type differences in that exposure treatment (simple effects from the ANCOVAs): *** $P < 0.001$, ** $P < 0.01$, * $P < 0.05$. Differences with a significance level between 0.05 and 0.09 indicated by actual P-value.

Table 1. Results from ANCOVA and ANOVA analyses testing the effects of forest type and exposure time on nine community measures of the beetle assemblage reared from 140 Norway spruce *Picea abies* bolts at four sites in central Sweden. The actual number of bolts used in each test differ (see text for details). Statistically significant differences ($P < 0.05$) are shown in bold

Source of variation	*df*	*F*	*P*	*df*	*F*	*P*	*df*	*F*	*P*
	Total beetle abundance			*Number of species*			*Evenness* (E_{var})		
Forest type	1,121	14.23	**<.001**	1,121	18.65	**<.001**	1,76	11.10	**0.001**
Exposure	4,121	13.58	**<.001**	4,121	20.32	**<.001**	4,76	3.14	**0.02**
Forest type x Exposure	4,121	3.87	**0.005**	4,121	5.90	**<.001**	4,76	2.47	0.05
Bark area	1,121	9.22	**0.003**	1,121	0.60	0.44	1,76	1.06	0.31
Simple effect: burnt forest	4,121	6.26	**<.001**	4,121	3.00	**0.02**	4,76	1.68	0.16
Simple effect: old-growth forest	4,121	12.76	**<.001**	4,121	23.03	**<.001**	4,76	3.86	**0.007**
Simple effect: exposure 0.5 weeks	1,121	1.29	0.26	1,121	0.07	0.79	1,76	0.48	0.49
Simple effect: exposure 1 week	1,121	0.31	0.58	1,121	0.10	0.75	1,76	0.12	0.73
Simple effect: exposure 2 weeks	1,121	6.43	**0.01**	1,121	1.50	0.22	1,76	0.98	0.33
Simple effect: exposure 4 weeks	1,121	23.81	**<.001**	1,121	21.25	**<.001**	1,76	22.06	**<.001**
Simple effect: exposure 8 weeks	1,121	4.96	**0.03**	1,121	19.36	**<.001**	1,76	3.03	0.09
	Taxonomic diversity (AvTD)			*Species composition similarity (Bray-Curtis)*			*Proportion cambivores*		
Forest type	1,74	20.18	**<.001**	1,191	6.43	**0.01**	1,106	76.15	**<.001**
Exposure	4,74	0.52	0.72	4,191	1.54	0.19	4,106	58.72	**<.001**
Forest type x Exposure	4,74	3.34	**0.01**	4,191	2.68	**0.03**	4,106	25.34	**<.001**
Bark area	1,74	0.25	0.62	n.t.	n.t.	n.t.	1,106	13.45	**<.001**
Simple effect: burnt forest	4,74	1.37	0.25	4,191	0.96	0.43	4,106	31.38	**<.001**
Simple effect: old-growth forest	4,74	2.30	0.07	4,191	4.24	**0.003**	4,106	50.17	**<.001**
Simple effect: exposure 0.5 weeks	1,74	18.08	**<.001**	1,191	4.49	**0.04**	1,106	14.50	**<.001**
Simple effect: exposure 1 week	1,74	2.86	0.09	1,191	0.00	0.97	1,106	0.00	0.99
Simple effect: exposure 2 weeks	1,74	3.67	0.06	1,191	0.88	0.35	1,106	66.94	**<.001**
Simple effect: exposure 4 weeks	1,74	0.01	0.93	1,191	12.98	**<.001**	1,106	104.47	**<.001**

Source of variation	*df*	*F*	*P*	*df*	*F*	*P*	*df*	*F*	*P*
	Taxonomic diversity (AvTD)			***Species composition similarity (Bray-Curtis)***			***Proportion cambivores***		
Simple effect: exposure 8 weeks	1,74	1.34	0.25	1,191	4.46	**0.04**	1,106	0.71	0.40
	Proportion fungivores			***Proportion predators***			***Proportion facultative predators***		
Forest type	1,85	110.39	**<.001**	1,106	4.22	**0.04**	1,85	5.53	**0.02**
Exposure	3,85	4.12	**0.009**	4,106	33.46	**<.001**	3,85	2.16	0.10
Forest type x Exposure	3,85	11.73	**<.001**	4,106	6.46	**<.001**	3,85	3.54	**0.02**
Bark area	1,85	24.80	**<.001**	1,106	0.11	0.74	1,85	0.48	0.49
Simple effect: burnt forest	3,85	17.27	**<.001**	4,106	13.61	**<.001**	3,85	2.72	**0.049**
Simple effect: old-growth forest	3,85	5.14	**0.003**	4,106	52.11	**<.001**	3,85	3.84	**0.01**
Simple effect: exposure 0.5 weeks	n.t.	n.t.	n.t.	1,106	0.50	0.48	n.t.	n.t.	n.t.
Simple effect: exposure 1 week	1,85	16.52	**<.001**	1,106	18.69	**<.001**	1,85	2.09	0.15
Simple effect: exposure 2 weeks	1,85	36.17	**<.001**	1,106	5.07	**0.03**	1,85	1.12	0.29
Simple effect: exposure 4 weeks	1,85	45.42	**<.001**	1,106	0.00	0.97	1,85	0.25	0.62
Simple effect: exposure 8 weeks	1,85	22.77	**<.001**	1,106	5.17	**0.02**	1,85	9.81	**0.002**

Notes: n.t. means that that effect is not tested for that particular community measure.

Table 2. Environmental variables at two burnt forest sites (B) and two old-growth forest (OG) sites in central Sweden. Sun exposure is a measure of percent sky (± SE) from photos taken at each of the bolt groups. Basal area, in m^2 / ha, of trees and CWD are means ± SE for five circular plots per study site (10 m radius). Tree species composition (percent basal area) are based on all five plots pooled within study site. Species are: Scots pine (*Pinus sylvestris*) / Norway spruce (*Picea abies.*) / deciduous species (mainly *Betula spp*.)

Variable	B1	OG1	B2	OG2
Area (ha)	30	25	15	25
Sun exposure	54 ± 4.6	48 ± 1.9	62 ± 2.4	65 ± 5.1
Basal area				
living trees	13.3 ± 3.6	29.9 ± 4.1	25.0 ± 3.4	18.5 ± 4.2
old CWD	2.6 ± 1.2	2.3 ± 1.5	4.3 ± 2.1	4.9 ± 1.4
Tree species composition				
living trees	92 / 3 / 5	42 / 44 / 14	100 / 0 / 0	67 / 27 / 6
old CWD	74 / 0 / 9	57 / 18 / 0	100 / 0 / 0	45 / 19 / 14

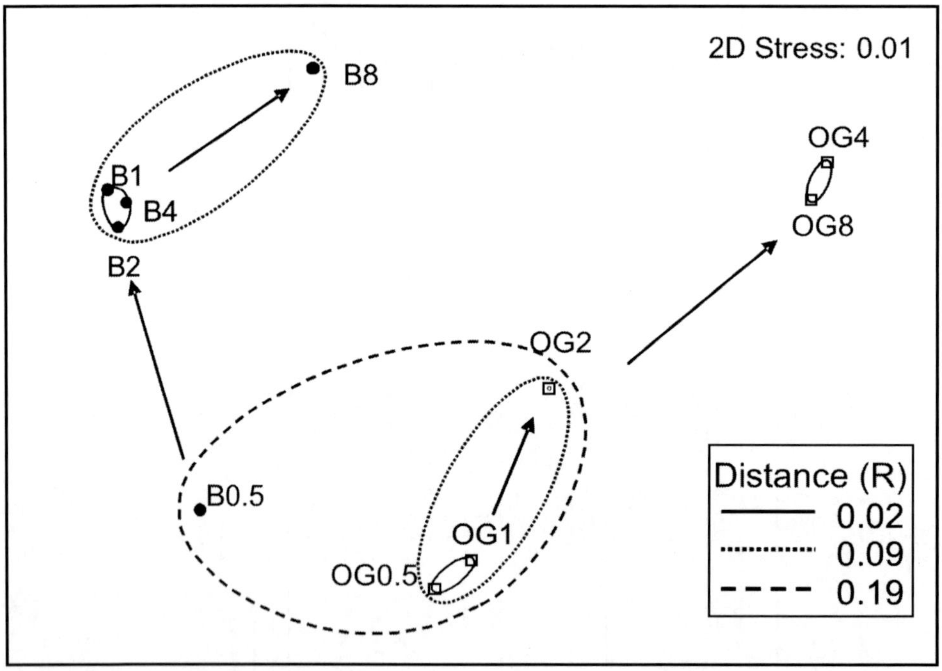

Figure 2. Beetle community assembly in species-composition space. Non-metric multi-dimensional scaling (NMDS) ordination of R-values from an analysis of similarity (ANOSIM) of all treatments (using Bray-Curtis similarity and 4th root transformed abundances). Arrows indicate assembly trajectory along the exposure-time sequence. Euclidian distances represent ANOSIM R-values (ranging from 0 to 1), a measure of how different treatments are from each other in species composition. Circles with R-value distances (from a cluster analysis on the same data) added to increase interpretability. B: burnt-forest bolts; OG: old-growth-forest bolts. Exposure time in weeks: 0.5, 1, 2, 4, 8.

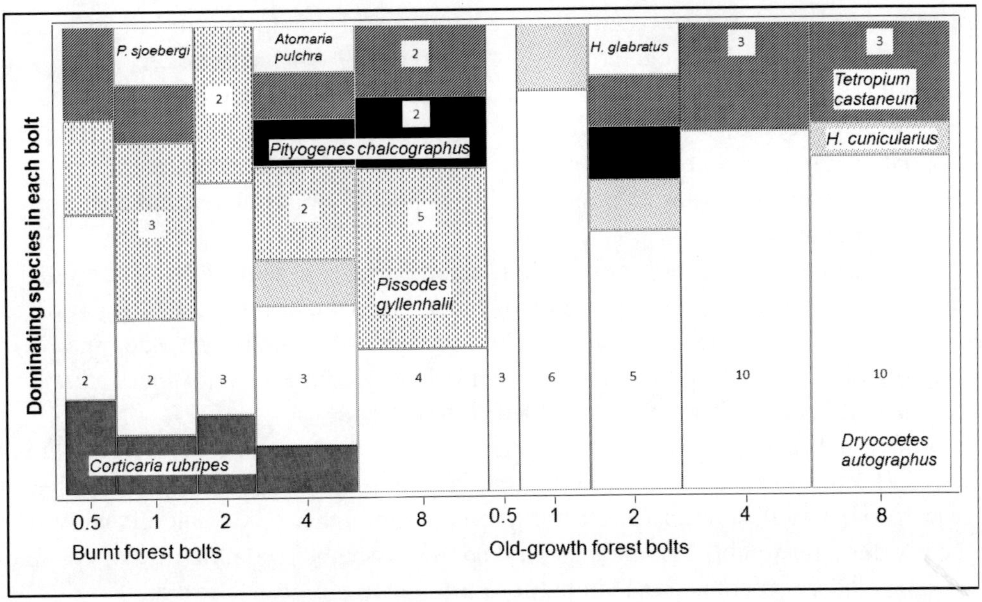

Figure 3. Bolt-dominants in each bolt. Mosaic plot showing proportion of bolts in each treatment that were dominated by a particular species. *Phloeonomus sjoebergi*, *Atomaria pulchra*, and *Hylastes glabratus* dominated one bolt each; all other bolts indicated in white are *Dryocoetes autographus*. Number of bolts indicated (no indication = 1 bolt). Exposure time in weeks: 0.5, 1, 2, 4, 8.

Discussion

Two Types of Community Assembly

Based on the two community types distinguished by Chase (2007), the community in burnt forest conformed to a dispersal-assembled community whereas that in old-growth forest conformed to a niche-assembled community. This is supported by five results.

Firstly, total abundance, the number of beetles emerging per bolt, reached the community-level carrying capacity (the increase levelled off) already after four weeks of exposure in the old-growth sites but not in the burnt sites. Resource competition is important in saproxylic beetles (Anderbrant et al. 1985, Rankin and Borden 1991, Schlyter and Anderbrant 1993, Schroeder and Weslien 1994, Weslien 1994, Denno et al. 1995, Victorsson 2012) and it is likely that larval resource competition created the levelling off in the number of emerging beetles in the old-growth bolts for two reasons: (1) Analogous results in the scolytines *Tomicus piniperda* (Sauvard 1989) and *Ips typographus* (Anderbrant et al. 1985), where the density also levelled off when the number of colonists increased, are interpreted as an effect of larval resource competition that offsets the increase in number of colonists (Dajoz 2000); (2) Competition is expected to be most intense within guilds (Pianka 1980) and in the old-growth 4- and 8-week bolts, where 95 % of all beetles were cambivores, intraguild resource competition among those cambivores probably created the levelling off in the number of emerging beetles. Density-dependent predation would be an alternative explanation but in both forest types predators were important only in the 1- and 2-week bolts whereas in the 4- and 8-week bolts, the predators could not follow the rapid increase in cambivore abundance and those treatments contained only 1-2 % predators.

Secondly, in the burnt forest, species similarity was low in replicate bolts during assembly which is consistent with a dispersal-assembled community where a high patch-to-patch variation in species composition is predicted. In the old-growth forest, on the other hand, species-composition varied less between replicate bolts as is predicted for niche-assembled communities.

Thirdly, the greater selection of bolt-dominating species in burnt forest shows that those assemblages had a less absolute competitive hierarchy, which is predicted for a dispersal-assembled community. The fact that most old-growth forest bolts were dominated by *D. autographus* indicates that the competitive hierarchy was quite strict in that forest type, an observation consistent with a niche-assembled community.

Fourthly, in 4- and 8-week bolts there were twice as many species per bolt in old-growth forest as in burnt forest. This indicates that the available niche space was more filled up in old-growth forest than in burnt forest. Furthermore, the higher taxonomic diversity in burnt forest early during assembly also indicates a dispersal-assembled community, where colonists come from a wider taxonomic range in burnt forest than in old-growth forest.

Fifthly, the higher evenness in burnt forest during assembly indicates an open community with colonization less constrained by species competitive traits since higher evenness means that species get a more equal share of resources. Buddle et al. (2000) also found increased evenness in burnt forest when studying spiders and conclude that this is an effect of lower levels of competition when spiders colonize a competition-relaxed habitat.

Two explanations for the forest-type specific community assembly could be important. Firstly, a wider selection of colonist species in burnt forest could explain the higher bolt-to-bolt variation in that forest type. Insects can colonize CWD from either within the stand, internal colonization, or from the surrounding forest, external colonization. In burnt forest, beetles present in bark-covered CWD (the source of potential colonists) would suffer high mortality during the fire since these trees often are destroyed by the fire (pers. obs. at site B1). Species residing deep inside old dead wood suffer less mortality whereas subcortical species are negatively affected by the fire (Ulyshen et al. 2010). Furthermore, Boulanger et al. (2010) show that the amount of internal CWD in burnt forest is unimportant as a colonist source. Insects are instead attracted to burnt forest from long distances by smoke or infrared radiation (Evans 1962, 1971, Schmitz et al. 1997, Schutz et al. 1999). For example the fire-dependent beetles in the genus *Melanophila* can detect fires at a distance of 130 km (Schmitz and Bousack 2012). The burnt sites therefore receive beetles from many CWD patches in the surrounding landscape. These patches differ in quality and this gives a wide selection of colonizing species which in turn could increase the variation in species composition and bolt dominants. In old-growth forest, internal colonists make up a larger proportion of all colonists since in that forest type the CWD still harbour subcortical colonists. This means that old-growth forest colonists are drawn from a narrower range of within-stand CWD types. The selection of colonizing species in this forest type should therefore be limited compared to that in burnt forest, as seen in the lower taxonomic richness in old-growth forest during assembly. The difference in tree-species composition between the two forest types, with more Norway spruce (the bolt species) in the old-growth sites than in the burnt-forest sites, is therefore not crucial for the results. Regardless of tree-species composition, the colonization mode for burnt forest is still external. (2) A high amount of CWD in the burnt sites could explain why the community-level carrying capacity was not reached in burnt forest. The colonizers arriving at the burnt sites would be spread out over the CWD created by the fire and the densities in individual CWD units would be low. The amount of fresh CWD is generally very high in newly burnt forest (Uotila et al. 2001, Stokland et al. 2012). The amount of CWD was most likely severely underestimated in the burnt sites in my study. A possible reason for this is that the CWD-inventory was done the same year as the fires. At this stage, it is difficult to estimate whether an injured tree will die or survive. Many of the trees classified as living at the time of inventory were probably sufficiently damaged by the fire (primarily by leaf-needle mortality) to be suitable as substrate for the focal species. Furthermore, Wikars and Schimmel (2001), who also did an experiment at Eggåsen, estimate that 50 % of the trees in that site were killed by fire. Also the high abundance of *Pissodes gyllenhalii* in burnt forest in the present study is important in this context as it is normally a rare and local species but can reach high abundance when there is locally a large amount of CWD (e.g. after stormfellings: Ehnström and Axelsson 2002). In my study it characterized burnt forest and was also bolt dominant in 31 % of the bolts in that forest type. *Pissodes gyllenhalii* could therefore indicate that the CWD amount was high in the burnt forest sites. However, further studies are clearly needed to establish whether the amount of CWD is the mechanism behind the forest-type specific community assembly.

Cambivore Specialization

The forest-type convergence in the proportion of cambivores and the forest-type difference in the species identity of those cambivores showed that the cambivore role offers opportunity for forest-type specialization. Patch quality, such as tree species, decay stage, and sun-exposure all affect saproxylic beetles (e.g. Vanderwel et al. 2006, Hjältén et al. 2007). Therefore, I used standardized bolts with standardized sun-exposure at the bolt group. However, the burnt sites were a more structurally open forest-habitat at the stand level, mainly because shrubs and small trees were consumed by the fire. The charred ground in the burnt forest could also increase heat retention. All this could increase temperatures on a forest-stand level in the burnt forest, a difference in microclimate that could be a basis for abiotic filtering. A pattern consistent with abiotic filtering was obvious for four cambivorous species. Old-growth forest had a higher abundance of *Dryocoetes autographus*, *Dryocoetes hectographus* and *Tetropium castaneum*, three species more common in closed forest than in sun-exposed forest (Ehnström and Axelsson 2002), whereas burnt forest had a higher abundance of *Pityogenes chalcographus*, a common species also in clear-cuts, a very open habitat (Ehnström and Axelsson 2002). *Pissodes gyllenhalii* also preferred burnt forest and the most important factor for this species could be the CWD amount as previously explained. Furthermore, as could be expected the fire-favoured cambivorous cerambycid *Rhagium inquisitor* was more abundant in burnt forest.

Erased Importance of Functional Groups

Fungivores, predators and facultative predators were all more important early during assembly than in the first-year final states. Their importance had become "erased" (as discussed in Drake 1991) and would not have been detected in a study of first-year final states only. This clearly points to the importance of community assembly studies in order to understand natural communities. In the present study fungivores made up 15-27 % of the individuals during the 1- to 4-week exposures in burnt forest whereas in old-growth forest they made up less than 1 %. Furthermore, the two fire-favoured fungivores, *Atomaria pulchra* and *Corticaria rubripes,* were even bolt dominants in burnt forest. This indicates that burnt forest is even more important for fungivores than previous studies on first-year final states indicate (Wikars 2002). Fungivorous beetles is a group that suffer from habitat fragmentation in managed forest (Komonen et al. 2000, Victorsson and Jonsell 2013) and prescribed burning of forest might be especially important for this group.

The pattern for the predators was driven by the staphylinids *Phloeonomus sjoebergi* and *Phloeonomus planus* since they were the most abundant predators. The prey choice for these species is scolytine eggs and larvae (Palm 1951, Nuorteva 1956). Likewise *Thanasimus formicarius* is a predator known to feed on scolytines (Kenis et al. 2004). The predator proportion was initially high in both forest types but later decreased as the number of scolytines increased. Such a pattern has been described by Aukema and Raffa (2004) as a predator-diluting effect contingent on scolytine biology. They conclude that the predators are unable to track the rapid, pheromone-mediated, increase of the scolytines. The pattern for the predators in the present study could have the same explanation. The pattern for the facultative predators was a scaled-down copy of the predator pattern. Among these species, the three *Rhizophagus* species as well as the *Epurea* species are associated with scolytines (Kenis et al. 2004) as predators and detritivores of dead scolytines and exuviae. It is noteworthy that the predator *P. sjoebergi* preferred old-growth forest (129 individuals) over burnt forest (37) even though it is fire-favoured (Wikars 2006). Predators must of course follow their prey and the much higher abundance of scolytines in old-growth forest compared to burnt forest could explain this result.

Fire-Dependent Species

The result that the community-level carrying capacity was not reached in burnt forest suggests that burnt forest can be a competition-reduced habitat for saproxylic beetles, as has previously been proposed (Esseen et al. 1992, Wikars 1992). Obviously, given an exposure time long enough, the community-level carrying capacity would be reached also in the burnt sites. However, the longest exposure time in the present study included most of the remaining season for the focal species (Vanderwel et al. 2006). In natural CWD, colonization would restart the following year (Boulanger and Sirois 2007) and the community-level carrying capacity might be reached also in burnt forest. However, even if larval resource competition becomes important in burnt forest over the years following the fire, my results suggest that it is unimportant during the fire-year.

Some beetle species are adapted to the burnt-forest habitat and depend on burnt forest for their reproduction. Such pyrophilous (fire loving) species are threatened by the lack of burnt forest in areas dominated by forestry (Wikars 1992, Esseen et al. 1997, Wikars 2002). Fire-adaptation in beetles can include long-range dispersal capacity, the ability to develop in sun-exposed CWD, specialized senses for fire-detection, a propensity to mate near still glowing fire, and possibly coloration (Evans 1964, Schmitz et al. 1997, Schutz et al. 1999, Allison et al. 2004, Evans 2010, Schmitz and Bousack 2012). I found two fire-dependent and nine fire-favoured species in burnt forest (sensu Wikars 2006). Such species generally prefer scorched CWD (Wikars 2002, Johansson et al. 2007) but the results presented here, where I used unscorched bolts, show that also factors other than scorching are important. The results in this chapter indicate that the dispersal-assembled community in burnt forest could be important. Pyrophilous species could invade these competition-reduced habitats and gain priority over more competitive, later-arriving species. In a niche-assembled community in old-growth forest such species might not be able to establish due to poor competitive abilities. Priority effects are important in saproxylic beetles (Weslien et al. 2011, Victorsson 2012) and arrival order could be crucial for pyrophilous species.

Conclusion

Community assembly differed in many respects between burnt and old-growth forest. The saproxylic beetle assemblage in old-growth forest conformed to a niche-assembled community (*sensu* Chase 2007) since the community-level carrying capacity was reached during assembly and most bolts were dominated by the same species, the cambivorous scolytine *D. autographus*. In contrast, the beetle assemblage in burnt forest conformed to a dispersal-assembled community (*sensu* Chase 2007) since the community-level carrying capacity was not reached and the bolts were dominated by a variety of species. Furthermore, there was a wider bolt-to bolt variation in species composition and a higher evenness in burnt forest. Additionally, the burnt forest bolts harboured only half the number of species per bolt as the old-growth bolts which could indicate that the available niche space was not filled up in burnt forest.

These results indicate that pyrophilous species could be adapted not only to the burnt wood *per se* but also to the dispersal-assembled community type developing in burnt forest. To conserve pyrophilous species it seems imperative that prescribed burning creates competition-reduced habitats where pyrophilous species can develop. In this respect, the practice of burning clear-cuts as a conservation measure seems less than perfect. If a large amount of CWD is created on the clear-cut, a prescribed burning might have some value but the importance of burning intact forest cannot be overemphasized.

Acknowledgments

Lars-Ove Wikars directed me towards the possibility of reduced competition in burnt forest. Mats Björklund, Stefan Ås, and Anders Ödeen offered valuable criticism on earlier drafts. Olof Hedgren, Anna Sandberg, Maria Forslund, and Lina Nittérus assisted in the field. Stig Lundberg and Gustav Carlander determined the beetles. StoraEnso provided bolts and study sites.

APPENDIX 1.

Abundance of all saproxylic beetles reared from 140 Norway spruce *Picea abies* bolts at two old-growth and two burnt forest sites in central Sweden. The bolts were exposed for colonization for 0.5, 1, 2, 4, or 8 weeks. The species biology, functional group and relation to fire, were taken from the literature (see text for details). Red list categories are from the Swedish national red list (threat category and publication year in parenthesis after the species name).

			Burnt forest					Old-growth forest					
Family	Species	Biology	0.5 w	1 w	2 w	4 w	8 w	0.5 w	1 w	2 w	4 w	8 w	Total
Ptiliidae	*Pteryx suturalis*	f					1					1	2
Staphylinidae	*Acrulia inflata*	f										3	3
	Hapalarea linearis	p										2	2
	Phloeonomus planus	p	2	2	1	1	1	6	21	12	11	10	67
	Phloeonomus lapponicus	p, FF		1	1	1						5	8
	Phloeonomus sjoebergi	p, FF	3	12	16	2	4	4	8	45	38	34	166
	Coryphium angusticolle									1			1
	Dadobia immersa	f					2						2
	Dinaraea aequata						2						2
	Dinaraea linearis										1		1
	Leptusa pulchella	f								1	1	7	9
	Placusa tachyporoides	p		1									1
	Placusa incompleta	p							5				5
	Placusa suecica	p		1	2				5	2	1		11
	Placusa atrata	p, FF		1						1			2
Cantharidae	*Malthodes brevicollis*	p						1					1
Anobiidae	*Ptinus villiger*		1										1
	Ptinus subpilosus	f							1			2	3

			Burnt forest					Old-growth forest					
Family	**Species**	**Biology**	**0.5 w**	**1 w**	**2 w**	**4 w**	**8 w**	**0.5 w**	**1 w**	**2 w**	**4 w**	**8 w**	**Total**
Cleridae	*Thanasimus formicariaus*	p								5	1	1	7
Nitidulidae	*Epuraea angustula*	fac			1	1			1			2	5
	Epuraea boreella	fac		1		1		3	3	5	7	11	31
	Epuraea pygmaea	fac									1	1	2
	Epuraea rufomarginata	fac		1									1
	Glischrochilus quadripunctatus	fac,FF		1									1
Monotomidae	*Rhizophagus ferrugineus*	fac					1				12	26	39
	Rhizophagus dispar	fac							2	2	1	4	9
	Rhizophagus parvulus	fac			1						1		2
Silvanidae	*Silvanus bidentatus* (NT 2005)	fac, FF			1								1
Cryptophagidae	*Henoticus serratus*	FD			1								1
	Cryptophagus corticinus	f, FD	1										1
	Atomaria morio	f							1				1
	Atomaria fuscipes	f					1						1
	Atomaria elongatula	f			1	1							2
	Atomaria pulchra	f, FF	3	10	12	15	11						51
Latridiidae	*Corticaria rubripes*	f, FF	13	23	11	18	3						68
	Corticaria ferruginea	f, FF	1	5			1						7
Salpingidae	*Salpingus ruficollis*	p							1				1
Cerambycidae	*Tetropium castaneum*	c	11	33		15	13			41	197	208	518
	Tetropium fuscum	c									1		1
	Rhagium inquisitor	c, FF		4	5	7	2		1			1	20
	Pogonocherus fasciculatus	c									2		2
Curculionidae	*Hylobius piceus*	c										1	1
	Pissodes gyllenhalii	c	10	128	29	67	210		13	45	29	51	582

(Continued)

Family	Species	Biology	Burnt forest					Old-growth forest					Total
			0.5 w	1 w	2 w	4 w	8 w	0.5 w	1 w	2 w	4 w	8 w	
Scolytinae	*Hylurgops glabratus*	c							1	16	25		42
	Hylurgops palliatus	c						1	1				2
	Hylastes brunneus	c		1			1				2		4
	Hylastes cunicularius	c	1	3	1	7	19	1	7	25	77	74	215
	Xylechinus pilosus	c								1			1
	Tomicus piniperda	c		1									1
	Dendroctonus micans	c									28		28
	Pityogenes chalcographus	c		2		16	702			67			787
	Pityogenes bidentatus	c					2						2
	Dryocoetes autographus	c	39	119	77	111	163	205	149	478	2148	1652	5141
	Dryocoetes hectographus	c							22		256	234	512
	Dryocoetes spp.	c	1	4	1	2	10	9	13	28	146	209	423
	Crypturgus pusillus	c								1			1
	Trypodendron lineatum	f										1	1
	Total		86	354	161	265	1149	230	255	776	2986	2540	8802

Abbreviations: w, weeks of exposure; NT, near threatened; c, cambivorous; f, fungivorous; p, predator; fac, facultative predator (also feeds as a detritivore or fungivore).FD, fire-dependent; FF, fire-favoured.

REFERENCES

Allison, J. D., Borden, J. H. & Seybold, S. J. (2004). A review of the chemical ecology of the Cerambycidae (Coleoptera). *Chemoecology*, *14*, 123-150.

Anderbrant, O., Schlyter, F. & Birgersson, G. (1985). Intraspecific competition affecting parents and offspring in the bark beetle *Ips typographus*. Oikos, *45*, 89-98.

Aukema, B. H. & Raffa, K. F. (2004). Does aggregation benefit bark beetles by diluting predation? Links between a group-colonisation strategy and the absence of emergent multiple predator effects. *Ecological Entomology*, *29*, 129-138.

Boulanger, Y. & Sirois, L. (2007). Postfire succession of saproxylic arthropods, with emphasis on coleoptera, in the north boreal forest of Quebec. *Environmental Entomology*, *36*, 128-141.

Boulanger, Y., Sirois, L. & Hébert, C. (2010). Distribution of saproxylic beetles in a recently burnt landscape of the northern boreal forest of Québec. *Forest Ecology and Management*, *260*, 1114-1123.

Buddle, C. M., Spence, J. R. & Langor, D. W. (2000). Succession of boreal forest spider assemblages following wildfire and harvesting. *Ecography*, *23*, 424-436.

Chase, J. M. (2007). Drought mediates the importance of stochastic community assembly. *Proceedings of the National Academy of Sciences of the United States of America*, *104*, 17430-17434.

Clarke, K. R. & Gorley, R. N. (2006). PRIMER v6: user manual/tutorial. PRIMER-E Ltd., Plymoth.

Clarke, K. R. & Warwick, R. M. (2001). Change in marine communities: an approach to statistical analysis and interpretation. PRIMER-E Ltd., Plymoth.

Dajoz, R. (2000). Insects and forests: the role and diversity of insects in the forest environment. Intercept, Paris.

Denno, R. F., Mcclure, M. S. & Ott, J. R. (1995). Interspecific interactions in phytophagous insects - competition reexamined and resurrected. *Annual Review of Entomology*, *40*, 297-331.

Diamond, J. M. (1975). Assembly of species communities. Pages 342-444 *in* M. L. Cody and J. M. Diamond, editors. Ecology and evolution of communities. Belknap Press of Harvard University Press, Cambridge, Massachusetts.

Drake, J. A. (1991). Community assembly mechanics and the structure of an experimental species ensemble. *American Naturalist*, *137*, 1-26.

Drake, J. A., Zimmerman, C. R., Puruker, T. & Rojo, C. (1999). On the nature of the assembly trajectory. Pages 233-250 *in* E. Weiher and P. Keddy, editors. Ecological assembly rules: perspectives, advances, retreats. Cambridge University Press, Cambridge.

Ehnström, B. & Axelsson, R. (2002). Insektsgnag i bark och ved. ArtDatabanken, SLU, Uppsala.

Esseen, P. A., Ehnström, B., Ericson, L. & Sjöberg, K. (1992). Boreal forests-the focal habitats of Fennoscandia. Pages 252-325 *in* L. Hansson, editor. Ecological principles of nature conservation. Elsevier Applied Science, Amsterdam.

Esseen, P. A., Ehnström, B., Ericson, L. & Sjöberg, K. (1997). Boreal forests. *Ecological Bulletins*, *46*, 16-47.

Evans, W. G. (1962). Notes on the biology and dispersal of *Melanophila*. The Pan-Pacific Entomologist, *38*, 59-63.

Evans, W. G. (1964). Infra-red receptors in *Melanophila acuminata* DeGeer. Nature, *202*, 211.

Evans, W. G. (1971). The attraction of insects to forest fires. Pages 115-127 *in* Tall Timbers conference on ecological animal control by habitat management, proceedings, Tall Timbers Research Station, Tallahasse, Florida.

Evans, W. G. (2010). Reproductive Role of Infrared Radiation Sensors of *Melanophila acuminata* (Coleoptera: Buprestidae) at Forest Fires. *Annals of the Entomological Society of America*, *103*, 823-826.

Fridman, J. & Walheim, M. (2000). Amount, structure, and dynamics of dead wood on managed forestland in Sweden. *Forest Ecology and Management*, *131*, 23-36.

Hjältén, J., Johansson, T., Alinvi, O., Danell, K., Ball, J. P., Pettersson, R., Gibb, H. & Hilszczanski, J. (2007). The importance of substrate type, shading and scorching for the attractiveness of dead wood to saproxylic beetles. *Basic and Applied Ecology*, *8*, 364-376.

Hjältén, J., Stenbacka, F. & Andersson, J. (2010). Saproxylic beetle assemblages on low stumps, high stumps and logs: Implications for environmental effects of stump harvesting. *Forest Ecology and Management*, *260*, 1149-1155.

Hubbell, S. P. (2001). The unified neutral theory of biodiversity and biogeography. Princeton University Press, Princeton, NJ.

Jenkins, D. G. & Buikema, A. L. (1998). Do similar communities develop in similar sites? A test with zooplankton structure and function. *Ecological Monographs*, *68*, 421-443.

Johansson, T., Hjältén, J., Gibb, H., Hilszczan´ski, J., Stenlid, J., Ball, J. P., Alinvi, O. & Danell, K. (2007). Variable response of different functional groups of saproxylic beetles to substrate manipulation and forest management: implications for conservation strategies. *Forest Ecology and Management*, *242*, 496-510.

Kenis, M., Wermelinger, B. & Grégoire, J. C. (2004). Research on parasitoids and predators of scolytidae, a review. Pages 237-290 *in* F. Lieutier, K. R. Day, A. Battisti, J.-C. Grégoire, and H. F. Evans, editors. Bark and wood boring insects in living trees in Europe, a synthesis. Springer, Dordrecht.

Komonen, A., Penttilä, R., Lindgren, M. & Hanski, I. (2000). Forest fragmentation truncates a food chain based on an old-growth forest bracket fungus. *Oikos*, *90*, 119-126.

Koskela, H. & Hanski, I. (1977). Structure and succession in a beetle community inhabiting cow dung. *Annales Zoologici Fennici*, *14*, 204-223.

Kuusela, S. & Hanski, I. (1982). The structure of carrion fly communities - the size and the type of carrion. *Holarctic Ecology*, *5*, 337-348.

Lieutier, F., Day, K. R., Battisti, A., Grégoire, J. C. & Evans, H. F. editors. (2004). Bark and wood boring insects in living trees in Europe, a synthesis. Springer, Dordrecht.

Littell, R., Milliken, G., Stroup, W., Wolfinger, R. & Schabenberger, O. (2006). SAS for Mixed Models. 2 edition. SAS Press, Cary.

Lundberg, S. & Gustafsson, B. (1995). Catalogus Coleoperorum Sueciae. Naturhistoriska riksmuseet & Entomologiska föreningen Stockholm, Stockholm.

Magurran, A. E. (2004). Measuring biological diversity. Blackwell Publishing, London.

Martikainen, P., Siitonen, J., Kaila, L., Punttila, P. & Rauh, J. (1999). Bark beetles (Coleoptera, Scolytidae) and associated beetle species in mature managed and old-growth boreal forests in southern Finland. *Forest Ecology and Management*, *116*, 233-245.

McCune, B. & Grace, J. B. (2002). Analysis of ecological communities. MJM Software Design, Gleneden Beach, Oregon.

Niklasson, M. & Granström, A. (2000). Numbers and sizes of fires: Long-term spatially explicit fire history in a Swedish boreal landscape. *Ecology*, *81*, 1484-1499.

Nuorteva, M. (1956). Über den Fichtenstamm-Bastkäfer, *Hylurgops palliatus* Gyll., und seine Insektenfeinde. *Acta Entomologica Fennica*, *13*, 1-116.

Palm, T. (1951). Die Holz- und Rinden-Käfer der nordschwedischen Laubbäume. Esselte, Stockholm.

Pianka, E. R. (1980). Guild structure in desert lizards. *Oikos*, *35*, 194-201.

Quinn, G. P. & Keough, M. J. (2002). Experimental design and data analysis for biologists. Cambridge University Press, Cambridge.

Rankin, L. J. & Borden, J. H. (1991). Competitive interactions between the Mountain pine-beetle and the Pine engraver in Lodgepole pine. *Canadian Journal of Forest Research-Revue Canadienne De Recherche Forestiere*, *21*, 1029-1036.

Sauvard, D. (1989). Multiplication capacities of *Tomicus piniperda* L. (Col, Scolytidae).1. Effects of attack density. *Journal of Applied Entomology-Zeitschrift Fur Angewandte Entomologie*, *108*, 164-181.

Schlyter, F. & Anderbrant, O. (1993). Competition and niche separation between two bark beetles - existence and mechanisms. *Oikos*, *68*, 437-447.

Schmitz, H., Bleckmann, H. & Murtz, M. (1997). Infrared detection in a beetle. *Nature*, *386*, 773-774.

Schmitz, H. & Bousack, H. (2012). Modelling a historic oil-tank fire allows an estimation of the sensitivity of the infrared receptors in pyrophilous *Melanophila* beetles. *Plos One 7*.

Schroeder, L. M. & Weslien, J. (1994). Interactions between the phloem-feeding species *Tomicus piniperda* (Col, Scolytidae) and *Acanthocinus aedilis* (Col, Cerambycidae), and the predator *Thanasimus formicarius* (Col, Cleridae) with special reference to brood production. *Entomophaga 39*, 149-157.

Schutz, S., Weissbecker, B., Hummel, H. E., Apel, K. H., Schmitz, H. & Bleckmann, H. (1999). Insect antenna as a smoke detector. *Nature*, *398*, 298-299.

Siitonen, J. (2001). Forest management, coarse woody debris and saproxylic organisms: Fennoscandian boreal forests as an example. *Ecological Bulletins*, *49*, 11-42.

Smith, B. & Wilson, J. B. (1996). A consumer's guide to evenness indices. *Oikos*, *76*, 70-82.

Stokland, J. N., Siitonen, J. & Jonsson, B. G. (2012). Biodiversity in dead wood. Cambridge University Press, Cambridge.

Ulyshen, M. D., Horn, S., Barnes, B. & Gandhi, K. J. K. (2010). Impacts of prescribed fire on saproxylic beetles in loblolly pine logs. *Insect Conservation and Diversity*, *3*, 247-251.

Uotila, A., Maltamo, M., Uuttera, J. & Isomäki, A. (2001). Stand structure in semi-natural and managed forests in eastern Finland and Russian Karelia. *Ecological Bulletins*, *49*, 149-158.

Vanderwel, M. C., Malcolm, J. R., Smith, S. A. & Islam, N. (2006). Insect community composition and trophic guild structure in decaying logs from eastern Canadian pine-dominated forests. *Forest Ecology and Management*, *225*, 190-199.

Warren, P. H., Law, R. & Weatherby. A. J. (2003). Mapping the assembly of protist communities in microcosms. *Ecology*, *84*, 1001-1011.

Warwick, R. M. & Clarke, K. R. (1995). New "biodiversity" measures reveal a decrease in taxonomic distinctness with increasing stress. *Marine Ecology-Progress Series*, *129*, 301-305.

Weiher, E. & Keddy, P. editors. (1999). Ecological assembly rules: perspectives, advances, retreats. Cambridge University Press, Cambridge, UK.

Weslien, J. (1994). Interactions within and between species at different densities of the bark beetle *Ips typographus* and its predator *Thanasimus formicarius*. Entomologia Experimentalis et Applicata, *71*, 133-143.

Weslien, J., Djupström, L. B., Schroeder, M. & Widenfalk, O. (2011). Long-term priority effects among insects and fungi colonizing decaying wood. *Journal of Animal Ecology*, *80*, 1155-1162. .

Victorsson, J. (2012). Semi-field experiments investigating facilitation: arrival order decides the interrelationship between two saproxylic beetle species. *Ecological Entomology*, *37*, 395-401.

Victorsson, J. & Jonsell, M. (2013). Effects of stump extraction on saproxylic beetle diversity in Swedish clear-cuts. *Insect Conservation and Diversity*, *6*, 483-493.

Wikars, L. O. (1992). Skogsbränder och insekter. *Entomologisk Tidskrift*, *113*, 1-11.

Wikars, L. O. (2002). Dependence on fire in wood-living insects: an experiment with burned and unburned spruce and birch logs. *Journal of Insect Conservation*, *6*, 1-12.

Wikars, L. O. (2006). Åtgärdsprogram för bevarande av brandinsekter i boreal skog.

Wikars, L. O. & Schimmel, J. (2001). Immediate effects of fire-severity on soil invertebrates in cut and uncut pine forests. *Forest Ecology and Management*, *141*, 189-200.

Zackrisson, O. (1977). Influence of forest fires on the North Swedish boreal forest. *Oikos*, *29*, 22-32.

In: Beetles: Biodiversity, Ecology and Role in the Environment ISBN: 978-1-63463-380-2
Editor: Camilla Stack

Chapter 3

ASSESSING THE POTENTIAL ROLE OF GROUND BEETLES (COLEOPTERA) AS ECOLOGICAL INDICATORS IN TROPICAL ECOSYSTEMS: A REVIEW

Reinaldo Lucas Cajaiba[1,*], Eduardo Périco[2], João Alexandre Cabral[3] and Mário Santos[3]

[1]Department of Environment-SEMMA, Uruará-PA, Brazil
[2]Ecology and Remote Sensing, University Center Univates, Lajeado-RS, Brazil
[3]Laboratory of Applied Ecology, CITAB - Centre for the Research and Technology of Agro-Environment and Biological Sciences, University of Trás-os-Montes e Alto Douro, Vila Real, Portugal

ABSTRACT

Emerging concerns with environmental issues has led researchers to look for indicators capable of providing information on the condition of ecosystems. Several animal species have been used for this role. Among these organisms, ground-beetles were found to be appropriate indicators, since they are adapted to all terrestrial habitats, being associated with ecosystem function and sensitive to ongoing changes. Several families of ground beetles, e.g., Carabidae, Cicindelidae, Scarabaeidae, Cerambycidae, Histeridae, Staphylinidae and Curculionidae were used in diverse ecological assessment studies due to its main characteristics, such as habitat fidelity and taxonomic diversity, and because they provide cheap and easy measurements from functionally diverse trophic chains.

Although the growing number of studies on ecological indicators little is known about the ecological patterns in the communities of ground beetles of tropical ecosystems. The present study provides a review on the use of ground beetles as ecological indicators in tropical ecosystems, with focus on the Brazilian Amazon region. We discuss its use for monitoring the effects of anthropogenic induced changes, namely the conversion of forests to agriculture and farming practices, among others on the functioning of tropical systems. We also highlight the major advances and applications employed, as well as proposals for future studies with this group.

* Corresponding author: reinaldocajaiba@hotmail.com.

Keywords: Coleoptera, ecological indicators, tropical ecosystems

INTRODUCTION

Human action has caused directly or indirectly, the reduction and loss of biodiversity (Halffter, 2005) due to the change of the natural ecological processes and the emergence of new combinations of species, mainly by the habitats transformation, fragmentation and damage (Marques et al., 2002). Several factors have been identified including the suppression of natural habitats, ecosystem fragmentation and overexploitation of natural resources, climate change and the introduction of alien species. However, all these factors have an intimate relation with a primary cause, population growth (Ibama, 2004). This biodiversity loss threatens the sustainability of the entire system, with the consequent reduction of the environmental services, such as pest control, nutrient cycling and the maintenance of soil structure. In many situations the changes are enhanced by the use of pesticides, fertilizers and inadequate soil management that can generate even more negative impacts on the biota (Giller et al., 1997).

Insects have been shown to be appropriate ecological indicators for evaluating changes in the environment, due to several characteristics that they possess (Holloway et al., 1981). An excellent group are beetles (Coleoptera), the largest group of insects, with about four hundred thousand known species, distributed in more than a hundred families occupying practically all ecosystems and feeding on a variety food sources (Audino et al., 2007).Our knowledge of the diversity and distribution of species in the Amazon biome is still at an early stage of development (Silva et al, 2005; Peres, 2005). Although Amazonia has welcomed a steady stream of scientists since the beginning of the 18th century and a huge number of books and articles on its diversity are available (e.g. Pitman et al., 2007), they give a very fragmented picture of the region. Much of the region is still little explored by scientists and many of the specimens that have been collected to date have not yet been studied (Silva et al., 2005; Verweij et al., 2009). Understanding the extent and importance of insect diversity in the humid tropics is one of the biggest challenges in modern ecology (Godfray et al., 1999). The fact that only a fraction of tropical insects (certainly less than 20%) has been properly described is certainly not the least of the problems in this respect (Verweij et al., 2009).

Currently there is much speculation about the effects of human generated disturbance on the biodiversity of tropical forests, in particular, the impacts on species composition and on the ecological services provided (Whitmore and Sayer, 1992; Laurance and Bierregaard, 1997).

This chapter describes the ecological importance of tropical forests (with an emphasis on the Brazilian Amazon), the main causes of deforestation and the impacts caused by such changes. We examine the use of soil invertebrates as ecological indicators, with a focus on the Soil Surface Dwelling Beetles (SSDB), considering their sensitivity to landscape changes and their abundance. We discuss the advantages and disadvantages of its use to monitor the effects of anthropogenic changes on the functioning of tropical systems. We also highlight the major advances and applications employed, as well as proposals for future studies with this group.

CAUSES AND EFFECTS OF AMAZON DEFORESTATION

The Amazonian has circa 40% of the world's remaining tropical rainforests (Laurance et al., 2001) being considered a hotspot in biodiversity but also important for carbon storage and regional/global regulation of the hydrological cycles and climate (Fernandes, 2009).

Located in northern South America, the Amazon rainforest occupies 5.5 million square kilometers, spread over territories of Brazil, Venezuela, Colombia, Peru, Bolivia, Ecuador, Suriname, Guyana and French Guiana (Rivero and Covas, 2013). In Brazil, this type of ecosystem, occurs in the states of Amazonas, Amapa, Rondônia, Acre, Pará, Roraima and part of Mato Grosso, Tocantins e Maranhão (Figure 1).

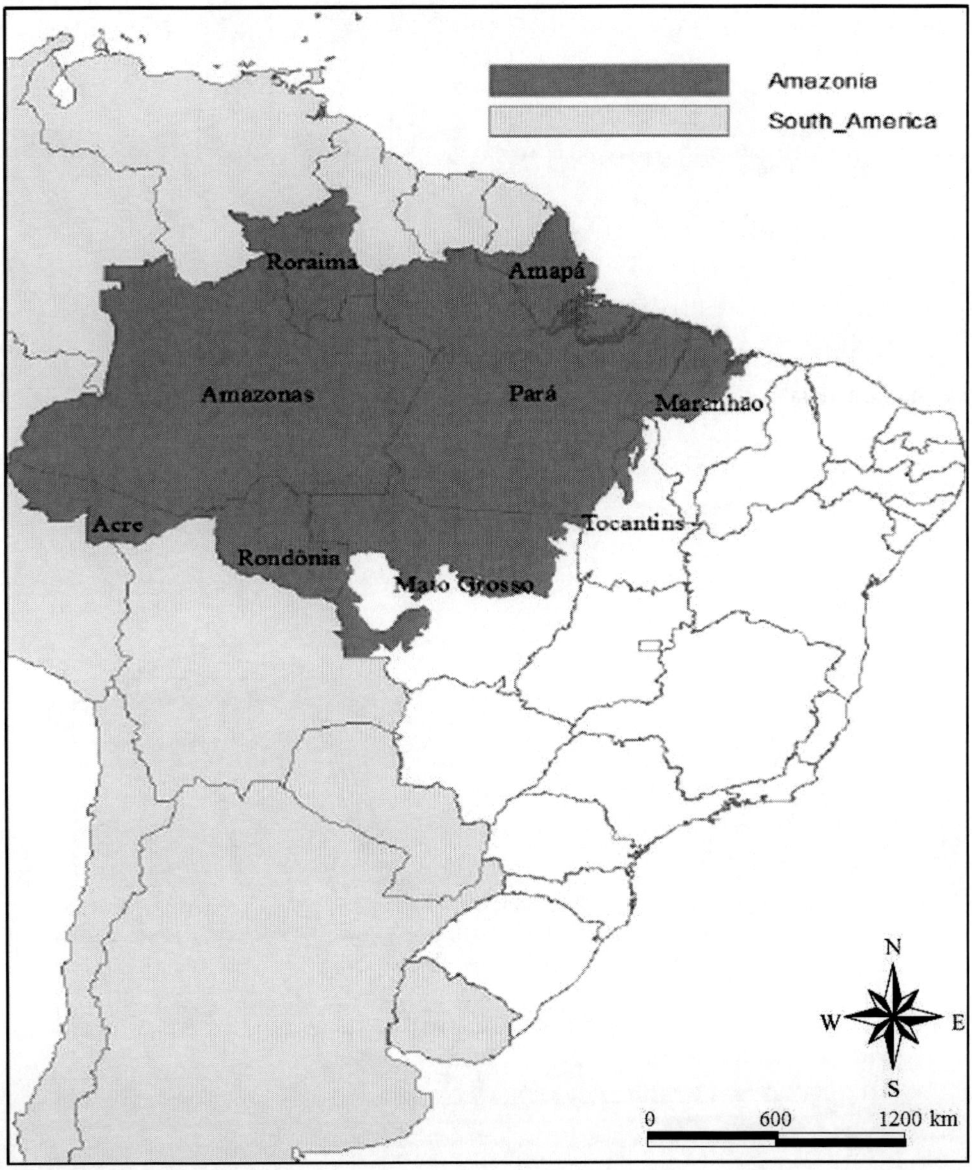

Figure 1. Location of the Brazilian Amazon (Amazonia) in South America (adapted from Souza-Rodrigues, 2011).

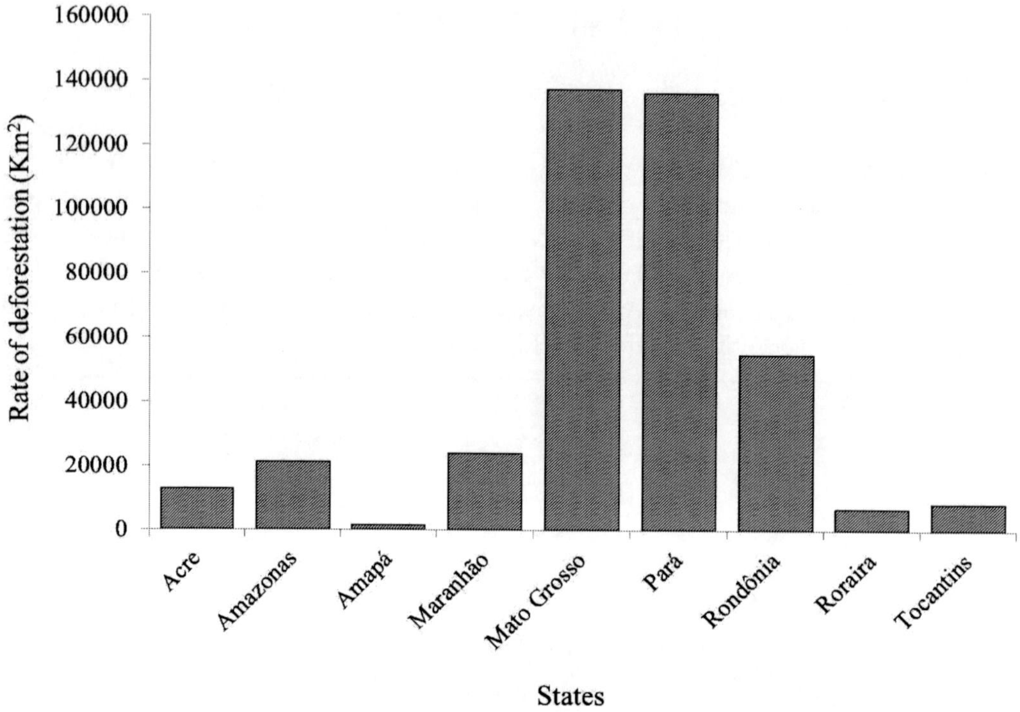

Figure 2.Cumulative rate of deforestation (Km^2) between 1988-2013 in the Brazilian Amazon (Inpe, 2014).

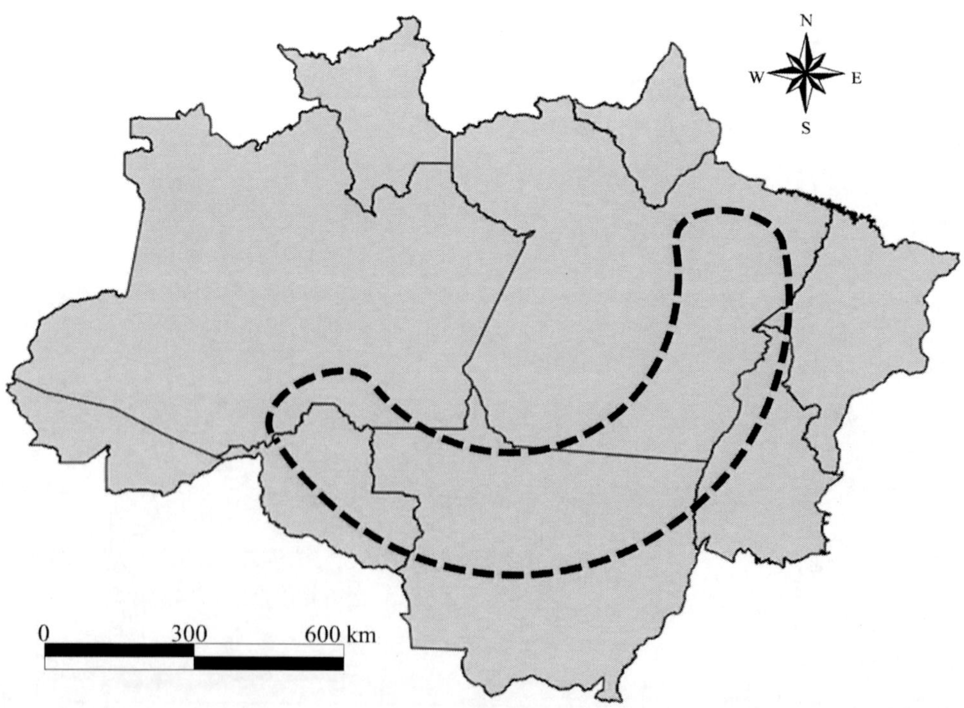

Figure 3. "Arc of deforestation" in the Brazilian Amazon - in dotted lines (adapted from Lui and Molina, 2009).

The deforestation of the Brazilian Amazon has attracted considerable attention over the last two decades (Souza-Rodrigues, 2011). According to satellite images, between 1988 to 2013 the accumulated rate of deforestation was 402.615 Km^2, divided by the states of Mato Grosso (137,251 Km^2), Pará (136,127 Km^2) and Rondônia (54,772 Km^2) (Inpe, 2014) (Figure 2). Although large areas remain intact, the rate of forest loss is dramatic and future projections simulate its increase (Fearnside 2005a; Soares-Filho et al., 2005).

The current deforestation has focused on the axis set to the "Arc of deforestation" (Becker, 2004) (that goes from the south of Pará, passing to the north of Mato Grosso, Rondônia and southwest of Amazonas until it reaches Acre) (Figure 3), where deforestation has been buoyed by land speculation and, more recently, by increasing the areas occupied by exports such as soy and beef (Fernandes, 2009).

The process of deforestation in the Amazon was molded by settlement patterns of the past and driven by different factors. There are, however, temporal and spatial variations in deforestation (Laurance et al., 2001b) and the causes and dynamics are very distinctive for different locations. In fact it is difficult to generalize what drives or accurately predict the extension of the phenomena (Fernandes, 2009).

The causes of deforestation change constantly, being mostly associated with socio-economic pressures. The main is the rise in the population, previously dominated by economic extractive activities and family farming. Indeed, the population of the Amazon has increased considerably, from 2.5 million in 1960 to more than 20 million in 2000 (Ibge, 2000) further increasing the pressure on forests (Fernandes, 2009). Additionally new economic activities begin to stand out: ranching, logging, mining and intensive agriculture, besides the construction of hydropower plants (Angelo and Sá, 2007).

The advance of soy plantations in the region represents a major threat, encouraged by massive government investments in infrastructures, such as waterways, railroads and highways (Fearnside, 2001). The Brazilian Amazon produces more than a third of the country's soy and the state of Mato Grosso accounts for more than 20% of the national area of soybean cultivation (Ibge, 2006). The expansion of soy in the Amazon states totaled 14.1% per year (1990-2005) with major increases in the last five years. The expansion of soybean production is causing serious deforestation, through the dynamics encompassed by forest burning/logging, implementation of livestock and subsequent conversion in intensive agriculture. With the increasing mechanization of production, several workers are excluded from these processes and tend to be displaced into marginal are as contributing to moving forest frontiers or expanding the areas occupied livestock and agriculture (Domingues and Bermann, 2012). According to Costa (2000) the cultivation of soy is only profitable if practiced in large areas, which demand large amounts of capital and technical expertise, generating a process of land and income concentration.

The logging of natural forests is another activity that plays a growing economic role in most of the Amazon region (Angelo and Sá, 2007). Forest is logged aiming first the trade of wood and possibly funding of economic activities that will be established next. When it comes to the implementation of another activity (Angelo and Sá, 2007), loggers acquire from ranchers the right to exploit logged areas (Veríssimo et al., 1995; Angelo and Sá, 2007). Most of the logging in the Amazon is illegal (Lentini et al., 2005), leading to the depletion of natural resources which could otherwise be used to generate sustainable socioeconomic development.

Due to the availability of cheap land to livestock, their numbers are expanding rapidly in the Amazon. Livestock is also present in small farms and it's the main land use associated with these properties. Increases in income expectations associated with livestock may have impacts on both types of agricultural systems of the Amazonia. That is, the increased expectation of profitability of livestock activity tends to have systemic effects with impacts on the rates of deforestation (Rivero et al., 2009). The expansion of livestock is associated with the increase of internal and external demands for beef, but is also influenced by other factors, such as reduction of transportation costs, increases in productivity and efficiency in management and the insignificant land price in the expansion regions of the Amazon (Rivero et al., 2009).

The construction of new roads and the rehabilitation of existing ones boost access to new areas, encouraging deforestation (Laurance et al., 2001; Nepstad et al., 2001; Soares-Filho et al., 2004) and in general anthropogenic pressures in areas adjacent to roads. The demographic distribution in the Amazon region shows a tendency of concentration of the rural population next to urbanized areas with higher road density (Angelo and Sá, 2007). Of all deforestation that occurred in the Brazilian Amazon, 85% took place within a radius of 30 km from roads (Verweij et al., 2009). Many of these projects create corridors between densely populated areas and the remote Amazonian frontier (Laurence et al., 2001), facilitating the process of colonization, which subsequently leads to deforestation and other irreversible environmental effects. According to Cattaneo (2002) the reduction of transport costs would imply a significant increase in return on arable land, thus increasing the incentive to deforestation (Verweij et al., 2009).

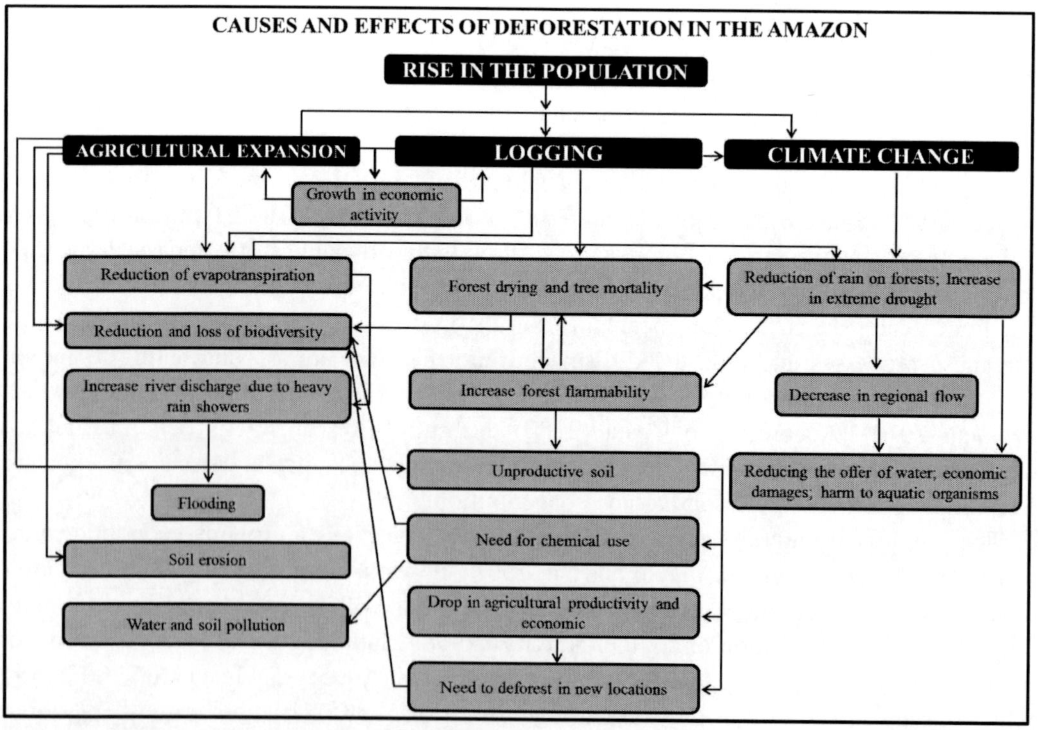

Figure 4. Flowchart showing the main causes and effects of Amazon deforestation in Brazil: The causes which lead to deforestation are black boxes; and consequences for human society are indicated by gray boxes (Adapted the Davidson et al., 2012).

Forests also are lost due to flooding from reservoirs associated with hydroelectric dams that usually tend to occupy vast areas (Fearnside, 1995, 2008). Besides the destruction of forests and riparian systems, hydroelectric projects have much wider impacts in encouraging road building, new migration, boosting deforestation (Fearnside 2005b, 2006, 2008). There are plans, in the near future, for more than 70 new dams that will submerge and destroy a total area of forest of more than 100,000 km² (Verweij et al., 2009). Figure 4 illustrates the main causes and effects of deforestation in the Amazon.

SURFACE SOIL INVERTEBRATES AS ECOLOGICAL INDICATORS OF ENVIRONMENTAL CHANGES

To assess the state of degradation or conservation of an ecosystem, biologists use ecological indicators (Fränzle, 2006; Heink and Kowarik, 2010). These indicators must be highly sensitive to changes in the structure of an ecosystem, if possible providing information concerning the alteration in the structure and functioning on the system according to their presence, absence and/or abundance (Majer, 1981).

In this context, Lindenmayer et al. (2000) defines seven types of attributes with relevance as ecological indicators at the level of species, guilds and communities: (1) species whose presence indicates the presence of a set of other species, (2) key species whose addition to or loss from an ecosystem leads to major changes, (3) species whose presence indicates a biotic conditions of human change as air pollution or water pollution, (4) dominant species that provide much of the biomass or number of individuals in an area, (5), species that indicate particular environmental conditions, such as soil or certain types of rocks; (6) species considered sensitive and thus serving as an early warning of environmental change (also called bioindicators) and (7) a management indicator species, reflecting the effects of a disturbance regime or the effectiveness of efforts to mitigate particular disturbance. These seven types of attributes can be effectively classified in three classes of ecological indicators in function of their main assessment vocation: (1) biodiversity, (2), environment or (3) ecosystem integrity (McGeoch, 1998).

Biodiversity indicators provides descriptive data with ecological relevance, such as species richness, native vs. exotic composition and/or degree of endemism in a given habitat (Noss, 1990, Gaston and Blackburn, 1995; Cano and Schuster, 2008). Species or groups of species that exhibit predictable responses to environmental degradation or to other changes in the environmental status, providing cheap and easy measurements if standard methodologies are applied, are termed environmental indicators (Venier and Pearce, 2004). Finally, ecological integrity indicators that demonstrate the effects of ecosystem changes, such as changes to habitats, fragmentation, climate change, pollution and other factors that generate impact on biota (McGeoch, 1998; Cano and Schuster, 2008). All three categories of ecological indicators cannot necessarily be exclusive and the most promising approach depends on their suitability for the investigation of key cause-effect relationships, as well as for the quality of the monitoring data and final diagnosis in supporting local management decisions (Van Straalen and Krivolutsky, 1996; Gerlach et al., 2013) (Figure 5).

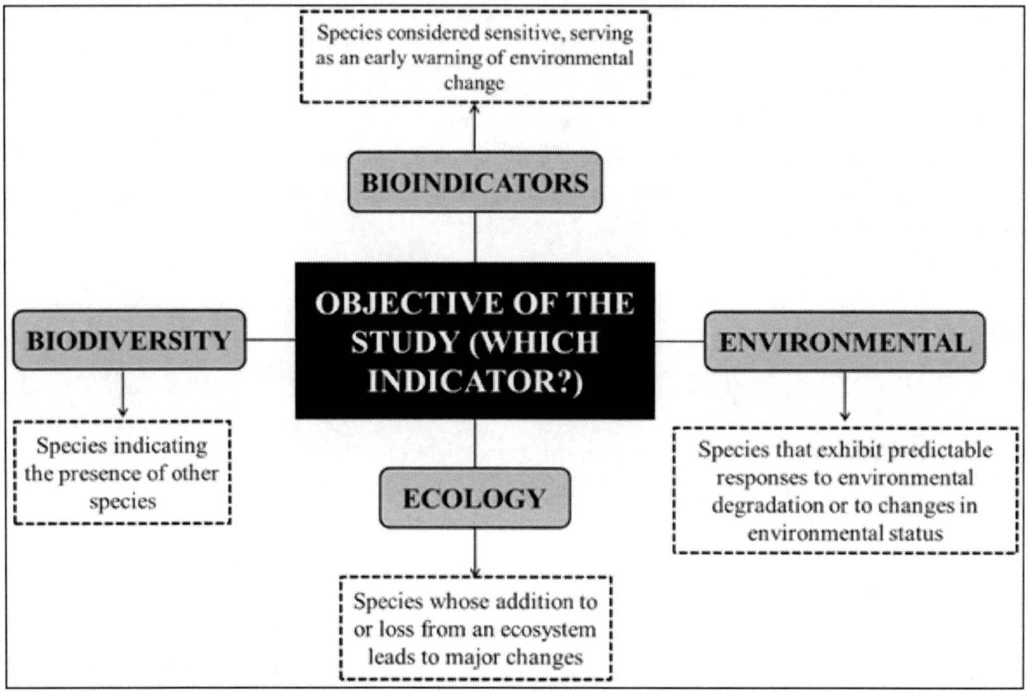

Figure 5. Types of ecological indicators.

The concept of bioindication was initially developed for aquatic environments while its application to terrestrial ecosystems gained momentum from 1980 (McGeoch, 1998). Subsequently, several criteria were proposed for the choice of suitable more generic ecological indicators (Figure 6) to be used also at the level of terrestrial systems (Pearson, 1994; Andreasen et al., 2001; Dale and Beyelerb, 2001; Freitas et al., 2006):

i. Taxonomy relatively well provided – groups with taxonomic problems provide low quality information, because of the difficulty or impossibility of defining the boundaries between the species;
ii. Knowledge about natural history, genetics, chemistry, and other aspects of biology – must have a recognized response to environmental change as well as responding clearly to the disturbance. In this way, must inform about the structure, functioning and composition of the ecological system and should be monitored in environmental disturbances of short- and long-term;
iii. Convenient diversity - tremendous diversity can derail the process of sorting and analysis, while a very low diversity may compromise the quality of the data;
iv. Short life cycle - in theory, the lower the time of generation, the more rapidly the effects of environmental changes are perceived.
v. Ecological diversity - a group that has different species with ecological associations provides information on different compartments of the habitat.
vi. Fidelity of habitat - species with little faithfulness of habitat do not provide specific information.

vii. Close association to resources or other species – is not essential, but species closely related to each other are more informative (represent another taxonomic group or resource).
viii. "Sedentary" relative - at one end, migratory species or very dispersive may be present in an environment without any relationship with the environmental conditions of the same.
ix. Ease in sampling, sorting and identification, including for non-specialists.
x. Little human use – groups whose species have a high economic value (for example, hunted for consumption or ornamentation) may disappear from the system regardless of their conservation status.

We highlight the surface soil invertebrates as good environmental and ecological indicators as their small size makes them sensitive to local conditions, while their mobility enables them to rapidly move in response to changing conditions (Samways et al., 2010), both at the habitat level and according to the gradients of environmental impact intensity (Oliver et al., 1998; Lewinsohn et al., 2005). Additionally as insects they tend to be very diverse and abundant when compared with the groups of vertebrates and plants (Figure 7). Although historically neglected in conservation and monitoring strategies, they are now becoming an important asset in the disciplines of landscape ecology and conservation (McGeoch et al., 2011; Gerlach et al., 2013).

Studies have been developed to evaluate the behavior of soil invertebrates in degraded environments and how these organisms respond to different restoration measures serving as a tool for monitoring (e.g. Correia, 2002; Rovedder et al., 2009; Cajaiba and Silva, 2014). Soil invertebrates composition reflects the functioning of the ecosystem due to its intimate association with litter fall-soil system processes and its great sensitivity to environmental changes (Ruf et al., 2003; Yi et al., 2006; Hoffmann et al., 2009; Oliveira and Souto, 2011).

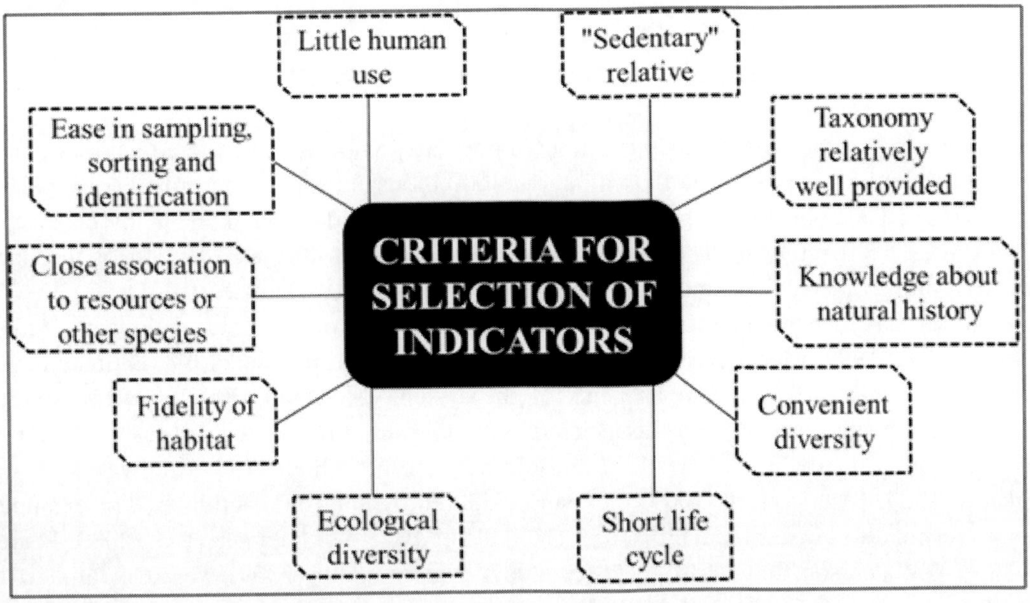

Figure 6. Criteria for selection of ecological indicators.

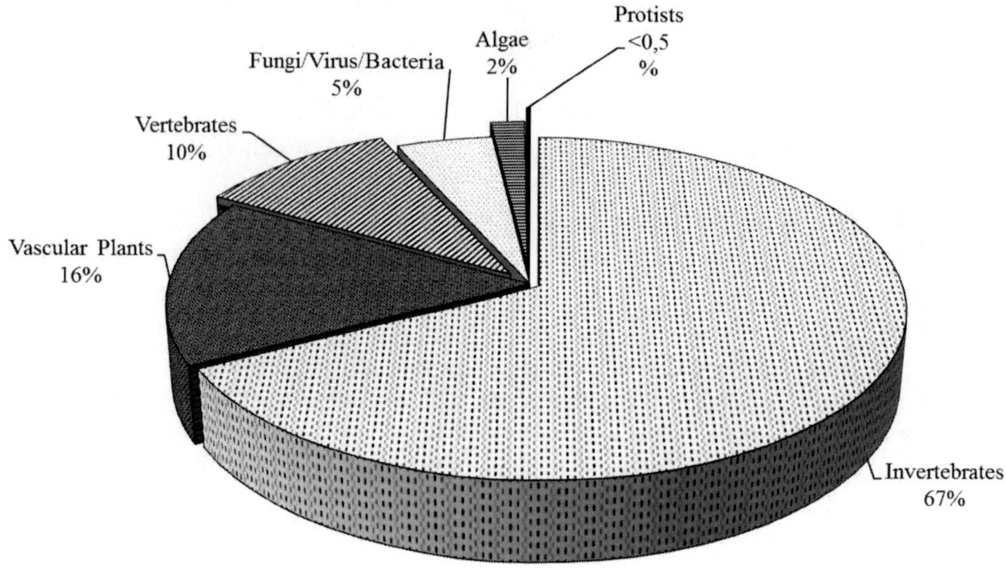

Figure 7. Diversity of invertebrate species compared to other indicators groups (Adapted the Calor, 2009).

Functional groups of soil fauna may disappear or be reduced (in their abundance and richness) as a result of soil degradation processes or be displaced by few opportunist organisms, highly adapted to disturbances (Lavelle, 1996; Loranger et al., 1998).Thus, environmental indicators must be highly sensitive organisms to changes in the structure of an ecosystem, featuring the integrity of the soil (Maluche et al., 2003; Lima et al., 2006).

Soil Surface Coleoptera as Indicators of Environmental Changes

The insects of the order Coleoptera, known popularly as beetles represent one of the most diverse animal life groups on Earth, with about 4000 thousand species described (Lara, 1992; Costa, 1999; Cajaiba, 2013), representing more than a third of all known insect species (Figure 8). They occupy most of the Earth's environments, with the exception of the open seas, exhibiting diverse diets and lifestyles (Grebennikov and Leschen, 2010).

Beetles have characteristics that qualify them as appropriate for ecological studies (Gaston et al., 1992; Marinoni and Dutra, 1997; Carlton and Robinson, 1998; Didham et al., 1998) monitoring different compartments of the system (Marinoni, 2001). Several studies suggest that beetles are strongly associated with the structure of the habitat and factors associated with microclimates (Halffter and Mathews, 1966; Thiele, 1977; Schwarzkopf and Rylands, 1989; Bishop et al., 2009) that can be altered by anthropic activities. For example, beetles respond to environmental disorder (Koivula et al., 1999; Niemelä et al., 2000; Pearce et al., 2003; Pearson and Cassola, 2005, 2007; Kaiser et al., 2009; Negro et al., 2010; Vasquez-Velez et al., 2010), soil depth (Bultman and Uetz, 1984), soil disturbance (Bergeron, 1991; Pearce et al., 2003), local humidity (Epstein and Kulman, 1990; Rykken et al., 1997), temperature (Li and Jackson, 1996), soil pH (Paje and Mossakowski, 1984) and pollution (Garcia et al., 2010). Beetles were also used for indicating success in managing and restoration of habitats (Jacobs et al., 2010; Paoletti et al., 2010; Gomez, 2010).

Carabidae, Scarabaeidae, Cerambycidae, Staphylinidae and Coccinellidae are groups that have species with high ecological fidelity, diversified taxonomically and ecologically, easily collectable in large samples, easily identifiable and functionally diverse (Oliver and Beattie, 1996; Langor and Spence, 2006; Pearce and Venier, 2006). These families also form groups relatively well known with other species and resources (Brown, 1991; Vulinec, 2000; Nichols et al., 2008).

Monitoring changes has been investigated with the use of the family Scarabaeidae, subfamily Scarabaeinae mainly (Spector, 2006) due to its relation with the forest environment and mammals that inhabit it (Gill, 1991; Halffter, 1991; Davis et al., 2001; Nichols et al., 2008). Scarabaeidae were also used as ecological indicator (McGeoch et al., 2002; Spector, 2006; Nichols et al., 2007, 2008), considering this group close relationship and dependency on the environment in which they live and of the resources available to the community (Halffter and Arellano, 2002; McGeoch et al., 2002; Andresen, 2003; Spector and Ayzama, 2003; Larsen and Forsyth, 2005; Scheffler, 2005; Nichols et al., 2007; Almeida et al., 2011). Several studies demonstrate that the composition and structure of Scarabaeinae communities are affected in modified areas by human action (Klein, 1989; Didham et al., 1998; Halffter and Arellano, 2002; Scheffler, 2005; Nichols et al., 2007), namely by habitat fragmentation and isolation (Klein, 1989; Andresen, 2003; Nichols et al., 2007; Escobar et al., 2008). In the tropical belt, their abundance and richness is heavily influenced by the climate, namely by precipitation that sets the advent of most species (Halffter and Matthews, 1966; Gill, 1991; Hernández, 2007; Andresen, 2008).

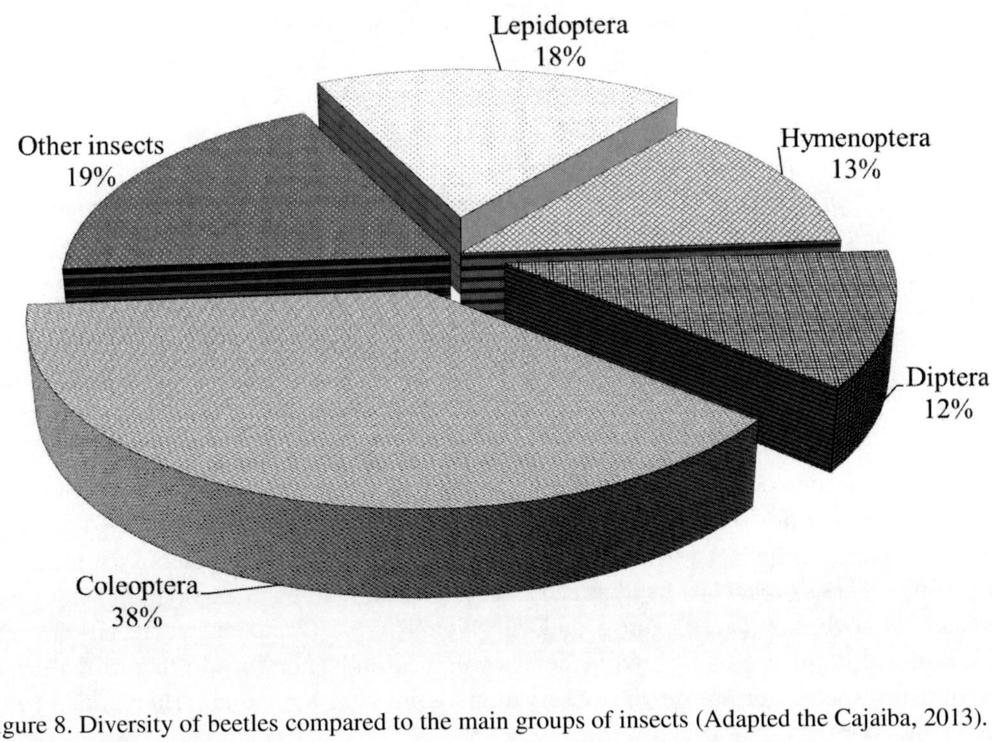

Figure 8. Diversity of beetles compared to the main groups of insects (Adapted the Cajaiba, 2013).

Kimberling et al. (2001), Escobar et al. (2007) highlight the Scarabaeidae sedentary nature and vulnerability to environmental changes (e.g. vegetation cover, vegetation type, fragmentation, habitat loss, physical structure, altitude of ecosystem, among others). The scavenger habit of these species may also indicate abundance/diversity of vertebrates in forest environments (Halffter, 1991). Size and biomass of Scarabaeidae are related to the presence of groups of vertebrates due to food scraps associated (Klein, 1989; Estrada et al., 1993; Scheffler, 2005).

The family Carabidae has been suggested to be used as indicator in studies with biodiversity and for evaluating habitat changes, for their diversity in morphological, behavioral, ecological and taxonomic traits, abundance and sensitivity to environmental changes (Heliovaara and Vaisanen, 1993; Niemelä et al., 2000; Koivula et al., 2002; Rainio and Niemelä, 2003; Niemelä and Kotze, 2009). This group is also used in assessing the sustainability of agricultural systems, being negatively affected by intensive agriculture, mechanized control of weeds and fire (Stork and Eggleton, 1992), urbanization, crop and forest management, overgrazing by domestic livestock, tourist flow in natural landscapes and soil pollution (Avgın and Luff, 2010). Their dominance is altered with the rhythm and phenology of the crop, as well as by the microclimate (Kromp, 1999). The group has also been used as indicator of metal contamination in soils (Paoletti et al., 1991; Avgın and Luff, 2010). Pearce and Venier (2006) demonstrate in his review that typical forest species tend to prefer lower temperatures and higher humidity, while open-habitat species are associated with for higher temperatures and lower humidity. Thus, large fluctuations in temperature and moisture levels can disrupt the assemblages of Carabidae (Pearce and Venier, 2006). Several disadvantages were also discussed, including seasonal variations, the uneven distribution and high number of generalist species. Also in tropical regions their taxonomy and ecology are little known (Rainio, 2009).

The family Staphylinidae, have been proposed as bioindicators or surrogates (Bohac, 1999; Anderson and Ashe, 2000; Pohl et al., 2007) given that they are abundant and diverse; occupy a great variety of habitats worldwide; are easy to collect, given that the majority of the species are found among leaf litter and fallen trunks, forming one of the commonest groups of insects and ecologically the most important of the soil-related fauna (Bohac, 1999; Navarrete-Heredia et al., 2002; Newton et al., 2005; Vasquez-Velez et al., 2010). Their abundance and diversity are dependent on the moisture of the soil foliage. For example, in studies conducted by García-Cárdenas et al. (2001) in fragments of forest, a positive relation was observed between richness of staphylinids and leaf litter moisture, as well as between richness and area of the fragments (García-Cárdenas and Ulloa, 2005). Preliminary evaluations suggest that richness and diversity depend, to a great extent, on the habitats with greater plant heterogeneity such as forests and riparian corridors (Bermúdez et al., 2006; Vasquez-Velez et al., 2010). This family is many times associated with locations containing important concentrations of potassium and phosphorus, which are frequent in soils containing organic material (Dunxião et al., 1999). Many species of Staphylinidae are also associated with the soil of agricultural crops (Sotherton, 1984), namely direct planting systems (Stinner and House, 1990). These organisms have already been cited in studies of population surveys on crops such as soybeans, cacao, guava and eucalyptus forest (Zanuncio et al., 1993). The biggest problem in the use of Staphylinidae as ecological indicators lies in the difficulties of identifying the species based on the observation of internal structures (Hammond, 1997; Freitas et al., 2006).

Another family of Coleoptera which presents good characteristics to be used as ecological indicators of environmental integrity/disturbance is the Coccinellidae. The microclimatic conditions and dietary factors, accordingly with Iperti (1999), are important in determining the habitat of beetles of this family, because each species shows preferences to the type of vegetation stratification and its seasonal variations. The specie *Semiadalia undecimnotata* was considered very sensible to the increase in chemical and physical pollution, increased cultivation of cereal (which create problems on trophic chain) and the impact of climatic fluctuations (Wink et al., 2005). Thomazini and Thomazini (2002) noted that the number of species as well as individuals was considerably less on pasture than in other areas, when compared with primary woods and secondary vegetation. Despite the already mentioned features, there are few studies developed with this group in tropical regions.

The Cerambycidae are important part of forest ecosystems, occupying from the ground up to the treetops and, in the process of co-evolution between the communities of insects and flora, have also played the important role of pollinators (Hequet, 1996). These insects have been used for planning of conservation and monitoring, because they maintain a close relationship with the natural resources they use, being sensitive to environmental changes (Brown, 1997; Lewinsohn et al., 2005). According Marinoni and Ganho (2003), the population dynamics of some species of Cerambycidae (e.g. *Nyssodrysina lignaria*) is influenced by high temperatures, humidity and daylight. This group is easily collectable by several types of traps (Monné and Hovore, 2005) and participates in the process of decomposition of dead wood, incorporating nutrients and allowing the renewal of forests through the opening of spaces and consequent seed germination. In Brazil, some areas such as the South and Southeast regions have important studies on the fauna of Cerambycidae, but in the Caatinga, Pantanal and especially the Amazon forest little is known about their natural history, behavior and ecology (Martins, 1999; Silva et al., 2005; Souza and Silva, 2012).

The beetles of the family Tenebrionidae are common, easily captured in pitfall traps, and straightforwardly identifiable when compared with other groups. Communities of these beetles are dependent on the availability of specific resources such as debris, vegetation cover and various environmental characteristics of soil such as humidity, hardness and granulometric composition (Henschel et al., 2010). They have also been considered sensitive biodiversity indicators due to its sensitivity to habitat loss and degradation. Their abundance and diversity is affected to forest fragmentation, fires, climatic factors, among others (Parenzee, 2001; Cheli, 2009; Fattorini et al., 2011; Thakare et al., 2012). The family also responds to vegetation post-fire recovery (Fattorini, 2010). Despite these advantages, deeper studies are needed to determine whether changes in diversity of Tenebrionidae are purely seasonal and bioecological results or whether they are influenced by land use changes (Henschel et al., 2010; Saji and Al Dhaheri, 2011).

Another group of beetles widely used in studies of conservation/degradation is the Histeridae family. These beetles are functional predators, both in larval stage and adulthood, using wide-ranging environments for their food, and debris attract the scavengers that comprise important portion of its diet (Summerlin et al., 1991). Theoretically the community of predatory insects that occurs in decomposed matter maintains is high diversity in environments with complex vegetation structure possibility by the existence of superior biomass of vertebrates (suppliers of resources) in these locations (Holloway and Schnell, 1997; Smith and Merrick, 2001). Lopes et al. (2005) studying restinga vegetation in the State of Espírito Santo, Brazil, found that species richness of Histeridae was practically equal in each of the four areas studied (Restinga arboreal, Restinga open from *Clusia* spp, Restinga burned and Pasture), but the pattern of diversity was very different, with great diversity in the pristine restinga arboreal environment. The same authors observed a tendency of decrease of diversity (H') and equity (J') with a reduction of the complexity of the vegetation. Summerlin (1989) mentions the trend of reduced abundance and high richness of Histeridae in forest environments, as opposed to greater abundance and reduced diversity of Histeridae in open habitats of grassland. For better understanding, studies with this family are necessary, because, as pointed out by Degallier et al. (2011) the number of known species in tropical areas is diminutive.

The Table 1 (the appendices) presents a summary of the main families of beetles used as indicators, as well as their advantages and disadvantages.

Soil Surface Beetlesas Indicators of Environmental Changes in the Tropics with a Focus on the Brazilian Amazon

Despite the recognized importance of using ground beetles as ecological indicators, its application in Brazil and more precisely in the Amazon, has been extremely limited. The effects of human actions (e.g. effects of deforestation, fire, grazing, logging and mining) on the functionality and structuring of ground beetles communities of the Amazonian forests are still poorly understood (Blanche et al., 2001; Dawes-Gromadzki, 2007; Andrade et al., 2011; Samu et al., 2011). Often, researcher's results are not comparable because of the lack of standardization in the sampling methodologies, hampering a direct comparison with other studies. In addition, there are few studies that examine long-term effects (Debinski and Holt, 2000; McGarigal and Cushman, 2002) owing practical difficulties and logistics of sampling and repetition in the same location and using the same methodology (Quintero and Halffter, 2009). As the number of forest areas to be searched or the total area under management increases, the number of researchers involved in surveys and the time required also increases, making surveys unfeasible (Aguilar-Amuchastegui and Henebry, 2007).

Studies with some families of Coleoptera such as Scarabaeidae, Carabidae, Staphylinidae have grown tremendously in the recent years, while for the other families most of the work are incipient. Amazonian Scarabaeidae beetles were studied in states Rondônia (Vulinec, 2000, 2002), Pará (Andresen, 2002, 2003; Scheffler, 2005; Gardner et al., 2008; Matavelli and Louzada, 2008; Vulinec et al., 2008; Barlow et al., 2010) and Amazonas (Vulinec, 2002; Quintero and Roslin, 2005; Radtke et al., 2007; Quintero and Halffter, 2009; Vulinec et al., 2009).

These studies highlight the effects of forest logging in the Amazon showing that abundance, species richness and composition of ground beetles is severely affected by forest fragmentation (Klein, 1989). According to the results of Imbrósio (2012), that performed studies in different soil uses (eucalyptus plantations, native forest and selective cutting of wood) and Gardner et al. (2008) in eucalyptus plantations, native and secondary forest, the system with original forest cover sustains a community with a different functional structure namely by the reduction in the variability in the more humanized systems. According to these studies, the fauna of dung beetles are severely depleted in eucalyptus plantations and secondary vegetation compared to areas of primary forest. According to the authors, dung beetle communities that are closely associated with tropical forest habitats are greatly influenced by differences in vegetation structure (Halffter & Arellano, 2002; Gardner et al., 2008), with individual species often having specific affinities for certain structural habitat properties (Davis et al., 2001; Gardner et al., 2008). Furthermore, secondary forests and Eucalyptus plantations, are typically characterized by low, relatively open canopies with hot and dry understory environments (Gardner et al., 2008). These microclimatic differences could help explain the observed impoverishment of dung beetle communities through a physiological intolerance to high temperatures (Chown, 2001; Campos and Hernández, 2013), the influence of solar radiation on adult activity patterns (Lobo et al., 1998; Moreira et al., 2009; Korasaki et al., 2013) or the influence of decreased soil moisture content on larval survival (Sowig, 1995; Gardner et al., 2008).

Teixeira (2006) analyzed the effect of isolation of forest islands and the effect of roads on the Scarabaeidae community noted that there was reduction of species in areas close to roads and in forest fragments when compared with intact and continuous regions. Vulinec (2002) in studies conducted in the States of Pará, Amazonas and Rondônia, noted that the diversity of most species of Scarabaeinae decreases in areas with high levels of impact (e.g. forest clearing and cutting), concluding that larger forest gaps affect the dynamics of this group. Scheffler (2005) examined the distribution and diversity of dung beetles (Coleoptera: Scarabaeidae) in intact forest and under three disturbance regimes (selective logging, clear-cutting and pasture), as a result, the author observed that Principal Components Analyses and the Sorensen's Index indicated that the intact and selectively logged areas were similar in terms of species composition but that the pasture areas and clear-cut areas had substantially different groups of species. Beetle size and diversity (as measured by species richness, Shannon index and Simpson's index) were lower in clear-cuts and pasture than in intact forest. While the fact that selective logging has a relatively minor impact on the dung beetle community is heartening, the projected increase in the amount of highly disturbed landscape in Amazonia is expected to have a severe impact on dung beetle biodiversity in the region (Scheffler, 2005).

Perspectives and Challenges

Despite already having a reasonable number of researchers and publications, the use of ecological indicators in Brazil is in its initial stage. However, these studies were restricted mostly to the South and Southeast of the country. The Amazon has still few studies with ecological indicators/ground beetles, and can be considered insignificant when compared with its biodiversity and environmental problems. Systematics, general inventories, and testing the effects of habitat disturbance should be considered a priority by research groups engaged in biology and natural resources managing (Freitas et al., 2006).

While recognizing the qualities of Coleoptera for ecological integrity studies, several works should also try to solve the highlighted difficulties/disadvantages identified. The key problem is the taxonomic challenge, considering that a low proportion of species are known or described (Samways et al., 2010; Cardoso et al., 2011; Gerlach et al., 2013). The rates of knowledge differ markedly, hindering the identification by non-specialists, and some of the taxa proposed for indicators can only be reliably identified by taxonomic specialists (Behan-Pelletier, 1993; Báldi, 2003). Thus, a group of invertebrates can show excellent characteristics to be used as ecological indicators, but their widespread practice can be impeded by taxonomic difficulties (Hodkinson and Jackson, 2005). A solution to overcome these challenges is the new advances in molecular identification techniques (particularly DNA barcoding) (Janzen et al., 2005). DNA barcoding (Hebert et al., 2003) has been particularly successful in the identification and delimitation of new species from various groups (Hebert et al., 2004; Ward et al., 2005; Cywinska et al., 2006; Hajibabaei et al.,2006a, b; Smith et al., 2007; Borisenko et al., 2008; Kerr et al., 2009). This method has received increased acceptance because it is simple and affordable (Padial and De La Riva, 2007; Pires and Marinoni, 2010). The advantages of the method would be the possibility of identifying individuals at any stage of development and the prospect of discriminating between morphologically identical species (Pires and Marinoni, 2010).

Another criterion that could "attenuate" this disadvantage would be sorting by morphospecies. This approach involves, above all, the creation of a reference collection of all new species found throughout the study. The fundamental key to this approach is that individuals of a species found in different samples are assigned the same species name (morphospecie). This allows the researcher to study the diversity and composition changes through a study, without having to wait for the species to be correctly identified (or described) (Gerlach et al., 2013), speeding the process until more accurate results are completed (Oliver and Beattie, 1996). Therefore, the approach by morphospecies is no substitute for taxonomy, just accelerate the studies and seems to provide good results concerning structuring and functioning of systems (Samways et al., 2010; Gerlach et al., 2013). Added to these factors, some impediments to the use of ground beetles as indicators (Cardoso et al., 2011):

i. Public dilemma – invertebrates and their ecological services are mostly unknown to the public;
ii. Political dilemma – policymakers and stakeholders are unaware the problems of invertebrate conservation;
iii. Scientific dilemma – the study of soil invertebrates is scarce and underfunded.
iv. Linnean shortfall – most species are not described;
v. Wallacean shortfall – the distribution of species described is virtually unknown;
vi. Prestonian shortfall – the abundance of species and their changes in time and space are unknown;
vii. Hutchinsonian shortfall – the ways of life of the species and their sensitivity to changes in habitat are largely unknown.

Besides the above-mentioned factors, many of the challenges of research applied to soil beetles could be significantly improved through a greater effort to collect standardized data sets on the species and patterns of functional response to disturbance. This could include the creation of a wide range of public available resources, including a comprehensive online catalog, new keys to genera and species and standardized survey protocols (Nichols and Gardner, 2011). Additionally, greater financial support to researchers and incentives for conducting research in more remote regions would help in solving these challenges.

FINAL REMARKS

In the last decade, there has been a growing recognition of the value of using ecological indicators for monitoring changes in terrestrial environments. Ground beetles are good indicators of biodiversity and environmental quality in a variety of spatial and temporal scales. This is because they are sensitive to climatic and ecological conditions, as well as to edaphic, physiognomic, trophic and microclimatic factors (Davis et al., 2004).

However, before choosing suitable ecological indicators, certain steps must be followed. It is necessary to define clearly the goals and endpoints of the study (McGeoch, 1998). It must be determined what to measure, how it will be measured, and why. Account must be taken in accordance with the nature of the problem, even if it is a response to a single pollutant in a restricted location or an attempt to compare the biodiversity of a wider area. The lack of accuracy at this stage will result in inadequate indication and results of little reliability (Hodkinson and Jackson, 2005). Only after establishing the groundwork it's possible to select the indicator group appropriate to support the rigorous protocols required (Hodkinson and Jackson, 2005).

Sometimes taxa that are good indicators at a spatial scale could lose their reliability at higher or lower scales (Allen et al., 1999). Instead of focusing on a few indicator species, more reliable information can be obtained from studies of a set of species/groups or at the community level, with measurements made not at the level of presence/absence but as numbers, biomass and dominance. The use of trophic guilds as detritivores, predators, parasitoids, decomposers, saprophagous, among others, may also reveal interesting differences in the landscape (Paoletti, 1999).

Selecting the appropriate indicator taxon is a compromise between advantages and drawbacks. The success of using ground beetles is based on cost-effective data collection and sensitivity to different environmental factors (Rainio and Niemelä, 2003). Despite the advantages and features already mentioned and the ecological importance of beetles, species identification is still very difficult for some groups, which added to the lack of systematic studies and natural history causes some beetles to be unsuitable for indicator studies (Gaston et al., 1992; Marinoni and Dutra, 1997; Freitas et al., 2006). However, these features are common to most invertebrate groups.

Finally, many existing studies on the change of habitat suffer from methodological limitations that may interfere with the interpretation of results (Gardner et al., 2007a) and such sampling problems are increasingly recognized in studies of biodiversity in general (Hamer and Hill, 2000; Hill and Hamer, 2004; Barlow et al., 2007a, b, c; Gardner et al., 2007b; Koh, 2007; Gardner et al., 2008). We stress here the importance of inventories well conducted, suggesting further discussion about general protocols and standardization of methods, taking into account the particularities of each group. The differences of method and the absence of general protocols, scientifically tested and feasible, can simplify complex realities, leading to wrong decisions (Silveira et al., 2010). However, more consistent studies it is necessary for understand the relationship between beetles and environment, and further explore the technique of beetles as ecological indicators, which might be applied to monitoring and evaluation of environment.

APPENDICES

Table 1. Advantages and limitations of ground beetles used with potential use as ecological indicators

Family	Main studies	Advantages	Limitations	Reference
Carabidae	Agriculture; Forest fire; Pollution by metals; Urban ecology; Insecticides; Climate change.	Each type of habitat has given set of species with generalist and specialist species. Thus, species of Carabidae individual or sets of species can be used as bioindicators.	Due to the high number of carabidae generalist species have been criticized as bioindicators.	Basedow (1990); Eyre and Luff (1990); Paoletti et al. (1991); Stork and Eggleton, (1992); Halme and Niemelä (1993); Butterfield (1996); Rykken et al. (1997); Venn (2000); Niemelä et al. (2000); Niemelä (2001); Maryański et al. (2002); Mozdzer et al. (2003); Ermakov (2004), Gongalsky et al. (2004); Schwerk et al. (2006); Belskaya and Zinoviev (2007); Cárdenas and Hidalgo (2007).
Scarabaeidae	Deforestation; Fragmentation; Logging; Agriculture; Reduction of rainfall; Livestock; Reduction of mammals.	Can represent to other animals (vertebrates); Easily collectable for several types of traps; Close relationship and dependency on the environment in which they live.	-	Howden and Nealis (1975); Wolda (1987); Klein (1989); Nummelin and Hanski (1989); Estrada et al. (1993); Hill et al (1995); Intachat et al. (1999); Nummelin (1996); Andresen (2003); Scheffler (2005); Nichols et al. (2007); Hernández

Family	Main studies	Advantages	Limitations	Reference
				(2007); Escobar et al. (2008); Andresen (2008).
Staphylinidae	Fragmentation; Agriculture; Fertilization of soil; Seasonality; Insecticides; Herbicides.	Easily collectable for several types of traps; Close relationship and dependency on the environment in which they live.	The biggest problem in the use of Staphylinidae as biological indicators lies in the difficulties of identifying the species, since this is often subject to the observation of internal structures.	Sotherton (1984); Stinner and House (1990); Zanuncio et al. (1993); Dunxião et al. (1999); Bohac (1999); Hunter (2002); Büchs (2003).
Coccinellidae	Climatic conditions; Food factors; Urbanization; Use of pesticides; Deforestation; Physical and chemical pollution.	Each species shows preferences to the type of vegetable strata and different seasonality.	Few studies developed with this group in tropical regions.	Iperti (1999); Thomazini and Thomazini (2002); Wink et al. (2005).
Cerambycidae	Planning of conservation and monitoring; Influence of temperatures and humidity	Easily collectable for several types of traps.	Little is known about their ecology, mainly in tropical forests.	Hequet, 1996; Brown, 1997; Marinoni and Ganho (2003)
Tenebrionidae	Climatic factors; Effects of fire; Influence of land use.	Are relatively plentiful, easily captured in pitfall traps and readily identifiable when compared with other groups. Sensitive to the change of biodiversity	Few studies developed with this group in tropical regions.	Parenzee (2001); Cheli (2009); Fattorini et al (2011); Fattorini (2010); Henschel et al. (2010); Saji and Al Dhaheri (2011); Thakare et al. (2012).
Histeridae	Effect of fire; Pasture; Replacement of natural landscape by agricultural activities.	Close relationship and dependency on the environment in which they live.	Few studies developed with this group in tropical regions.	Holloway and Schnell (1997); Smith and Merrick (2001); Lopes et al. (2005)

REFERENCES

Aguilar-Amuchastegui, N; Henebry, GM. (2007). Assessing sustainability indicators for tropical forests: Spatio-temporal heterogeneity, logging intensity, and dung beetle communities. *Forest Ecology and Management, 253*, 56-67.

Allen, AP; Whittier, TR; Larsen, DP; Kaufmann, PR; Connor, RJO; Hughes, RM; Stemberger, RS; Dixit, S; Brinkhurst, RO; Herlihy, AT; Paulsen, SG. (1999). Concordance of taxonomic composition patterns across multiple lake assemblages: effects of scale, body size, and land use. *Canadian Journal of Fisheries and Aquatic Science, 56*, 2029-2040.

Almeida, S; Louzada, J; Sperber, C; Barlow, J. (2011). Subtle land-use change and tropical biodiversity: dung beetle communities in cerrado grasslands and exotic pastures. *Biotropica, 43*, 704-710.

Anderson, RS; Ashe, JS. (2000). Leaf litter inhabiting beetles as surrogates for establishing priorities for conservation of selected tropical montane cloud forest in Honduras, Central America (Coleoptera: Staphylinidae and Curculionidae). *Biodiversity and Conservation, 9*, 617-653.

Andrade, RB; Barlow, J; Louzada, J; Vaz-de-Mello, FZ; Souza, M; Silveira, JM; Cochrane, MA. (2011). Quantifying responses of dung beetles to fire disturbance in tropical forests: The importance of trapping method and seasonality. *PLoS ONE, 6*, 1-9.

Andreasen, JK; O'Neill, RV; Noss, R; Slosser, NC. (2001). Considerations for the development of a terrestrial index of ecological integrity. *Ecological Indicators, 1*, 21-35.

Andresen, E. (2003). Effect of forest fragmentation on dung beetle communities and functional consequences for plant regeneration. *Ecography, 26*, 87-97.

Andresen, E. (2008). Short-term temporal variability in the abundance of tropical dung beetles. *Insect Conservation and Diversity, 2*, 120-124.

Angelo, H; Sá, SPP. (2007). O desflorestamento na Amazônia brasileira. *Ciência Florestal, 17*, p. 217-227.

Audino, LD; Nogueira, JM; Silva, PG; Neske, MZ; Ramos, AHB; Moraes, LP; Borba MFS. (2007). *Identificação dos coleópteros (Insecta: Coleoptera) das regiões de Palmas (município de Bagé) e Santa Barbinha (município de Caçapava do Sul), RS*. Embrapa: Pecuária Sul.

Avgın, SS; Luff, ML. (2010). Ground beetles (Coleoptera: Carabidae) as bioindicators of human impact. *Munis Entomology & Zoology, 5*, p. 209-215.

Báldi, A. (2003). Using higher taxa as surrogates of species richness: a study based on 3700 Coleoptera, Diptera, and Acari species in Central-Hungarian reserves. *Basic and Applied Ecology, 4*, 589-593.

Barlow, J; Gardner, TA; Araujo, IS; Ávila-Pires, TC; Bonaldo, AB; Costa, JE; Esposito, MC; Ferreira, LV; Hawes, J. Hernandez, MIM; Hoogmoed, MS; Leite, RN; Lo-Man-Hung, NF; Malcolm, JR; Martins, MB; Mestre, LAM; Miranda-Santos, R.; Nunes-Gutjahr, AL; Overal, WL; Parry, L; Peters, SL; Ribeiro-Junior, MA; Silva, MNF; Motta, CS; Peres, CA. (2007c). Quantifying the biodiversity value of tropical primary, secondary and plantation forests. *Proceedings of the National Academy of Sciences of the United States of America, 104*, 18555-18560.

Barlow, J; Louzada, J; Parry, L; Hernández, MIM; Hawes, J; Peres, CA; Vaz-de-Mello, FZ; Gardner, T. (2010). Improving the design and management of forest strips in human-dominated tropical landscapes: a field test on Amazonian dung beetles. *Journal of Applied Ecology*, *47*, 779-788.

Barlow, J; Mestre, LAM; Gardner, TA; Peres, CA. (2007a). The value of primary, secondary and plantation forests for Amazonian birds. *Biological Conservation, 126*, 212-231.

Barlow, J; Overal, WL; Araujo, IS; Gardner, TA; Peres, CA. (2007b). The value of primary, secondary and plantation forests for fruit-feeding butterflies in the Brazilian Amazon. *Journal of Applied Ecology, 44*, 1001-1012.

Basedow, T. (1990). Effects of insecticides on Carabidae and the significance of these effects for agriculture and species number.In E Stork (Ed.). *The Role of Ground Beetles in Ecological and Environmental Studies* (pp. 115-125). Intercept Ltd Andover, UK.

Becker, BK. (2004). *Amazônia*: geopolítica na virada do III milênio. Rio de Janeiro: Garamond.

Behan-Pelletier, VM. (1993). Diversity of soil arthropods in Canada: systematic and ecological problems. *Memoirs of the Entomological. Society of Canada, 165*, 11-50.

Belskaya, EA; Zinoviev, EV. (2007). Structure of the complexes of carabid beetles (Coleoptera, Carabidae) in natural and industry-disturbed forest ecosystems in the south-west of the Sverdlovsk region. *Siberian Journal of Ecology, 4*, 533-543.

Bergeron, Y. (1991). The influence of island and mainland lakeshore landscape on boreal forest fire regime. *Ecology, 72*, 1980-1992.

Bermúdez, C; Gutiérrez, C; Chacón, P. (2006). *Estimación de la diversidad de coleópteros Staphylinidae en el paisaje del bosque seco tropical*. In: Libro de Resúmens II Congreso Colombiano de Zoología, Santa Marta, Colombia, p 407.

Bishop, DJ; Majka, CG; Bondrup-Nielsen, S; Peck, SB. (2009). Dead-wood and saproxylic beetle diversity in naturally disturbed and managed spruce forests in Nova Scotia. *Zookeys, 22*, 309-340.

Blanche, KR; Andersen, NA; Ludwig, JA. (2001). Rainfall-contingent detection of fire impacts: Responses of beetles to experimental fire regimes. *Ecological Applications, 11*, 86-96.

Bohac, J. (1999). Staphylinid beetles as bioindicators. *Agriculture, Ecossystems and Environment, 74*, 357-372.

Borisenko, AV; Burton, KL; Ivanova, NV; Hanner, RH; Hebert, PDN. (2008). DNA barcoding in surveys of small mammal communities: a field study in Suriname. *Molecular Ecology Resources, 8*, 471-479.

Brown, KS. (1991). Conservation of neotropical environments: Insects as Indicators. In: NM Collins; JA Thomas (Eds.). *The conservation of insects and their habitats*. (pp. 349-404). Academic Press, London.

Brown, KS. (1997). Diversity, disturbance, and sustainable use of Neotropical forests: insects as indicators for conservation monitoring. *Journal of Insect Conservation, 1*, 25-42.

Büchs, W. (2003). Biodiversity and agri-environmental indicators-general scopes and skills with special reference to the habitat level. *Agriculture, Ecosystems and Environment, 98*, 35-78.

Bultman TL; Uetz GW. (1984). Effect of structure and nutritional quality of litter on abundance of litter-dwelling arthropods. *The American Midland Naturalist, 111*, 165-172.

Butterfield, J. (1996). Carabid life-cycle strategies and climate change: a study on an altitude transect. *Ecological Entomology, 21*, 9-16.

Cajaiba, RL. (2013). *Introdução à Entomologia*. Pará de Minas, MG: VirtualBooks Editora.

Cajaiba, RL; Silva, WB. (2014). Mirmecofauna (Hymenoptera, Formicidae) em fragmento florestal urbano no município de Uruará-Pa. *Enciclopédia Biosfera, Centro Científico Conhecer, 10*, 2226-2238.

Calor, AR. (2009). *Insecta*. Disponível em: Museu de Zoologia Virtual, Universidade Federal da Bahia, http://www.mzufba.ufba.br/insetos.html). Acessed in 01 jun 2014.

Campos, RC; Hernández, MIM. (2013). Dung beetle assemblages (Coleoptera, Scarabaeinae) in Atlantic forest fragments in southern Brazil. *Revista Brasileira de Entomologia, 57*, 47-54.

Cano, EB; Schuster, JC. (2008). *Beetles as indicators for forest conservation in Central America*. Encyclopedy of Life Support Systems (UNESCO-EOLSS).

Cárdenas, AM; Hidalgo, JM. (2007). Application of the mean individual biomass (MIB) of ground beetles (Coleoptera, Carabidae) to assess the recovery process of the Guadiamar Green Corridor (southern Iberian Peninsula). *Biodiversity and Conservation, 16*, 4131-4146.

Cardoso, P; Erwin, TL; Borges, PAV; New, TR. (2011). The seven impediments in invertebrate conservation and how to overcome them. *Biological Conservation, 144*, 2647-2655.

Carlton, CE; Robinson, HW. (1998). Diversity of litter-dwelling beetles in the Ouachita highlands of Arkansas, USA (Insecta: Coleoptera). *Biodiversity and Conservation, 7*, 1589-1605.

Cattaneo, A. (2002). *Balancing agricultural development and deforestation in the Brazilian Amazon*.Research Report 129, International Food Policy Research Institute, Washington DC.

Cheli, GH; Corley, J; Castillo, LD; Martinez, F. (2009). Una aproximación experimental a la preferencia alimentaria de *Nyctelia circumundata* (Coleoptera: Tenebrionidae) en el Noreste de la Patagonia. *Interciencia, 34*, 771-776.

Chown, SL. (2001). Physiological variation in insects: hierarchical levels and implications. *Journal of Insect Physiology, 47*, 649-660.

Correia, MEF. (2002). *Potencial de utilização dos atributos das comunidades de fauna de solo e de grupos chave de invertebrados como bioindicadores do manejo de ecossistemas*. Seropédica: Embrapa Agrobiologia.

Costa, C. (1999). Coleoptera. In CRF Brandão; EM Cancello (Eds.). *Invertebrados Terrestres*: Síntese do conhecimento ao final do século XX (pp. 225-243). São Paulo, FAPESP.

Costa, FG. (2000). *Avaliação do potencial de expansão da soja na Amazônia Legal: uma aplicação do modelo de Von Thünen*. MSc. Thesis. Universidade de São Paulo-USP, Piracicaba, SP.

Cywinska, A; Hunter, FF; Hebert, PDN. (2006). Identifying Canadian mosquito species through DNA barcodes. *Medical and Veterinary Entomology, 20*, 413-424.

Dale, VH; Beyelerb, SZ. (2001). Challenges in the development and use of ecological indicators. *Ecological Indicators, 1*, 3-10.

Davidson, EA; Araújo, AC; Artaxo, P; Balch, JK; Brown, IF; Bustamante, MMC; Coe, MT; DeFries8, RS; Keller, M; Longo, M; Munger, JW; Schroeder, W; Soares-Filho, BS;

Souza Jr, CM ; Wofsy, SC. (2012). The Amazon basin in transition. *Nature*, 481, p. 321-329.

Davis, AJ; Holloway, JD; Huijbregts, H; Krikken, J; Kirk-Spriggs, AH; Sutton, SL. (2001). Dung beetles as indicators of change in the forests of northern Borneo. *Journal of Appllied Ecology, 38*, 593-616.

Davis, ALV; Scholtz, CH; Dooley, PW; Bham, N; Kryger, U. (2004). Scarabaeinae dung beetles as indicators of biodiversity, habitat transformation and pest control chemicals in agro-ecosystems. *South African Journal of Science, 100*, 415-424.

Dawes-Gromadzki, TZ. (2007). Short-term effects of low intensity fire on soil macroinvertebrate assemblages in different vegetation patch types in Australian tropical savanna. *Austral Ecology, 32*, 663-668.

Debinski, DM; Holt, RD. (2000). A survey and overview of habitat fragmentation experiments. *Conservation Biology, 14*, 342-355.

Degallier, N; Leivas, FWT; Moura, DP. (2011). Histerid beetles of French Guiana. V. Revision of the genus Ebonius Lewis (Coleoptera, Histeridae, Omalodini). *Zootaxa, 2824*, 44-52.

Didham, RK; Hammond, PM; Lawton, JH; Eggleton, P; Stork, NE. (1998). Beetles species responses to Tropical Forest fragmentation. *Ecological Monographs, 68*, 295-323.

Domingues, MS; Bermann, C. (2012). O arco de desflorestamento na Amazônia: da pecuária à soja. *Ambiente & Sociedade, 15*, 2, 1-22.

Dunxião, H; Chunru, H; Yaling, X; Banwang, H; Liyuan, H; Paoletti, MG. (1999). Relationship between soil arthropods and soil properties in a Suburb of Qianjiang City, Hubei, China. *Critical Rewiews in Plant Sciences, 18*, 467-473.

Epstein, ME; Kulman, HM. (1990). Habitat distribution and seasonal occurrence of Carabid beetles in East Central Minnesota. *The American Midland Naturalist, 123*, 209-225.

Ermakov, AI. (2004). Structural changes in the carabid fauna of forest ecosystems under a toxic impact. *Russian Journal of Ecology, 35*, 403-408.

Escobar, F; Halffter, G; Arellano, L. (2007). From forest to pasture: an evaluation of the influence of environment and biogeography on the structure of dung beetle (Scarabaeinae) assemblages along three altitudinal gradients in the Neotropical region. *Ecography, 30*, 193-208.

Escobar, F; Halffter, G; Solís, A; Halffter, V; Navarrete, D. (2008). Temporal shifts in dung beetle community structure within a protected area of tropical wet forest: a 35-year study and its implications for long-term conservation. *Journal of Appllied Ecology, 45*, 1584-1592.

Estrada, A; Halffter, G; Estrada, RC; Meritt, A. (1993). Dung beetles attracted to mammalian herbivore (Alouattapalliata) and omnivore (Nasuanarica) dung in the tropical rain forest of Los Tuxtlas, Mexico. *Journal of Tropical Ecology, 9*, 45-54.

Eyre, MD; Luff, ML. (1990). A preliminary classification of European grassland habitats using carabid beetles. In NE Stork (Ed.). *The Role of Ground Beetles in Ecological and Environmental Studies* (pp. 227-236). Intercept, Andover, UK.

Fattorini, S. (2010). Effects of fire on tenebrionid communities of a Pinus pinea plantation: a case study in a Mediterranean site. *Biodiversity and Conservation, 9*, 1237-1250.

Fattorini, S; Dennis, RLH; Cook, LM. (2011). Conserving organisms over large regions requires multi-taxa indicators: one taxon's diversity-vacant area is another taxon's diversity zone. *Biological Conservation, 144*, 1690-1701.

Fearnside, PM. (1995). Hydroelectric dams in the Brazilian Amazon as sources of 'greenhouse' gases. *Environmental Conservation, 22*, 7-19.

Fearnside, PM. (2001). Soybean cultivation as a threat to the environment in Brazil. *Environmental Conservation, 28*, 23-38.

Fearnside, PM. (2005a). Deforestation in Brazilian Amazonia: History, rates and consequences. *Conservation Biology, 19*, 680-688.

Fearnside, PM. (2005b). Brazil's Samuel Dam: Lessons for hydroelectric development policy and the environment in Amazonia. *Environmental Management, 35*, 1-19.

Fearnside, PM. (2006). Dams in the Amazon: Belo Monte and Brazil's Hydroelectric Development of the Xingu River Basin. *Environmental Management, 38*, 16-27.

Fearnside, PM. (2008). Amazon Forest Maintenance as a Source of Environmental Services. *Anais da Academia Brasileira de Ciências, 80*, 101-114.

Fernandes, AL. (2009). The social and environmental impacts of industrial agriculture in the Legal Amazon. In *Anais XIV Simpósio Brasileiro de Sensoriamento Remoto, Natal, Brasil*, (pp. 159-165). INPE.

Fränzle, O. (2006). Complex bioindication and environmental stress assessment. *Ecological Indicators, 6*, 114-136.

Freitas, AVL; Leal, IR; Uehara-Prado, M; Iannuzzi, L. (2006). Insetos como indicadores de conservação da paisagem. In CFD Rocha; M, Van Sluys; HG Bergallo; MAS Alves. (Org.). *Biologia da Conservação* (pp. 357-385). São Carlos, Rima Editora.

Garcia, M; Ortego, F; Castanera, P; Farinos, GP. (2010). Effects of exposure to the toxin Cry1Ab through Bt maize fed-prey on the performance and digestive physiology of the predatory rove beetle Atheta coriaria. *Biological Control, 55*, 225-233.

García-Cárdenas, R; Armbrecht, I; Ulloa, CP. (2001). Staphylinidae (Coleoptera): Composición y mirmecofilia en bosques secos relictuales de Colombia. *Folia Entomológica Mexicana, 40*, 1-10.

García-Cárdenas, R; Ulloa, CP. (2005). Estafilínidos (Coleoptera: Staphylinidae) en fragmentos de bosque seco del valle geográfico del río Cauca. *Revista Colombiana de Entomología, 31*, 43-50.

Gardner, TA; Barlow, J; Parry, LTW; Peres, CA. (2007a). Predicting the uncertain future of tropical forest species in a data vacuum. *Biotropica, 39*, 25-30.

Gardner, TA; Hernández, MIM; Barlow, J; Peres, CA. (2008). Understanding the biodiversity consequences of habitat change: the value of secondary and plantation forests for neotropical dung beetles. *Journal of Applied Ecology, 45*, 883-893.

Gardner, TA; Ribeiro, MA; Barlow, J; Ávila-Pires, TAS; Hoogmoed, M; Peres, CA. (2007b). The value of primary, secondary and plantation forests for a neotropical herpetofauna. *Conservation Biology, 21*, 775-787.

Gaston, KJ; Blackburn, TM. (1995). Mapping biodiversity using surrogates for species richness: macro-scales and New World birds. *Proceedings of the Royal Society of London B, 262*, 335-341.

Gaston, KJ; Warren, PH; Hammond, PM. (1992). Predator: non-predator ratios in beetle assemblages. *Oecologia, 90*, 417-421.

Gerlach, J; Samways, M; Pryke, J. (2013). Terrestrial invertebrates as bioindicators: an overview of available taxonomic groups. *Journal of Insect Conservation, 17*, 831-850.

Gill, BD. (1991). Dung beetle in Tropical American Forest. In I Hanski; Y Cambefort (Eds.). *Dung beetle ecology*. (pp. 211-229). Princeton, Princeton University Press.

Giller, KE; Beare, MH; Lavelle, P; Izac, AMN; Swift, MJ. (1997). Agricultural intensification, soil biodiversity and agroecosystem function. *Applied Soil Ecology 6*, 3-16.

Godfray, HCJ; Lewis, OT; Memmott, J. (1999). Studying insect diversity in the tropics. *Philosophical Transactions of the Royal Society of London, Series B, 354*, 1811-1824.

Gomez, AD. (2010). Activity density versus biomass distribution patterns among ground-beetle species as bioindicator for conservation planning over a wide range of altitudes in Tenerife (Canary Islands). *Ecological Indicator, 10*, 1213-1217.

Gongalsky, KB; Chudnyavtseva, II; Pokarzhevskii, AD; Samonov, AE; Slobodyan, VY. (2004). Arsenic bioaccumulation by beetles in an arsenic-rich region. *Bulletin of Environmental Contamination and Toxicology, 72*, 1115-1121.

Grebennikov, VV; Leschen, RAB. (2010). External exoskeletal cavities in Coleoptera and their possible mycangial functions. *Entomological Science,13*, 81-98.

Hajibabaei, M; Janzen, DH; Burns, JM; Hallwachs, W; Hebert, PDN. (2006a). DNA barcodes distinguish species of tropical Lepidoptera. *Proceedings of the National of Sciences, 103*, 968-971.

Hajibabaei, M; Smith, MA; Janzen, DH; Rodriguez, JJ; Whitfiels, JB; Hebert, PDN. (2006b). A minimalist barcode can identify a specimen whose DNA is degraded. *Molecular Ecology Notes, 6*, 959-964.

Halffter, G. (1991). Historical and ecological factors determining the geographical distribution of beetles (Coleoptera: Scarabaeidae: Scarabaeinae). *Folia Entomológica Mexicana, 82*, 195-238.

Halffter, G. (2005). Towards a culture of biodiversity conservation. *Acta Zoológica Mexicana, 21*, 133-153.

Halffter, G; Arellano, L. (2002). Response of dung beetle diversity to human-induced changes in a tropical landscape. *Biotropica, 34*, 144-154.

Halffter, G; Matthews, EG. (1966). The natural history of dung beetles of the subfamily Scarabaeinae (Coleoptera: Scarabaeidae). *Folia Entomológica Mexicana, 12*, 1-312.

Halme, E; Niemelä, J. (1993). Carabid beetles in fragments of coniferous forest. *Annales Zoologici Fennici, 30*, 17-30.

Hamer, KC; Hill, JK. (2000). Scale-dependent effects of habitat disturbance on species richness in tropical forests. *Conservation Biology, 14*, 1435-1440.

Hammond, HEJ. (1997). Arthropod biodiversity from Populus coarse woody material in North-central Alberta: a review of taxa and collection methods. *The Canadian Entomologist, 129*, 1009-1033.

Hebert, PDN; Cywinska, A; Ball, SL; Waard, JR. (2003). Biological identifications through DNA barcodes. *Proceedings of the Royal Society B, 270*, 313-321.

Hebert, PDN; Stoeckle, MY; Zemlak, TS; Francis, CM. (2004). Identification of birds through DNA barcodes. *Plos Biology, 2*, 1657-1663.

Heink, U; Kowarik, I. (2010). What are indicators? On the definition of indicators in ecology and environmental planning. *Ecological Indicators, 10*, 584-593.

Heliovaara, K; Vaisanen, R. (1993). *Insects and Pollution*. Boca Raton: CRC Press.

Henschel, JR; Grohmann, C; Siteketa, V; Linsenmair, KE. (2010). Monitoring Tenebrionid Beetle Biodiversity in Namibia. *African Study Monographs, 40*, 91-102.

Hequet V. (1996). *Longicornes de Guyane*. Cayenne, Orstom: Silvolab.

Hernández, MIM. (2007). Besouros escarabeíneos (Coleoptera: Scarabaeidae) da Caatinga paraibana, Brasil. *Oecologia Brasiliensis, 11*, 356-364.

Hill, JK; Hamer, KC. (2004). Determining impacts of habitat modification on diversity of tropical forest fauna: the importance of spatial scale. *Journal of Applied Ecology, 41*, 744-754.

Hill, JK; Hamer, KC; Lace, LA; Banham, WMT. (1995). Effects of selective logging on tropical forest butterflies on Buru, Indonesia. *Journal of Applied Ecology, 32*, 754-760.

Hodkinson, IA; Jackson, JK. (2005). Terrestrial and Aquatic Invertebrates as Bioindicators for Environmental Monitoring, with Particular Reference to Mountain Ecosystems. *Environmental Management, 35*, 649-666.

Hoffmann, RB; Nascimento, MSV; Diniz, AA; Araújo, LHA; Souto, JS. (2009). Diversidade da mesofauna edáfica como bioindicadora para o manejo do solo em Areia, Paraíba, Brasil. *Revista Caatinga, 22*, 121-125.

Holloway, AK; Schnell, GD. (1997). Relationship between numbers of the endangered American burying beetle *Nicrophorus americanus* Oliver (Coleoptera: Silphidae) and available food resourses. *Biological Conservation, 81*, 145-152.

Holloway, JD; Bradley, JD; Carter, JD. (1981). *CIE guides to insects of importance to man.* Lepidoptera, 1.C.A.B. International: Wallinford.

Howden, HF; Nealis, VG. (1975). Effects of clearing in a tropical rain forest on the composition of the coprophagous scarab beetle fauna (Coleoptera). *Biotropica*,**7**, 77-83.

Hunter, MD. (2002). Lanscape structure, habitat fragmentation, and the ecology of insects. *Agricultural and Forest Entomology, 4*, 159-166.

Imbrósio, WB. (2012). *Resposta da diversidade funcional de scarabaeinae (Coleoptera: Scarabaeidae) aos diferentes usos de solo na Amazônia*. MSc. Thesis. Universidade Federal de Lavras, Minas Gerais, Brasil.

Instituto Brasileiro de Geografia e Estatística - IBGE. (2000). *Censo demográfico do Brasil.* Available on: www.ibge.gov.br. [Accessed in: 20/01/2014].

Instituto Brasileiro de Geografia e Estatística-IBGE. (2006). *Soja bate novo recorde de produção em 2006.* http://saladeimprensa.ibge.gov.br/noticias?view=noticia&id=1&busca=1&idnoticia=931. [Accessed in: 15/04/2014].

Instituto Brasileiro do Meio Ambiente e dos Recursos Naturais Renováveis-IBAMA. (2004). *Contextualização de Unidades de Conservação*: Curso de Capacitação de Analistas Ambientais. [Palestra]. Brasília.

*Instituto Nacional de Pesquisas Espaciais-INPE. (2014). Monitoramento da Floresta Amazônica brasileira por satélite.*Available on: http://www.obt.inpe.br/prodes/prodes_1988_2013.htm [Accessed in: 21/05/2014).

Intachat, J; Holloway, JD; Speight, MR. (1999). The impact of logging on geometroid moth populations and their diversity in lowland forests of Peninsular Malaysia. *Journal of Tropical Forest Science, 11*, 61-78.

Iperti, G. (1999). Biodiversity of predaceous coccinellidae in relation to bioindication and economic importance. *Agriculture, Ecosystems and Environment, 74*, 323-342.

Jacobs, CT; Scholtz, CH; Escobar, F; Davis, ALV. (2010). How might intensification of farming influence dung beetle diversity (Coleoptera: Scarabaeidae) in Maputo Special Reserve (Mozambique)? *Journal of Insect Conservation, 14*, 389-399.

Janzen, DH; Hajibabaei, M; Burns, JM; Hallwachs, W; Remigio, E; Hebert, PDN. (2005). Wedding biodiversity inventory of a large and complex Lepidoptera fauna with DNA barcoding. *Philosophical Transactions of the Royal Societ B, 360*, 1835-1845.

Kaiser, W; Avenant, NL; Haddad, CR. (2009). Assessing the ecological integrity of a grassland ecosystem: the applicability and rapidity of the SAGraSS method. *African Journal of Ecology, 47*, 308-317.

Kerr, KCR; Lijtmaer, DA; Barreira, AS; Hebert, PDN; Tubaro, PL. (2009). Probing evolutionary patterns in neotropical birds through DNA barcodes. *Plos One* 4:e4379.

Kimberling, DN; Karr, JR; Fore, LS. (2001). Measuring human disturbance using terrestrial invertebrates in the shrub-steppe of eastern Washington (USA). *Ecological Indicators, 1*, 63-81.

Klein, BC. (1989). Effects of forest fragmentation on dung and carrion beetles communities in Central Amazonia. *Ecology, 70*, 1715-1725.

Koh, LP. (2007). Impacts of land use change on south-east Asian forest butterflies: a review. *Journal of Applied Ecology, 44*, 703-713.

Koivula, M; Kukkonen, J; Niemelä, J. (2002). Boreal carabid-beetle (Coleoptera, Carabidae) assemblages along the clear-cut originated succession gradient. *Biodiversity and Conservation, 11*, 1269-1288.

Koivula, M; Punttila, P; Haila, Y; Niemela, J. (1999). Leaf litter and the smallscale distribution of carabid beetles (Coleoptera, Carabidae) in the boreal forest. *Ecography, 22*, 424-435.

Korasaki, V; Braga, RF; Zanetti, R; Moreira, FMS; Vaz-de-Mello, FZ; Louzada, J. (2013). Conservation value of alternative land-use systems for dung beetles in Amazon: valuing traditional farming practices. *Biodiversity and Conservation, 22*, 1485-1499.

Kromp, B. (1999). Carabid beetles in sustainable agriculture: a review on pest control efficacy, cultivation impacts and enhancement. *Agriculture, Ecosystem & Environment, 74*, 187-228.

Langor, DW; Spence, JR. (2006). Arthropods as ecological indicators. *The Forestry Chronicle, 82*, 344-350.

Lara, FM. (1992). *Princípios de Entomologia*. Ícone: São Paulo, SP.

Larsen, TH; Forsyth, A. (2005). Trap spacing and transect design for dung beetle biodiversity studies. *Biotropica, 37*, 322-325.

Laurance, WF; Albernaz, AKM; Costa, C. (2001b). Is deforestation accelerating in the Brazilian Amazon? *Environmental Conservation, 28*, 4, 305-311.

Laurance, WF; Bierregaard, RO. (1997). *Tropical Forest Remnants*: Ecology, Management, and Conservation of Fragmented Communities. The University of Chicago Press: Chicago, IL.

Laurance, WF; Cochrane, MA; Bergen, S; Fearnside, PM; Delamônica, P; Barber, C; D'Angelo, S; Fernandes, T. (2001). The future of the Brazilian Amazon. *Science, 291*, 438-439.

Lavelle, P. (1996). Diversity of soil fauna and ecosystem function. *Biology International, 33*, 3-15.

Lentini, M; Pereira, D; Celentano, D; Pereira, R. (2005). *Fatos Florestais da Amazônia*. Belém, Brazil: Imazon.

Lewinsohn, TH; Freitas, AVL; Prado, PI. (2005). Conservação de invertebrados terrestres e seus habitats no Brasil. *Megadiversidade, 1*, 62-69.

Li, D; Jackson, RR. (1996). Prey preferences of Portia fimbriata, naaraneophagic, web-building jumping spider (Araneae: Salticidae) from Queensland .*Insect Behavior, 9*, 613-642.

Lima, AA; Lima, WL; Berbara, RLL. (2006). Diversidade da mesofauna de solo em sistemas de produção agroecológica. *Revista Brasileira de Agroecologia, 1*, p. 1199-1202.

Lindenmayer, DB; Margules, CR; Botkin, DB. (2000). Indicators of biodiversity for ecologically sustainable forest management. *Conservation Biology, 14*, 941-950.

Lobo, JM; Lumaret, JP; Jay-Robert, P. (1998). Sampling dung beetles in the French Mediterranean area: effects of abiotic factors and farm practices. *Pedobiologia, 42*, 252-266.

Lopes, PP; Louzada, JNC; Oliveira-Rebouças, P L; Nascimento, LM; Santana-Reis, VPG. (2005). Resposta da Comunidade de Histeridae (Coleoptera) a Diferentes Fisionomias da Vegetação de Restingas no Espírito Santo. *Neotropical Entomology, 34*, 025-031.

Loranger, G; Ponge, JF; Blanchart, E; Lavelle, P. (1998). Influence of agricultural practices on arthropod communities in a vertisol (Martinique). *European Journal of Soil Biology, 34*, 157-165.

Lui, GH; Molina, SMG. (2009). Ocupação humana e transformação das paisagens na Amazônia brasileira. *Amazônica - Revista de Antropologia, 1*, 200-229.

Majer JD. (1981). *The role of invertebrates in bauxite mine rehabilitation*. Bulletin, 93.Peth: Forests Departament of Western Australia.

Maluche, CRD; Santos, JCP; Sinhorati, D; Amarante, CVT; Baretta, D. (2003). *Fauna edáfica como bioindicadora da qualidade do solo em pomares de macieiras conduzidos nos sistemas orgânico e convencional* (pp. 1-4). In: Congresso Brasileiro de Agroecologia. Porto Alegre: EMATER/RS-ASCAR.

Marinoni, RC. (2001). Os grupos tróficos em Coleoptera. *Revista Brasileira de Zoologia, 18*, 205-224.

Marinoni, RC; Dutra, RRC. (1997). Famílias de Coleoptera capturados com armadilhas Malaise em oito localidades do estado do Paraná, Brasil. Diversidades alfa e beta. *Revista Brasileira de Zoologia,14*, 751-770.

Marinoni, RC; Ganho, NG. (2003). Sazonalidade de *Nyssodrysina lignaria* (Bates) (Coleoptera, Cerambycidae, Lamiinae) no Estado do Paraná, Brasil. *Revista Brasileira de Zoologia, 20*, 141-152.

Marques, AAB; Fontana, CS; Vélez, E; Bencke, GA; Schneider, M; Reis, RE. (2002). *Lista de Referência da Fauna Ameaçada de Extinção no Rio Grande do Sul*. Decreto nº 41.672, de 10 de junho de 2002. Porto Alegre: FZB/MCTPUCRS/PANGEA.

Martins, UR. (1999). Cerambycidae. In CRF Brandão; EM Cancello (Eds.). *Biodiversidade do Estado de São Paulo, Brasil*. (pp. 125-132). São Paulo, FAPESP.

Maryański, M; Kramarz, P; Laskowski, R; Niklińska, M. (2002). Decreased energetic reserves, morphological changes and accumulation of metals in carabid beetles (*Poecilus cupreus* L.) exposed to Zinc-and Cadmium-contaminated food. *Ecotoxicology, 11*, 127-139.

Matavelli, RA; Louzada, JNC. (2008). Invasion of intra-Amazonian savannas by the dung beetle *Digitontophagus gazelle* (Fabricius, 1787) (Insecta: Coleoptera: Scarabaeidae). *Acta Amazonica, 38*, 153-158.

McGarigal, K; Cushman, S. (2002). Comparative evaluation of experimental approaches to the study of habitat fragmentation effects. *Ecological Applications*, *12*, 335-345.

McGeoch, MA. (1998). The selection, testing and application of terrestrial insects as bioindicators. *Biology Review, 73*, 181-201.

McGeoch, MA; Rensburg, BJV; Botes, A. (2002). The verification and application of bioindicators: a case study of dung beetles in a savanna ecosystem. *Journal of Applied Ecology, 39*, 661-672.

McGeoch, MA; Sithole, H; Samways, MJ; Simaika, JP; Pryke, JS; Picker, M; Uys, C; Armstrong, AJ; Dippenaar-Schoeman, AS; Engelbrecht, IA; Braschler, B; Hamer, M. (2011). Conservation and monitoring of invertebrates in terrestrial protected areas. *Koedoe, 53*, 1-13.

Monné, MA; Hovore FT. (2005). *Electronic Checklist of the Cerambycidae of the Western Hemisphere*.<http://www.cerambycids.com/checklist>. [Accessed in 08/05/2014].

Moreira, FMS; Nóbrega, RSA; Jesus, EC; Ferreira, DF; Pérez, DV. (2009). Differentiation in the fertility of inceptisols as related to use in the upper Solimões river region, western Amazon. *Scince of the Total Environment, 408*, 349-355.

Mozdzer, T; Kramarz, P; Piúkiewicz, A; Niklińska, M. (2001). *Accumulation of cadmium and zinc in the larval stage of the ground beetle, Pterostichus oblongopunctatus*. In Abstracts of the SECOTOX World Congress: Ecotoxicology and Environmental Safety on the Verge of Third Millennium: Trends, Threats and Challenges, Kraków, Poland.

Navarrete-Heredia, JL; Newton, AF; Thayer, MK; Ashe, JS; Chandler, DS. (2002). *Guía ilustrada para los géneros de Staphylinidae (Coleoptera) de México*. Universidad de Guadalajara and CONABIO: Guadalajara, Mexico.

Negro, M; Isaia, M; Palestrini, C; Schoenhofer, A; Rolando, A. (2010). The impact of high-altitude ski pistes on ground-dwelling arthropods in the Alps. *Biodiversity and Conservation, 19*, 1853-1870.

Nepstad, DC; Carvalho, G; Barros, AC; Alencar, A; Capobianco, JP; Bishop, J; Moutinho, P; Lefebvre, P; Silva, UL; Prins, E. (2001). Road paving, fire regime feedbacks, and the future of Amazon forests. *Forest Ecolology an Management, 154*, 395-407.

Newton, AF; Gutiérrez-Chacón, C; Chandler, DS. (2005). Checklist of the Staphylinidae (Coleoptera) of Colombia. *Biota Colombiana, 6*, 1-72.

Nichols, E; Larsen, T; Spector, S; Davis, AL; Escobar, F; Favila, M; Vulinec, K. (2007). Global dung beetle response to tropical forest modification and fragmentation: A quantitative literature review and meta-analyses. *Biological Conservation, 137*, 1-19.

Nichols, E; Spector, S; Louzada, J; Larsen, T; Amezquita, S; Favila, ME. (2008). Ecological functions and ecosystem services provided by Scarabaeinae dung beetles. *Biological Conservation, 141*, 1461-1474.

Nichols, ES; Gardner, TA. (2011). Dung Beetles as a Candidate Study Taxon in Applied Biodiversity Conservation Research. In LW Simmons; TJ Ridsdill-Smith (Eds.).*Ecology and Evolution of Dung Beetles*. (pp. 267-290). John Wiley & Sons, Ltd, Chichester, UK.

Niemelä, J. (2001). Carabid beetles (Coleoptera, Carabidae) and habitat fragmentation: a review. *European Journal of Entomology, 98*, 127-132.

Niemelä, J; Kotze, DJ. (2009). Carabid beetle assemblages along urban to rural gradients: A review. *Landscape and Urban Planning, 92*, 65-71.

Niemelä, J; Kotze, J; Ashworth, A; Brandmayr, P; Desender, K; New, T; Penev, L; Samways, M; Spence, J. (2000). The search for common anthropogenic impacts on biodiversity: a global network. *Journal of Insect Conservation, 4*, 3-9.

Noss, RF. (1990). Indicators for Monitoring biodiversity: A hierarchical approach. *Conservation Biology, 4*, 354-364.

Nummelin, M. (1996). The community structure of arthropods in virgin and managed sites in the Kibale Forest, Western Uganda. *Tropical Ecology, 37*, 203-213.

Nummelin, M; Hanski, I. (1989). Dung beetles of the Kibale Forest, Uganda: comparison between virgin and managed forests. *Journal of Tropical Ecology, 5*, 349-352.

Oliveira, EM; Souto, JS. (2001). Mesofauna edáfica como indicadora de áreas degradadas. *Revista Verde de Agroecologia e Desenvolvimento Sustentável, 6*, 01-09.

Oliver, I; Beattie, A; York, A. (1998). Spatial fidelity of plant, vertebrate, and invertebrate assemblages in multipleuse forest in eastern Australia. *Conservation Biology, 12*, 822-835.

Oliver, I; Beattie, AJ. (1996). Invertebrate morphospecies as surrogates for species: a case study. *Conservation Biology, 10*, 99-109.

Padial, JM; De La Riva, I. (2007). Integrative taxonomists should use and produce DNA barcodes. *Zootaxa, 1586*, 67-68.

Paje, F; Mossakowski, D. (1984).pH-preferences and habitat selection in Carabid beetles. *Oecologia, 64*, 41-46.

Paoletti, MG. (1999). Using bioindicators based on biodiversity to assess landscape sustainability. *Agriculture, Ecosystems and Environment, 74*, 1-18.

Paoletti, MG; D'Inca, A; Tonin, E; Tonon, S; Migliorini, C; Petruzzelli, G; Pezzarossa, B; Gomiero, T; Sommaggio, D. (2010). Soil invertebrates as bio-indicators in a natural area converted from agricultural use: the case study of Vallevecchia-Lugugnana in north-eastern Italy. *Journal of Sustainable Agriculture, 34*, 38-56.

Paoletti, MG; Favretto, MR; Stimer, BR; Purrigton, FF; Bater, JE. (1991). Invertebrates as bioindicators of soil use.*Agriculture, Ecosystems and Environment, 34*, 341-362.

Parenzee, L. (2001). *Use of Tenebrionid Beetles as Indicators of Habitat Quality*.MSc. Thesis. University of the Witwatersrand, Johannesburg.

Pearce, JL; Venier, LA. (2006). The use of ground beetles (Coleoptera: Carabidae) and spiders (Araneae) as bioindicators of sustainable forest management: A review. *Ecological Indicators, 6*, 780-793.

Pearce, JL; Venier, LA; McKee, J; Pedlar, J; Mckenney, D. (2003). Influence of habitat and microhabitat on Carabid (Coleoptera: Carabidae) assemblages in four stand types. *Canadian Entomologist, 135*, 337-357.

Pearson, DL. (1994). Selecting indicator taxa for the quantitative assessment of biodiversity. *Philosophical Transactions of the Royal Society of London, Series B, 345*, 75-79.

Pearson, DL; Cassola, F. (2005). A quantitative analysis of species descriptions of tiger beetles (Coleoptera: Cicindelidae), from 1758 to 2004, and notes about related developments in biodiversity studies. *The Coleopterists Bulletin, 59*, 184-193.

Pearson, DL; Cassola, F. (2007). Are we doomed to repeat history? A model of the past using tiver beetles (Coleopter: Cicindelidae) and conservation biology to anticipate the future. *Journal of Insect Conservation, 11*, 47-59.

Peres, CA. (2005). Why we need mega reserves in Amazonia. *Conservation Biology, 19*, 728-733.

Pires, AC; Marinoni, L. (2010). DNA barcoding and traditional taxonomy unified through Integrative Taxonomy: a view that challenges the debate questioning both methodologies. *Biota Neotropica, 10*, 339-346.

Pitman, NCA; Azáldegui, MCL; Salas, K; Vigo, GT; Lutz, DA. (2007). Written accounts of an Amazonian landscape over the last 450 years. *Conservation Biology, 21*, 253-262.

Pohl, GR; Langor, DW; Spence, JR. (2007). Rove beetles and ground beetles (Coleoptera: Staphylinidae, Carabidae) as indicators of harvest and regeneration practices in western Canadian foothills forests. *Biological Conservation, 37*, 294-307.

Quintero, I; Halffter, G. (2009). Temporal changes in a community of dung beetles (Insecta: Coleoptera: Scarabaeinae) resulting from the modification and fragmentation of tropical rain forest. *Acta Zoológica Mexicana, 25*, 625-649.

Quintero, I; Roslin, T. (2005). Rapid recovery of dung beetle communities following habitat fragmentation in Central Amazonia. *Ecology, 86*, 3303-3311.

Radtke, MG; Fonseca, CRV; Williamson, GB. (2007). The old and young Amazon: dung beetle biomass, abundance, and species diversity. *Biotropica, 39*, 725-730.

Rainio, J. (2009). *Carabid beetles (Coleoptera, Carabidae) as indicators of environmental change in Ranomafana National Park, Madagascar*. MSc. Thesis. University of Helsinki, Finland.

Rainio, J; Niemelä, J. (2003): "Ground beetles (Coleoptera: Carabidae) as bioindicators". *Biodiversity and Conservation, 12*, 487-506.

Rivero, S; Almeida, O; Ávila, S; Oliveira, W. (2009). Pecuária e desmatamento: uma análise das principais causas diretas do desmatamento na Amazônia. *Nova Economia, 19*, 41-66.

Rivero, S; Covas, T. (2013). The Processing and Marketing of Seeds from the Amazon City of Porto Velho, RO. *Journal of Technology Management & Innovation, 8*, 321-331.

Rovedder, APM; Eltz, FLF; Drescher, MS; Schenato, RB; Antoniolli, ZI. (2009). Organismos edáficos como bioindicadores da recuperação de solos degradados por arenização no Bioma Pampa. *Ciência Rural, 39*, 1061-1068.

Ruf, A; Beck, L; Dreher, P; Hund-Rinke, K; Römbke, J; Spelda, J. (2003). A biological classification concept for the assessment of soil quality: "Biological soil classification scheme" (BBSK). *Agriculture, Ecosystems & Environmental, 98*, 263-271.

Rykken, JJ; Capen, DE; Mahabir, SP. (1997). Ground beetles as indicators of land type diversity in the green mountains of Vermont. *Conservation Biology, 11*, 522-530.

Saji, A; Al Dhaheri, S. (2011). Ecological Distribution and Seasonality of Darkling Beetles (Coleoptera: Tenebrionidae) in the Western Region of Abu Dhabi, UAE Middle-East. *Journal of Scientific Research, 9*, 704-710.

Samu, F; Kadar, F; Onodi, G; Kertesz, M; Sziranyi, A; Szita, E; Fetyko, K; Botos, E; Altbacker, V. (2010). Differential ecological responses of two generalist arthropod groups, spiders and carabid beetles (Araneae, Carabidae), to the effects of wildfire. *Community Ecology, 11*, 129-139.

Samways, MJ; McGeoch, MA; New, TR. (2010). *Insect conservation*: a handbook of approaches and methods. Oxford University Press: Oxford.

Scheffler, PY. (2005). Dung beetle (Coleoptera: Scarabaeinae) diversity and community structure across three disturbance regimes in eastern Amazonia. *Journal of Tropical Ecology, 21*, 9-29.

Schwarzkopf, L; Rylands, AB. (1989). Primate species richness in relation to habitat structure in Amazonian rainforest fragments. *Biological Conservation, 48*, 1-12.

Schwerk, A; Salek, P; Duszczyk, M; Szyszko, J. (2006). Variability of Carabidae in time and space in open areas. *Entomologica Fennica, 17*, 258-268.

Silva, CF; Fernandes, GW; Santos, JC. (2005). *Comportamento de Dorcadocerus barbatus (Olivier) (Coleoptera: Cerambycidae) em Lantana camara L. (Verbenaceae)*. In: Congresso de Ecologia do Brasil: Caxambu. Resumo n. 616.

Silva, JMC; Rylands, AB; Fonesca, GAB. (2005). The fate of the Amazonian areas of endemism. *Conservation Biology, 19*, 689-694.

Silveira, LF; Beisiegel, BM; Curcio, FF; Valdujo, PH; Dixo, M; Verdade,VK; Mattox, GMT; Cunningham, PTM. (2010). Para que servem os inventários de fauna? *Estudos Avançados, 24*, 173-207.

Smith, MA; Wood, DM; Janzen, DH; Hallwachs, W; Hebert, PD. (2007). DNA barcodes affirm that 16 species of apparently generalist tropical parasitoid flies (Diptera, Tachinidae) are not all generalists. *Proceedings of the National of Sciences, 104*, 4967-4972.

Smith, RJ; Merrick, MJ. (2001). Resource availability and population dynamics of Nicrophorus investigator, an obligate carrion breeder. *Ecological Entomology, 26*, 173-180.

Soares-Filho, BS; Alencar, A; Nepstad, D; Cerqueira, G; Diaz, M; Del, CV; Rivero, S; Solórzano, L; Voll, E. (2004). Simulating the response of landcover changes to road paving and governance along a major Amazon highway: The Santarém-Cuiabá corridor.*Global Change Biology, 10*, 745-764.

Soares-Filho, BS; Nepstad, DC; Curran, L; Cerqueira, GC; Garcia, RA; Ramos, CA; Voll, E; McDonald, V; Lefebvre, P; Schlesinger, P; McGrath, D. (2005). Cenários de desmatamento para a Amazônia. *Estudos Avançados, 19*, 137-152.

Sotherton, NW. (1984). The distribution and abundance of predatory arthropods overwintering on farmland. *Annals of Applied Biology,105*, 423-429.

Souza, DS; Silva, AA. (2012). Cerambycidae (Insecta: Coleoptera) of the Parque Natural Municipal de Porto Velho, Rondônia, Western Amazon, Brazil. *Biota Neotropica, 12*, 237-240.

Souza-Rodrigues, EA. (2011). Demand for Deforestation in the Amazon. Job Market Paper. Availabe in: http://cbey.yale.edu/uploads/Environmental%20Economics%20 Seminar/Souza-Rodrigues_Deforestation.pdf. [Accessed in 05/02/2014].

Sowig, P. (1995). Habitat selection and offspring survival rate in three paracoprid dung beetles: the influence of soil type and soil moisture. *Ecograpahy, 18*, 147-154.

Spector, S. (2006). Scarabaeine dung beetles (Coleoptera: Scarabaeidae: Scarabaeinae): an invertebrate focal taxon for biodiversity research and conservation. *The Coleopterists Bulletin, 5*, 71-83.

Spector, S; Ayzama, S. (2003). Rapid turnover and edge effects in dung beetle assemblages (Scarabaeidae) at a Bolivian Neotropical forest-savanna ecotone. *Biotropica, 35*, 394-404.

Stinner, BR; House, GJ. (1990). Arthropods and other invertebrates in conservation-tillage agriculture. *Annual Review of Entomology, 35*, 299-318.

Stork, NE; Eggleton, P. (1992). Invertebrates as determinants and indicators of soil quality. *American Journal of Alternative Agriculture, 7*, 38-47.

Summerlin, JW. (1989). Techniques for collecting, rearing and handling histerid beetles. *Southwester Entomology, 14*, 127-132.

Summerlin, JW; Fincher, GT; Roth, JP; Meola, SM. (1991). Laboratory observations on the life history and habitats of *Phelisterhae morrhous*. *Southwester Entomology 16*, 311-315.

Teixeira, FM. (2006). *A composição de Scarabaeidae (Coleoptera) coprófagos na Região de Alter do Chão, Pará*: A influência dos biomas Amazônia e Cerrado e da sazonalidade e os efeitos de tamanho de área, isolamento e proximidade de estradas. MSc. Thesis. Universidade Federal do Pará, Brasil.

Thakare, VG; Zade, VS; Hegde, VD. (2012). Darkling Beetles (Coleoptera: Tenebrionidae) of Melghat Tiger Reserve, Central India. *Journal on New Biological Reports, 1*, 29-32.

Thiele, HU. (1977). *Carabid Beetles in Their Environments*. Berlin: Springer Verlag.

Thomazini, MJ; Thomazini, APBW. (2002). *Levantamento de Insetos e Análise Entomofaunística em Floresta, Capoeira e Pastagem no Sudeste Acreano*. Boletim de Pesquisa e Desenvolvimento (Embrapa Acre).

Van Straalen, NM; Krivolutsky, DA. (1996). *Bioindicator Systems for Soil Pollution*. Kluwer Academic Publishers: Dordrecht.

Vasquez-Velez, LM; Bermudez, C; Chacon, P; Lozano-Zambrano, FH. (2010). Analysis of the richness of Staphylinidae (Coleoptera) on different scales of a sub-Andean rural landscape in Colombia. *Biodiversity and Conservation, 19*, 1917-1931.

Venier, LA; Pearce, JL. (2004). Birds as indicators of sustainable forest management. *Forestry Chronicle, 80*, 61-66.

Venn, S. (2000). *The effects of urbanization on boreal forest ecosystems*. MSc. Thesis. University of Helsinki, Helsinki, Finland.

Veríssimo, A; Barreto, P; Tarifa, R; Uhl, C. (1995). Extraction of a high-value natural resource in Amazonia: the case of Mahogany. *Forest Ecology and Management, 72*, 39-60.

Verweij, V; Schouten, M; Van Beukering, P; Triana, J; Van der Leeuw, K; Hess, S. (2009). *Keeping the Amazon forests standing: a matter of values*. WWF: Netherlands.

Vulinec, K. (2000). Dung beetles (Coleoptera: Scarabaeidae), monkeys and conservation in Amazonia. *Florida Entomologist, 83*, 229-241.

Vulinec, K. (2002). Dung beetle communities and seed dispersal in primary forest and disturbed land in Amazonia. *Biotropica, 34*, 297-309.

Vulinec, K; Fonseca, CRV; Mellow, DJ. (2009). Scarabaeidae (Coleoptera: Lamellicornia) coprófagos da Reserva Florestal Ducke, Manaus, Brasil. In: CRV Fonseca; C Magalhães; JA Rafael; E Franklin (Org.). *A fauna de Artrópodes da Reserva Florestal Ducke. Parte II - Biologia e Ecologia* (pp. 249-261). Manaus, INPA.

Freitas, AVL; Leal, IR; Uehara-Prado, M; Iannuzzi, L. (2006). Insetos como indicadores de conservação da paisagem. In CFD Rocha; M Van Sluys; HG Bergallo; MAS Alves. (Org.). *Biologia da Conservação* (pp. 357-385). São Carlos, Rima Editora.

Vulinec, K; Lima, AP; Carvalho-Jr, EAR; Mellow, DJ. (2008). Dung beetles and long-term habitat fragmentation in Alter do Chão, Amazônia, Brazil. *Tropical Conservation Science, 1*, 111-121.

Ward, RD; Zemlak, ST; Innes, BH; Last, PR; Hebert, PDN. (2005). DNA barcoding Australia´s fish species. *Philosophical Transactions of the Royal Society B 360*, 1847-1857.

Whitmore, TC. (1997). Tropical forest disturbance, disappearance, and species loss.In WF Laurance; RO Bierregaard (Eds.). *Tropical Forest Remnants: Ecology, Management, and Conservation of Fragmented Communities*. (pp. 3-12). The University of Chicago Press, Chicago, IL.

Wink, C; Guedes, JVC; Fagundes, CK; Rovedder, AP. (2005). Insetos edáficos como indicadores da qualidade ambiental. *Revista de Ciências Agroveterinárias, 4*, 60-71.

Wolda, H. (1987). Altitude, habitat and tropical insect diversity. *Biological Journal of the Linnean Society, 30*, 313-323.

Yi, L; You, W; Song, Y. (2006). Soil animal communities at five succession stages in the litter of the evergreen broad-leaved forest in Tiantong, China.*Frontiers of Biology in China, 1*, 142-150.

Zanuncio, JC; Bragança, MAL; Laranjeiro, AJ; Fagundes, M. (1993). Coleópteros associados à eucaliptocultura nas regiões de São Mateus e Aracruz, Espírito Santo. *Revista Ceres, 40*, 584-590.

In: Beetles: Biodiversity, Ecology and Role in the Environment ISBN: 978-1-63463-380-2
Editor: Camilla Stack

Chapter 4

TWIG-GIRDLER BEETLES OF THE ATLANTIC RAINFOREST

Alberto Arab[1,*], Claudia Moreno Paro[2], Gustavo Quevedo Romero[2], Gustavo Schiffler[2] and João Vasconcellos-Neto[2]

[1]Centro de Ciências Naturais e Humanas (CCNH), Universidade Federal do ABC – UFABC, Santo André-SP,

[2]Instituto de Biologia, Departamento de Biologia Animal, Universidade Estadual de Campinas, Unicamp, Campinas-SP, Brazil

ABSTRACT

Cerambycids constitute one of the most diverse groups of insects in the world, with approximately 35,000 described species. They are commonly cited as models in several areas of Entomology, especially regarding their interactions with host plants and their economic impacts on agriculture. Females of the tribe Onciderini (Cerambycidae: Lamiinae) are twig-girdlers. They cut the branches of healthy plants, perforate the bark and insert their eggs. Usually, the branch falls to the ground and larvae develop inside the branch on the forest floor. Although some information is available for non tropical species, the biology of twig-girdler beetles and their influence on the floral communities in the neotropics remains largely unknown. This chapter will cover the current available data on the diversity of twig-girdler beetles (Onciderini) of the Atlantic rainforest of Brazil, with special emphasis on community composition, population dynamics, host plants range and ecological interactions. The first two sections are an introduction of this chapter and description of the study site. The third section will focus on diversity and host-plant preference. The community structure and the phenology will be also presented in this sections. In the fourth section the authors will discourse on the effects of twig-girdler beetles on host-plants population dynamics and community structure in their natural habitats. In the fifth and final section the authors will discuss the host utilization and microhabitat specialization of these beetles.

* Corresponding author: E-mail: albertoarab@gmail.com.

1. Introduction

Cerambycid beetles are known by their high diversity (an estimate of 35,000 species) and polyphagous behavior [Linsley 1959]. The success of these group of insects has been directly associated with the radiation of angiosperms and the proliferation of plant-history traits [Farrell, 1998]. In most cerambycids, females search for cracks and bark crevices of host plants branches [Hanks 1999], but a more specialized behavior occurs in the tribe Onciderini (Lamiinae). Adult females from several genera of Onciderini are twig-girdlers [Dillon and Dillon, 1945]. They cut the branches of healthy plants, perforate the bark and insert their eggs. Usually, the branch falls to the ground and larvae develop inside the branch on the forest floor (Figure 1). These beetles are essentially homogeneous in their behaviors, with adults often feeding, mating, and ovipositing on the same host-plant individual [Hanks 1999]. Depending on the intensity of the attack, twig-girdlers can cause plant mortality, reduction of reproductive capacity, penetration of pathogens, as well as alterations in the recruitment and age structure of the host-plant populations [Caraglio et al. 2001; Romero et al. 2005; Uribe-Mú and Quesada 2006].

Onciderini beetles are widely distributed in the Nearctic and Neotropical regions, consisting of approximately 480 species and 80 genera. In general, cerambycids play a crucial role in ecosystem processes, such as nutrient cycling and succession [Hanks 1999], since they initiate and are primary agents of wood degradation in forests [Edmonds and Eglitis 1989]. Despite the importance of understanding the association with their host plants, this issue has received little attention. In Brazil, vast areas of tropical forests have been converted into farmland in the last decades [Gibbs et al., 2010]. As an example, the Atlantic rain forest remains on only 7% of its original covered area. This ecosystem is known as one of the world's biodiversity hotspots where forests containing myriad endemic and threatened species are being substituted by large scale farming of commodity crops (such as coffee, sugarcane and logging) [Prance et al. 2000]. Over time, these large regions have been transformed into small fragments due to the expansion of agricultural and urban areas [Viana and Tabanez 1996].

Figure 1. *Oncideres saga* and *O. dejeani* females girdling host-plant branches at the study site.

Considering this scenario and the fact that knowledge on the biodiversity of the Atlantic rain forest is a keystone in any conservation effort, in the present chapter we discuss different approaches of the biology and ecology of twig-girdler beetles (Onciderini) of the Atlantic rainforest of Southeastern Brazil, with special emphasis on community composition, population dynamics, host plants range and ecological interactions. First, we describe the diversity and host-plant preference and how climatic factors influence beetle phenology. Second, we provide data about host-plant utilization and we discuss the effects of twig-girdler beetles on host-plants population dynamics and community structure in their natural habitat.

2. Study Site and Data Collection

This study was conducted at the Biological Reserve of the Serra do Japi (23°14' S; 46°56' W, 35.400 ha, altitude 700 - 1300 m), located at Jundiaí city, São Paulo State, Southeastern Brazil. The Serra do Japi is one of the last continuous fragments of the Atlantic rain forest in Brazil, representing an important natural corridor containing high diversity of plant and animal species. The climate of the region is seasonal and characterized by a temperature range from 11.8 to 15.3^{o} C in the winter (June to August) and from 18.4 °C to 22.2 °C in the summer (December to February) and an annual accumulated precipitation of approximately 1500 mm [Pinto 1992]. The region experiences a single dry season from June to September. The vegetation of the Serra do Japi is considered a transitional forest between Altitude Semi-Deciduous Mesophyll and Semi-Deciduous Mesophyll Forests and the flora was estimated in 864 species of vascular plants from 141 families, with a maximum canopy height of 15 m [Leitão-Filho 1992; Rodrigues and Shepherd 1992; Lombardi et al. 2012]. Field work was carried out for four consecutive years from December 2002 to December 2006 along 5.2 Km of trails in areas 4 m wide on both sides. The structure of the vegetation was similar along the trails. Data collection were conducted on host-plants (summarized in Table 1) with over 1 cm of diameter at breast height (DBH) [Paro et al. 2011]. Sampling included direct observation and manual capture of adult beetles and girdled branches when there was evidence of girdling activity on the accesible parts of plants. Since some Onciderini girdle apical or secondary branches of host-plants, the collection of these insects was possible by using ropes to climb trees in the forest after detecting girdled branches on the floor.

3. Diversity of Twig-Girdlers Beetles of the Atlantic Rainforest

3.1. Community Structure and Host Plants

Twelve species of Onciderini beetles were found associated to 35 native species of plants in the the Atlantic rain forest of the Serra do Japi: *Oncideres bucki* Melzer 1934; *Oncideres captiosa* Martins 1981; *Oncideres cervina* Thomson 1968; *Oncideres dejeani* Thomson 1968; *Oncideres humeralis* Thomson 1968; *Oncideres impluviata* Germar 1824; *Oncideres irrorata* Melzer 1934; *Oncideres macra* Thomson 1968; *Oncideres miniata* Thomson 1968; *Oncideres saga* Dalman 1823; *Oncideres ulcerosa* Germar 1824 and *Psyllotoxus griseocinctus* Thomson

1968. Of the beetles, five species had only one host-plant species (*O. bucki*, *O. cervina*, *O. miniata*, *O. irrorata* and *O. macra*); *O. ulcerosa* had two, *O. captiosa* had four, *O. humeralis* had five, two beetle species (*O. impluviata* and *O. dejeani*) had seven host plants and *P. griseocinctus* had eight (Table 1).

Table 1. Host-plants species associated with Onciderini beetles at the Serra do Japi

Plant family	Species	Number of available plants	Beetle species
Mimosaceae	*Acacia foliolosa*	16	*Oncideres impluviata*
	Acacia polyphyla	26	*O. saga*
	Albizia polycephala	22	*O. saga*
	Inga edulis	105	*O. dejeani; O. impluviata; O. saga*
	Inga sessilis	35	*O. impluviata*
	Leucochloron incuriale	36	*O. saga*
	Piptadenia gonoacantha	42	*O. impluviata*
	Senna multijuga	23	*O. saga*
Euphorbiaceae	*Alchornea glandulosa*	7	*O. dejeani; O. saga; O. ulcerosa*
	Croton floribundus	8	*O. impluviata*
	Croton urucurana	9	*O. dejeani*
Bombacaceae	*Ceiba speciosa*	11	*O. dejeani*
	Pseudobombax grandiflorum	21	*O. dejeani*
Myrtaceae	*Psidium cattleianum*	8	*P. griseocinctus*
	Psidium guajava	39	*O. saga; P. griseocinctus*
	Eugenia pyriformis	12	*P. griseocinctus*
	Eugenia uniflora	64	*P. griseocinctus*
Melastomataceae	*Miconia jucunda*	84	*O. humeralis*
	Miconia latecrenata	3	*O. humeralis*
	Miconia sellowiana	96	*O. humeralis*
	Leandra gracilis	57	*O. humeralis*
	Leandra scabra	18	*O. humeralis*
Lauraceae	*Nectandra rigida*	72	*O. captiosa*
	Ocotea puberula	81	*O. captiosa*
	Persea pyrifolia	133	*O. captiosa; O. cervina; P. griseocinctus*
Meliaceae	*Cedrella fissilis*	24	*O. dejeani*
Urticaceae	*Cecropia glaziovii*	9	*O. dejeani*
Myrsinaceae	*Myrsina ferruginea*	5	*O. impluviata*
Anacardiaceae	*Lithraea molleoides*	92	*O. ulcerosa*
Fabaceae	*Cajanus cajan*	63	*O. impluviata*
Thymelaeaceae	*Daphnopsis fasciculata*	149	*O. cervina*
Vochysiaceae	*Vochysia tucanorum*	13	*O. irrorata*

Total abundance was highly variable across the sampling period, ranging from 58 to 272 individuals. Abundance was also unevenly distributed across species, and the the study site was dominated by three species: *O. saga*, *O. captiosa* and *O. humeralis*, that accounted for 59 % of all sampled individuals. The most abundant species was *O. saga* with 238 individuals, followed by *O. captiosa* with 231, *O. humeralis* with 188, *O. impluviata* with 117, *O. ulcerosa* with 92, *O. dejeani* with 81, *P. griseocinctus* with 73, *O. irrorata* with 58, and *O. cervina* with 29. In contrast, *O. miniata* was represented by three individuals, *O. bucki* by two and *O. macra* by only one individual. The number of species was almost constant from 2002 to 2006; however, *O. bucki* was found only in November 2003 and December 2004 and *O. macra* in November 2003 (both species). For the period between December 2002 and May 2003, we recorded 10 different Onciderini species; from October 2003 to May 2004, 11 species; from October 2004 to May 2005, 10 species; from October 2005 to May 2006, 9 species; and from October 2006 to December 2006, only 9 species were collected.

Of the native host-plants available in the sampled area, 29% belonged to the Melastomataceae, 25% to the Mimosaceae, 10% to the Myrtaceae and 10% to the Euphorbiaceae. The remaining 26% were distributed as follows: Lauraceae 5%, Anacardiaceae 5%, Meliaceae and Bombacaceae 3%, Fabaceae, Cecropiaceae, Myrsinaceae and Vochysiaceae 2%, Thymelaeaceae and Lecythidaceae 1%. In total, 1046 of the 1436 sampled plants were girdled by Onciderini beetles as follows: Vochysiaceae (represented by *Vochysia tucanorum*) constituted 45% of the records of girdled plants (though only 2% of the sampled plants are in this family) 15% were Mymosaceae, 12% were Melastomataceae, 9% were Lauraceae, 4% were Anacardiaceae. The remaining 15% were distributed among and Meliaceae, Euphorbiaceae, Bombacaceae, Fabaceae, Thymelaeaceae, Cecropiaceae, Myrtaceae, Lecythidaceae, and Myrsinaceae [Paro et al. 2011] (Table 1).

Despite the wide range of available plants in the study site, Onciderini beetles did not select plants randomly. Twelve species of these beetles were found girdling plants from 14 different families, which represent 24% of the total plant families and 13% of the tree species found in the Serra do Japi [Leitão-Filho 1992]. Eight of the associations with host-plants were species-specific (monophagy) or family-specific (oligophagy). One such interaction in the study site was *Oncideres irrorata* found associated exclusively with *Vochysia tucanorum* (Vochysiaceae). Trees of *V. tucanorum* are elements typical of the transition zone between the Brazilian Cerrado and the Atlantic rain forest. Adults of *O. irrorata* were observed girdling the apical branches of the tallest individuals (over seven meters tall) of *V. tucanorum* and this is the first report of the interaction between this plant and *O. irrorata*. Other species (*O. miniata*, *O. bucki*, and *O. macra*) were also found to be species-specific to host plants (*Lecythis* sp, *Sebastiana* sp and *Ocotia* sp, respectively) in the study site; however, caution should be taken before considering these beetle species as monophagous due to the low frequency of girdled plants by these beetles. Four species were oligophagous or family specialists: *O. captiosa*, *O. humeralis*, *O. ulcerosa*, and *Psyllotoxus griseocinctus*. *O. captiosa* is a specialist on Lauraceae plants, as reported in other studies [Link and Costa 1993; Monné 2005]. *O. humeralis* is a specialist on Melastomataceae. This beetle has been reported on four species of melastoma plants (*Miconia sellowiana*, *M. jucunda*, *M. latecrenata* and *Leandra scabra*) in Serra do Japi [Paulino-Neto et al. 2005; Romero et al. 2005] with preference for *Miconia* species. In this study, however, we found another melastoma species (*L. gracilis*) as the most preferred host plant by *O. humeralis* during the sampling period. *O. ulcerosa* and *P. griseocinctus* are specialists on Anacardiaceae and Myrtaceae, respectively. Previous reports of *O. ulcerosa* also included Myrtaceae, Lauraceae, Myrsinaceae and Tiliaceae in the host-plant repertory of this beetle. Nevertheless, we did not observe any interactions of *O. ulcerosa* with those families in the study site.

Polyphagous beetles were responsible by most of the associations with host-plants in Serra do Japi. *O. dejeani* showed the widest plant range among the Onciderini beetles found in the study site. This beetle is the most polyphagous Onciderini found in Southeastern Brazil and has been reported to use over 50 plants from 17 families [Link and Costa 1994]. Other polyphagous beetles showed preferences that diverge from specimens sampled in other ecosystems. *O. saga* and *O. impluviata* preferred Leguminosae, specifically those of the Mimosaceae. In Brazil, *O. saga* and *O. impluviata* have been found to girdle more than twenty plant species, mainly Mimosaceae, and are thus considered important pests of forestry [Costa and Marques 1988; Costa et al. 1988; Link and Costa 1993; Monné 2005]. Finally, *O. cervina* was found associated to two plant species belonging to two families (Lauraceae and Thymelaeaceae) with preference for Thymelaeaceae. *O. cervina* has been reported to be a specialist on Lauraceae plants [Monné 2001] and to our knowledge this is the first report of this beetle on Thymelaeaceae.

The findings reported for Onciderini beetles in the Atlantic rain forest of the Serra do Japi are consistent with the assumptions proposed by Berkov and Tavakilian [1999] that proposed that many tropical insects may be polyphagous in their host-plant range and, therefore, monophagy in the Cerambycidae from tropical ecosystems could be considered rare since these communities have a high diversity of plant species that may be utilized by the beetles. According to Janzen [1980, 1981], some coleopterans can appear to be locally monophagous, but polyphagous over their geographic range. In this case, *O. dejeani* and *O. ulcerosa* showed a restricted range in the study site compared to other localities [Link and Costa 1994], whereas *O. cervina* was found on a new association with *Daphnopsis fasciculata* (Thymelaeaceae). Differences in the local degree of utilization of host-plants have been found in other cerambycidae and appear to be associated with plant density [Janzen 1970] and size [Forcella 1984; Romero et al. 2005], plant chemistry, or seasonal changes [Berkov and Tavakilian 1999]. However, further studies will be necessary to explore which factors are associated with plant range of Onciderini beetles in the Atlantic rain forest.

3.2. Population Dynamics

The population of adult Onciderini found in the Serra do Japi was seasonal, reaching their highest abundance in the months of December and January [Paro et al., 2012]. Although the Onciderini species showed slight differences in relation to their abundance throughout the year, all presented the same pattern of population dynamics, i.e. the adult phase always occurred in the rainy season (October to May) and the larval phase of some species occurred in the dry season (June to September) while for others the larval cycle could take up to three years [J. Vasconcellos-Neto - pers. observation). The abundance was restricted from November to May, excepting for *O. irrorata* where the adults emerged in October. Five species emerged in November (*O. saga*, *O. impluviata*, *O. ulcerosa*, *O. dejeani* and *P. griseocinctus*) and three species emerged in December (*O. captiosa*, *O. humeralis* and *O. cervina*). *O. saga*, *O. captiosa*, *O. humeralis*, and *O. irrorata* reached their highest abundances from October to January (Figure 2A), whereas *O. ulcerosa*, *O. dejeani*, *O.*

impluviata, *O. cervina* and *P. griseocinctus* showed a uniform abundance through the months of activity (Figure 2B).

The seasonal pattern of the Onciderini beetles of the Serra do Japi was found to be related to the strongly seasonal climate, i.e., the abundance of twig-girdler beetles increased within two months of rainfall peak and increase in temperature. Consequently, the emergence, dispersal and mating of adults occurred over the months of highest rainfall (Figures 3A and B). These pattern could be also associated with different factor as the phenology of host-plants and reproductive behavior of beetles. Climatic changes are probably the primary factors molding the phenology of plants in the Serra do Japi. Flowering and leaf flush patterns of atlantic rain forest trees are significantly seasonal concentrated at the beginning of the wettest season, and are significantly correlated with temperature [Morellato et al. 1990, 2000; Morellato and Leitão-Filho 1990, 1992]. Onciderini females girdle branches of healthy plants and then lay eggs on the distal portion of the girdled branch. According to Forcella [1984], girdling elevate the content of nitrogen in the twigs and may also circumvent host resistance [Hanks 1999; Martínez et al. 2009]. During this phase, adults also feed on the flowers and leaves of host-plants [Paulino-Neto 2006; Paro 2008]. In this way, the reproductive activity of the adults of this group would be also synchronized when the plants are more nutritious for larval development.

Following eclosion, larvae penetrate and feed in the internal cortex of the girdled twigs during the dry season (from June to September) [Hanks 1999; Paulino-Neto 2005] and depending on the species, they would reach the adult stage during the subsequent rainy season. In some cerambycids, egg eclosion is reduced when exposed to high moisture [Coulson 1979] and therefore, phloem-boring larvae may be especially vulnerable to moisture conditions because they are embedded in tissues of their host plant [Hanks et al. 1999], or when moist conditions increase the presence of predators, parasitoids [Martínez et al. 2009] or pathogens. It is possible that natural selection leads the Onciderini of the atlantic rain forest to adopt a synchrony with climate and host-plant phenology, because the development of larvae could be affected negatively if the twigs on the forest floor are exposed to high levels humidity. Therefore, rain and high temperatures are probably a stimulus for the different species of adult Onciderini to emerge, search for host-plant, mate and lay their eggs.

3.3. Sex Ratios

During the period of adult activity, females were more abundant than males in all species except for *O. saga*, *O. captiosa* and *O. irrorata* in which the sex ratio was 1:1. Males and females of *O. irrorata* were found always mating during the collection period, suggesting a sex ratio of 1:1. Due the low number of collected individuals, *O. bucki*, *O. macra* and *O. miniata* were not included in the analysis. According to the literature, sex ratios of Lamiinae differed among species [Lu et al. 2007; Rodriguez-del-Bosque and Garza-Cedillo, 2008] and the reason for female or male-biased sex ratios is unknown. Female-biased sex ratios could be explained, in part, by their sexual behavior, especially in species in which females emerge from twigs and look for host-plants to feed and lay their eggs whereas males migrate in the search for females, possibly attracted by sex pheromones or plant volatiles. After emergence, females of the species *O. humeralis*, a specialist of Melastomataceae, arrived at the host plants before the males [Paulino-Neto et al. 2006]. Males of this species could become more vulnerable to natural enemies while searching for the females that remain on the host plants, thus resulting in populations containing more females than males. Moreover, males could spend more energy on courtship and/or defense against other males [Hanks, 1999], and wound hence die earlier than females. This phenomenon could explain why the sex ratios of *O. humeralis*, *O. impluviata*, *O. ulcerosa* and *O. dejeani* (which presented a 1:1 sex ration at the initial phase of the population) were favorable to females at declining phase of the population.

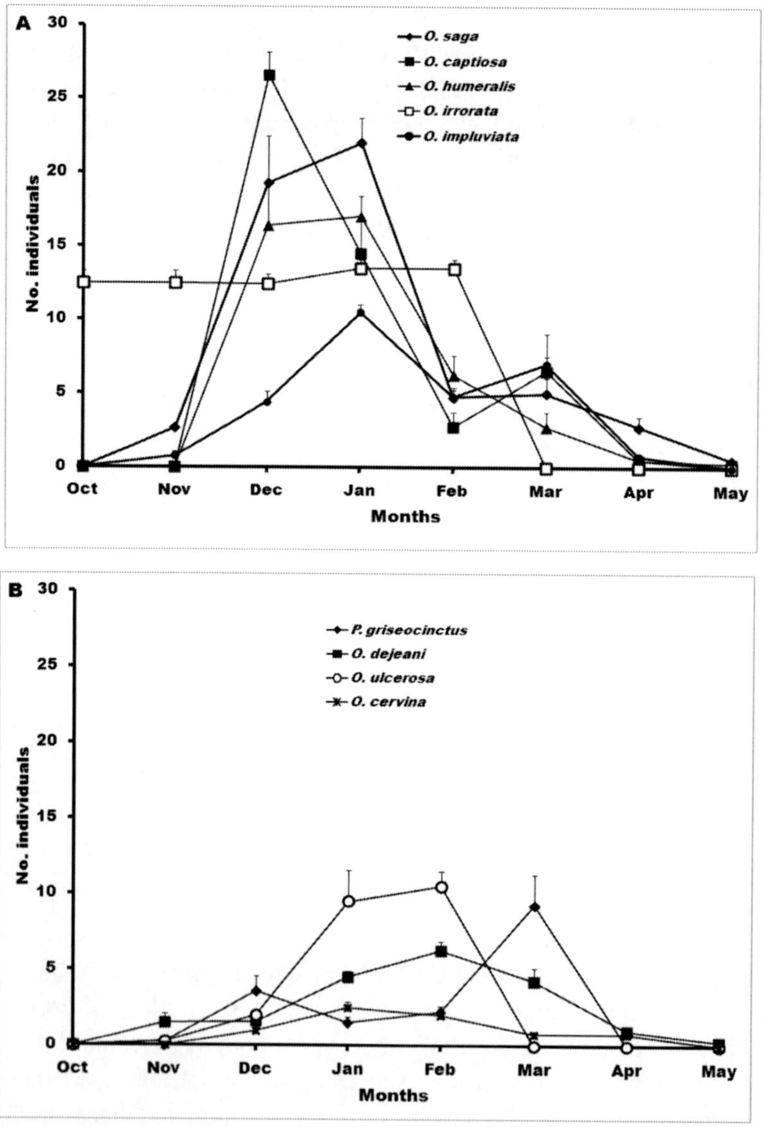

Figure 2. Annual pattern of abundance for nine species of Onciderini during the period of adult activity (between October and May). (A) Species with highest abundance between December and January. (B) Species with relatively uniform abundance throughout the activity period. Values indicate the mean (± SE).

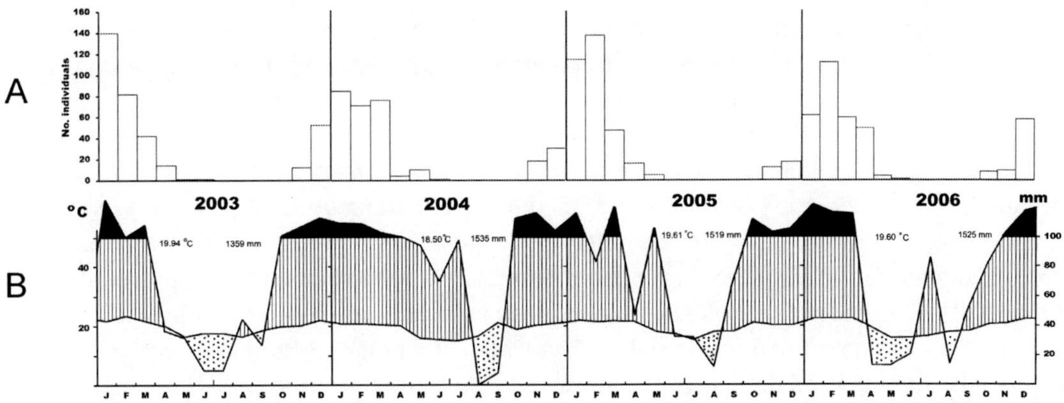

Figure 3. (A) Population fluctuation of Onciderini beetles during the sampling period. (B) Monthly rainfall and temperature variation during the study period (data from a meteorological station approximately 8 km from the study site). Values in the climatic graph indicate mean temperature and accumulated rainfall for each sampling year.

4. Impacts of Onciderini on Plant Community Structure

Herbivorous insects can damage hosts in any phase of their life cycle and decrease plant fitness or even kill them [Marquis 1992, Coley & Barone 1996, Price et al. 2011]. As a consequence, herbivores alter plant population dynamics by affecting recruitment [Crawley 1989, Ehrlén 1995, Louda & Potvin 1995, Clarke & Kerrigan 2002, Sullivan 2003]. Herbivores can also affect plant community structure by attacking dominant species, providing competitor-free space and favouring the establishment of rare species thereby increasing local plant diversity [Janzen 1970, Harper 1977, Hunter 1992, Rao et al. 2001]. Although many studies describe the effects of herbivores on plant density, population dynamics and plant community structure, most of them consider only seed predation, seedling meristem herbivory or leaf defoliation.

Adult females of Onciderini saw stems with their mandibles and prepare the oviposition site on the cut portion by perforating the bark before laying eggs [Rice 1989, 1995, Caraglio et al. 2001]. By clipping a huge quantity of plant biomass, they can stress, and sometimes kill, host plants [Caraglio et al. 2001, Feller 2002, Paulino-Neto et al. 2006, Uribe-Mú & Quesada 2006, Martínez et al. 2009]. In addition, larvae of these beetles bore live or dead stems of their host plants, and when they use active tissues (e.g., xylem), can affect the development and survival of their hosts [e.g., Paulino-Neto et al. 2006, Uribe-Mú & Quesada 2006, Martínez et al. 2009]. Consequently, woodboring beetles can cause changes in population dynamics and community structure of their host plants [Caraglio et al. 2001, Romero et al. 2005].

In this section we report the effects of *Oncideres humeralis* Thomson, on the mortality, number, size, and community structure of its host plants at the Serra do Japi. This species seems to be specialized in Melastomataceae [Paulino-Neto et al. 2005, 2006, Paro et al. 2011] Females oviposit at the base of the secondary branches and girdle the principal trunk; all the plants attacked by this beetle lose the tree crown and the main trunk appears transversely sectioned, as if cut with a handsaw [Paulino-Neto et al. 2006, Paro et al. 2011].

We quantified the effects of *O. humeralis* on its host plants over three years of beetle outbreaks by censusing undamaged Melastomataceae host and non-host species of this beetle along permanent 400 m x 3 m forest-edge transects (see details of material and methods in Romero et al 2005). The Melastomataceae host species in these transects were *Miconia sellowiana* (the most common host plant), *M. jucunda*, *M. latecrenata*, and *Leandra scabra*; the non-host species were *Leandra* sp. 1, 2 and 3. Of 296 *M. sellowiana* censused in the first year of outbreak, 23% had the trunk girdled by *O. humeralis* and had lost their crown. Of the 126 individuals censused in the third year of outbreak, 11 (9%) had their canopy cut down by the beetle. Of 35 individuals of *Miconia sellowiana* freshly girdled by *O. humeralis*, which were marked and monitored in the field, 15 (43%) died and the remaining produced 2-5 new shoots near the point of the cut in the following year [Romero et al. 2005]. Caraglio et al. [2001] found similar results in the system *Dicorynia guianensis* (Caesalpiniaceae) – *Oncideres* spp. (Cerambycidae) in French Guiana.

Whereas two of four host species decreased in abundance over three years of outbreak, one non-host species increased in abundance in this period. And more importantly, one non-host species appeared in the area for the first time in the third year of outbreak, after the decrease of the dominant host species (Figure 4).

The beetles had contrasting effects on the size structure of the non-host and host species over the three years of outbreak (Figure 5). Whereas the non-host species increased in size, occupying higher classes of trunk diameter, some host species decreased in size over the three years (e.g., *M. sellowiana*).

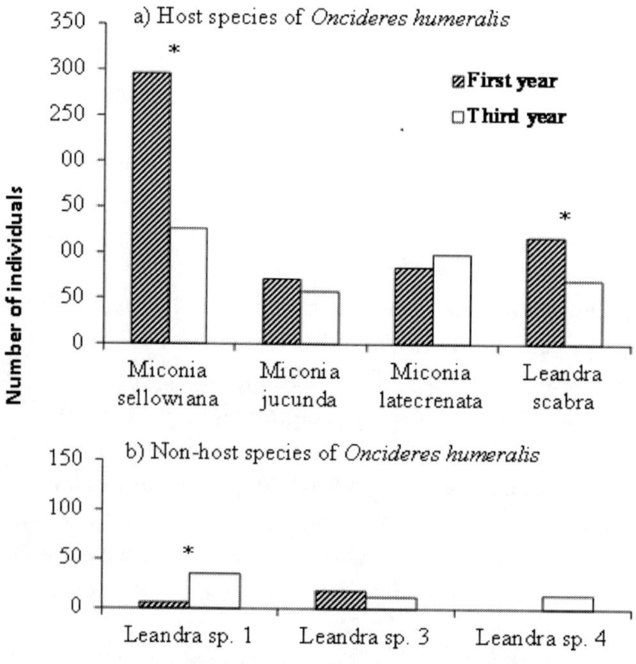

Figure 4. Number of individuals (not girdled) of Melastomataceae (a) host species of *O. humeralis* and (b) non-host species, in the first and third year of beetle outbreak, in the Serra do Japi. * significant differences (G-test, P<0.05).

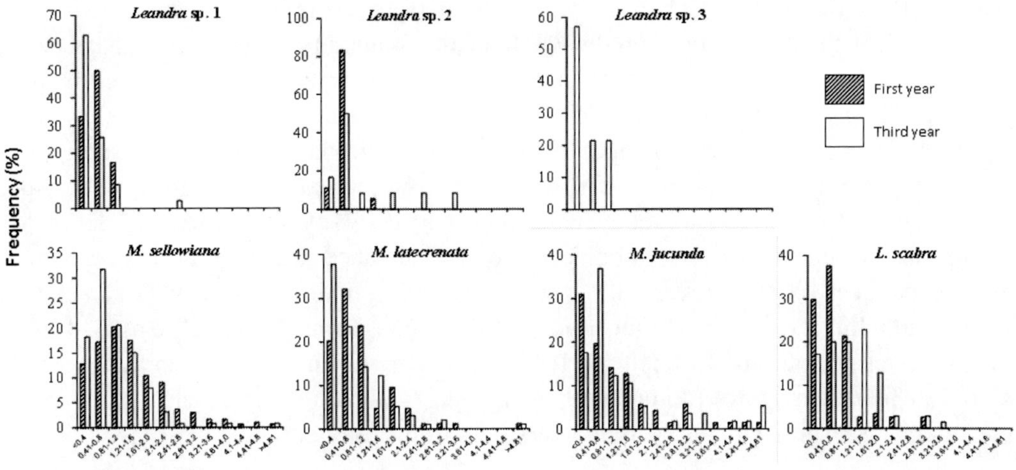

Figure 5. Size structure (trunk diameter at 10 cm above ground) of the Melastomataceae non-host and host species of *O. humeralis*, in the first and third year of beetle outbreak, at the Serra do Japi. The number of individuals is indicated in the graphs of the Figure 4. Kolmogorov-Smirnov two sample test: *M. sellowiana*, P = 0.038; other host species, P ≥ 0.8; tests not performed for non-host species.

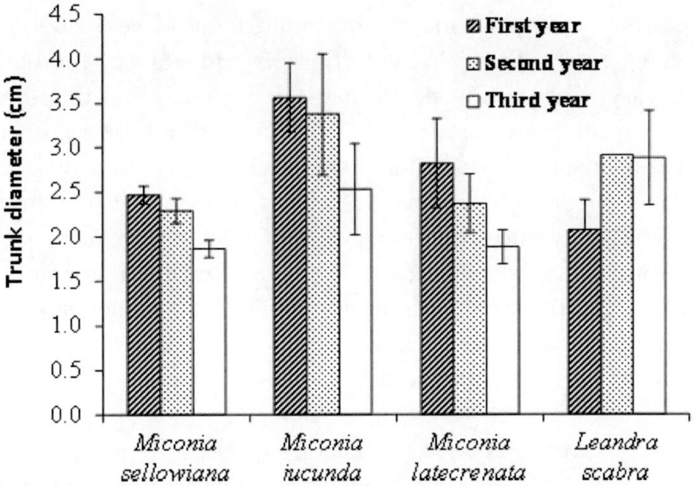

Figure 6. Mean diameter (± 1 SE) of the trunk (at 10 cm above ground) of Melastomataceae host-species girdled by *O. humeralis* during 3 successive years of beetle outbreak at the Serra do Japi. Only one individual of *L. scabra* was sampled in the second year of outbreak. A significant difference among years occurred only for *M. sellowiana*.

Consequently, *O. humeralis* tended gradually to attack smaller plants (trunk diameter) of host plants in successive years (Figure 6).

These findings highlighted the importance of stem girdling cerambycids on the mortality, size and community structure of host trees in the tropics. These beetles attacked and killed many adult and immature plants. As a short-term consequence, this beetle had a strong, rapid effect on the size structure of some of its host plant species. As a long-term consequence, the beetle might decrease seedling recruitment of its host plants by diminishing the seed output of the population. Further studies are needed to test this hypothesis.

We believe that the impact of *O. humeralis* may be even stronger than reported here, since we underestimate its effects on the whole plant community (only Melastomataceae were evaluated). Taken into account that this tropical forest (Serra do Japi) is inhabited by at least 12 stem girdling cerambycids (Onciderini), which attack many other plant families [Paro et al. 2011], we predict that competitor-free space provided by this beetle to rare Melastomataceae species (e.g., *Leandra* sp. 3) may also be extended to many other species belonging to other families, thereby increasing plant diversity throughout an entire forest. And the basic mechanism seems to be decrease of dominance of certain host species through density-dependent control over the population, thus indirectly providing space and microhabitats for rare species. Our findings were consistent with the density-dependent Janzen-Connell hypothesis [Connell 1971, Janzen 1970], which occurs when herbivores limit the local abundance of the dominant host plants, facilitating coexistence of subordinate competitors and maintaining alpha tree diversity in tropical forests.

5. Host-Plant Utilization and Microhabitat Specialization

Host-specialization has been considered a characteristic of leaf-chewing insects [Erwin 1982] due to the co-evolutionary process between defensive plant secondary metabolites (which can be specific or appear more concentrated on leaves) and the adaptation to overcome plant defenses by these insects [Ehrlich and Raven 1964]. However, host-specialization may vary substantially between feeding guilds, non-folivorous insect guilds appeared to be more specialized than leaf-chewing insects and their level of specialization varied between microhabitats and/or feeding guilds in the tropics [Novotny and Basset 2005]; however, information about xylophagous insects is scarce. Despite the wide range of available plants in the study site, adult females of Onciderini do not select microhabitats randomly and microhabitat specialists are present in the stem-borer community. Indeed, generalist and family-specialist species showed preferences to girdle some host-plant species over other, depending on the availability and morphological characteristics of hosts as found for *O. humeralis* and its Melastomataceae hosts in the study site (see section 3 and 4 of this chapter).

Three species were specialists on upper canopy trees (*O. irrorata*, *O. captiosa*, and *O. ulcerosa*) and two were specialists on young trees and shrubs (*O. humeralis* and *P. griseocinctus*). Additionally, all beetle species showed preferences for girdling particular types of branches. Of these, *O. irrorata* and *O. cervina* were specialists on apical branches and the main trunk, respectively, whereas *O. impluviata*, *O. captiosa*, *O. ulcerosa* and *Psyllotoxus griseocinctus* were specialists on removing secondary branches of host plants (Table 2) [Paro 2008].

This evidence suggests that Onciderini beetles are a specialized group of herbivores and that some microhabitats in the Atlantic rain forest trees support a relatively specific twig-girdler beetle fauna. According to Novotny and Basset [2005], non-folivorous insect guilds appeared to be more specialized than leaf-chewing insects and their level of specialization varied between microhabitats and/or feeding guilds in the tropics; however, information about xylophagous insects is scarce.

Table 2. Host-specificity of Onciderini beetles for each microhabitat category at the Serra do Japi

Microhabitat category	Species-specialist	Family-specialists	Generalists
Upper canopy trees	*O. irrorata*	*O. ulcerosa*	–
	–	*O. captiosa*	–
	–	–	*O. cervina*
	–	–	*O. dejeani*
	–	–	*O. saga*
	–	–	*O. impluviata*
Lower canopy trees		–	*O. dejeani*
	–	–	*O. cervina*
	–	–	*O. saga*
Young trees and shrubs	–	*O. humeralis*	–
	–	*P. griseocinctus*	–
	–	–	*O. impluviata*
	–	–	*O. saga*
Apical branches	*O. irrorata*	–	–
Secondary branches	–	*O. captiosa*	–
	–	*O. ulcerosa*	–
	–	*P. griseocinctus*	–
	–	–	*O. impluviata*
	–	–	*O. saga*
	–	–	*O. dejeani*
	–	*O. humeralis*	–
Principal branches	–	–	*O. cervina*
	–	*O. humeralis*	–
	–	–	*O. dejeani*
	–	–	*O. saga*

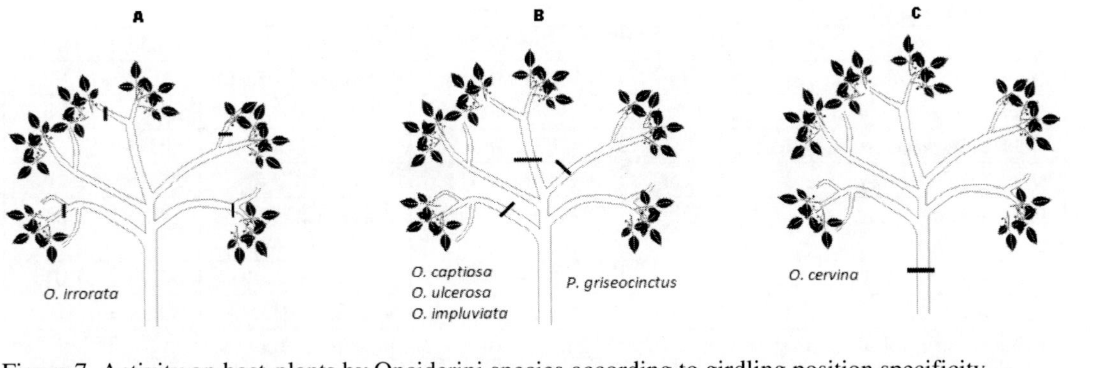

Figure 7. Activity on host-plants by Onciderini species according to girdling position specificity.

We also found an interaction between host-plant preferences and microhabitat specificity. Canopy trees supported the entire assemblage of generalist beetle species. Additionally, of the six species identified as specific to a particular type of branch, the species-specialist *O. irrorata* girdled only apical branches; three family-specialist species were secondary branch specialists (*O. captiosa, O. ulcerosa, P. griseocinctus*) whereas the generalist species *O. cervina* and *O. impluviata* were main trunk and secondary branch specialists, respectively (Figure 7). We are aware of only two previous studies focused on stem-borer specialization and our findings were consistent with both. In the rain forests of French Guiana, more than 40% of the described stem-borer species displayed a higher degree of host specificity [Tavakilian et al.,1997] and microhabitat specialization [Berkov and Tavakilian 1999].

Functional size of girdled branches varied significantly among Onciderini, and species that showed preference for young trees and shrubs removed shorter and slender branches, whereas those species specialist on canopy trees girdled larger branches and made more egg incisions, indicating that twig-girdler beetles have an optimum host-plant size as we expected (Table 3). Girdling branches serves to circumvent plant defenses and provides food for larvae [Forcella 1982; Rice 1985] and is expected that beetle species that remove larger branches of larger trees will laid more eggs since branch size serves as an estimate of the nutritional value of the host [Mitchell 1981].

At this time, we do not know why microhabitat specialization and host-plant preference among Onciderini adult females in the Atlantic rain forest of the Serra do Japi is so remarkable. We suggest that microhabitat selection by females could be related to different factors as feeding ecology, host-plant age or density [Caraglio et al. 2001]. Since larvae of Onciderini beetles cannot move to find suitable food sources, searching behavior is presumed to be critical in the location of an appropriate site for mating and oviposition [Linsley 1959; 1961], requiring that adults have good dispersal abilities and effective host recognition mechanisms [Hanks et al. 1993]. On the other hand, females are attracted by host-specific factors that could indicated a relative fitness advantage for their larvae that is greater than that provided by alternative hosts [Mitchell 1981].

Table 3. Functional size of girdled twigs (mean ± SD) and number of egg incisions made by Onciderini species on host-plants at the Serra do Japi

Beetle species	Diameter (cm)	Length (cm)	Volume (cm^3)	Number of egg incisions
Oncideres irrorata	1.81 ± 0.16d	39.10 ± 4.19d	28.90 ± 7.99c	3.35 ± 1.63d
Oncideres humeralis	2.17 ± 0.04c	118.50 ± 28.26c	28.33 ± 11.7b	13.33 ± 56.45b
Oncideres saga	4.56 ± 1.14a	160 ± 45.36b	290.81 ± 170.91a	62.77 ± 12.92a
Oncideres impluviata	1.86 ± 0.12c	140.48 ± 60.24c	49.76 ± 42.65b	21.40 ± 1.61b
Oncideres captiosa	4.25 ± 0.83ac	271.17 ± 45.81a	362.72 ± 184.90a	57.74 ± 5.3a
Oncideres ulcerosa	2.78 ± 0.12ab	197.95± 41.31c	62.17 ± 12.16b	28.46 ± 2.51b
Psyllotoxus griseocinctus	1.73 ± 0.32bc	92.67 ± 38.33c	22.19 ± 13.34b	19.53 ± 4.63b
Oncideres dejeani	5.15 ± 0.92a	197.68 ± 39.06ab	25.66 ± 33.19b	64.94 ± 24.01a
Oncideres cervina	4.79 ± 1.23ab	215 ± 22.68a	413.57 ± 238.28a	52.84 ± 4.01a

Means within columns followed by different letters indicate significant difference at $p < 0.05$ between functional host-plants size utilized by Onciderini species

CONCLUSION

Stem-borers constitute an insect group that play a crucial role in ecosystems processes, since they are primary agents of plant degradation in forests. Most of thespecies begin their life cycle as eggs laid under bark; the larvae feed on the wood inside stems and eventually emerge as adults to repeat the cycle. Twig-girdlers in the tribe Onciderini (Cerambycidae: Lamiinae) are a large group of stem borers in the tropical regions. Although twig-girdler beetles have been studied in other ecosystems, there is little information about the biology of these insects and their implications on the floral communities in the tropics. By studying Onciderini in a rain forest of Southeastern Brazil, we found that twig-girdling beetles are seasonal and did not girdle host-plants randomly. It could be associated to plant phenology and suitability of larval food. Additionally, the evidences has shown that twig-girdlers can modified plant community by creating a competitor-free space to rare plant species, thereby increasing plant diversity throughout the forest. This work constitutes an important step in understanding the host-plant interactions of an insect guild poorly studied and further investigations on the population fluctuations and impact of these beetles on natural forests could help evaluate the dynamics of this insect–plant community.

REFERENCES

Berkov, A. & Tavakilian, G. (1999). Host utilization of the Brazil nut family (Lecythidaceae) by sympatric wood-boring species of *Palame* (Coleoptera, Cerambycidae, Lamiinae, Acanthocinini). *Biological Journal of the Linnean Society*, *67*, 181-198.

Caraglio, Y., Nicolini, E. & Petronelli, P. (2001). Observations on the links between the architecture of a tree (*Dicorynia guianensis* Amshoff) and Cerambycidae activity in French Guiana. *Journal of Tropical Ecology*, *17*, 459-463.

Clarke, P. J. & Kerrigan, R. A. (2002). The effects of seed predators on the recruitment of mangroves. *Journal of Ecology*, *90*, 728-736.

Coley, P. D. & Barone, J. A. (1996). Herbivory and plant defenses in tropical forests. *Annual Review of Ecology and Systematics*, *27*, 305-335.

Connell, J. H. (1971). On the role of natural enemies in preventing competitive exclusion in some marine animals and in rain forest trees. In P. J. de Boer, & G. R. Gradwell (Eds.), Dynamics of Populations: Proceedings of the Advanced Study Institute on Dynamics of Numbers in Populations (298-312). Wageningen, The Netherlands: Centre for Agricultural Publishing and Documentation.

Costa, E. C., Marques, E. N. & Link, D. (1988). Periodo pupal, emergência e distribuição dos galhos cortados por *Oncideres impluviata* (Germar 1824) em povoamentos de bracatinga. Revista do Centro de Ciencias Rurais, *Santa Maria*, *18*, 229-237.

Costa, E. C. & Marques, E. N. (1988). Aspectos etológicos de *Oncideres impluviata* (Germar 1824) em bracatinga. Revista do Centro de Ciências Rurais, *Santa Maria*, *18*, 219-228.

Coulson, R. N. (1979). Population dynamics of bark beetles. *Annual Review of Entomology*, *24*, 417-447.

Crawley, M. J. (1989). Insect Herbivores And Plant Population Dynamics. *Annual Review of Entomology*, *34*, 531-564.

Dillon, L.S. & Dillon, E.S. (1945). The tribe Onciderini (Coleoptera: Cerambycidae). *Part I. Scientific Publications of the Reading Public Museum*, *5*, 1-186.

Edmonds, R. L. & Eglitis, A. (1989). The role of the Douglas-fir beetle and wood borers in the decomposition of and nutrient release from Douglas-fir logs. *Canadian Journal of Foresty Research*, *19*, 853-859.

Ehrlén, J. (1995). Demography of the perenial herb *Lathyrus vernus*. II. Herbivory and population dynamics. *Journal of Ecology*, *83*, 297-308.

Ehrlich, P. R. & Raven, P. H. (1964). Butterflies and plants: a study in coevolution. *Evolution*, *18*, 586-608.

Erwin, T. R. (1982). Tropical forests: their richness in Coleoptera and other arthropod species. *The Coleopterist Bulletin*, *36*, 74-75

Farrel, B. D. (1998). "Inordinate fondness" explained: Why are there so many beetles? *Science*, *281*, 555-558.

Feller, I. (2002). The role of herbivory by wood boring insects in mangrove ecosystem in Belize. *Oikos*, *97*, 167-176.

Forcella, F. (1982.) Why twig-girdling beetles girdle twigs. *Naturwissenschaften*, *69*, 398-400

Forcella, F. (1984). Tree size and density affect twig-girdling intensity of *Oncideres cingulata* (Say) (Coleoptera: Cerambycidae). *The Coleopterist Bulletin*, *38*, 37-42.

Gibbs, H. K., Ruesch, A. S., Achard, F., Clayton, M. K., Holmgren, P., Ramankutty, N., & Foley, J. A. (2010). Tropical forests were the primary sources of new agricultural land in the 1980s and 1990s. *Proceedings of the National Academy of Sciences*, *107*, 16732-16737.

Hanks, L. M. (1999). Influence of the larval host pant on reproductive strategies of cerambycid beetles. *Annual Review of Entomology*, *44*, 483-505.

Hanks, L. M., Paine, T. D. & Millar, J. G. (1993). Host species preference and larval performance in the wood-boring beetle *Phoracantha semipunctata* F. *Oecologia*, *95*, 22-29

Harper, J. L. (1977). *Population Biology of Plants*. London, UK: Academy Press,.

Hunter, M. D. (1992). Interactions within herbivore communities mediated by the host plant: The Keystone Herbivore Concept. In M. D Hunter, T. Ohgushi, & P.W. Price (Eds.), *Effects of resource distribution on Animal-Plant Interactions*, (287-325). San Diego, California: Academic Press, Inc.

Janzen, D. H. (1970). Herbivores and the number of tree species in tropical forest. *American Naturalist*, *104*, 501-528.

Janzen, D. H. (1980). Specificity of seed-attacking beetles in a Costa Rican deciduous forest. *Journal of Ecology*, *68*, 929-952.

Janzen, D. H. (1981). Patterns of herbivory in a tropical deciduous forest. *Biotropica*, *13*, 271-282.

Leitão– Filho, H. F. (1992). A flora arbórea da Serra do Japi. In: L. P. C. Morellato (Ed.), História natural da Serra do Japi: ecologia e preservação de uma área florestal no Sudeste do Brasil (40-62). Campinas, São Paulo: Editora da Unicamp.

Link, D. & Costa, E.C. (1993). Aspectos da biologia de *Oncideres impluviata* e de *O. captiosa* em Santa Maria, RS. (Coleoptera: Cerambycidae). *Ciência Florestal*, *3*, 77-84.

Link, D. & Costa, E. C. (1994). Diâmetro de galhos cortados por *Oncideres* spp. (Coleoptera, Cerambycidae) na região Central do Rio Grande do Sul. *Revista de Agricultura-Piracicaba*, *69*, 183-192.

Linsley, E. G. (1959). Ecology of Cerambycidae. *Annual Review of Entomolology*, *4*, 99-138.

Linsley, E. G. (1961). The Cerambycidae of North America, Part I. Introduction. California, USA: University of California publications in entomology.

Lombardi, J. A., Carvalho, C. S., Biral, L., Saka, M. N. & Hieda, S. M. (2012). Vascular flora of Serra do Japi Biological Reserve, Jundiaí, southeastern Brazil. *Rodriguesia*, *63*, 333–340

Louda, S. M. & Potvin, M. A. (1995). Effect Of Inflorescence-Feeding Insects On The Demography And Lifetime Fitness Of A Native Plant. *Ecology*, *76*, 229-245.

Lu, W., Wang, Q., Tian, M. Y., He, X. Z., Zeng, X. L. & Zhong, Y. X. (2007). Mate location and recognition in *Glenea cantor* (Fabr.) (Coleoptera: Cerambycidae: Lamiinae): roles of host plant health, female sex pheromone, and vision. *Environmental Entomology*, *36*, 864-870.

Marquis, R. J. (1992). The Selective Impact of Herbivores. In R. S. Fritz, & Simms, E. L. (Eds.), Pp Plant resistence to herbivores and pathogens: ecology, evolution, and genetics. (301-325). Chicago, USA: University Of Chicago Press.

Martínez, A. J., López-Portillo, J., Eben, A. & Golubov, J. (2009). Cerambycid girdling and water stress modify mesquite architecture and reproduction. *Population Ecology*, *51*, 533-541.

Mitchell, R. (1981). Insect behavior, resource exploitation, and fitness. *Annual Review of Entomology*, *26*, 373-396

Monné, M. A. (2001). Catalogue of the Neotropical Cerambycidae (Coleoptera) with known host plant. Part III: Subfamily Lamiinae, Tribes Acanthocinini to Apomecynini. *Publicações Avulsas do Museu Nacional*, *92*, 1-94.

Monné, M. A. (2005). Catalogue of the Cerambycidae (Coleoptera) of the Neotropical Region. Part II. Subfamily Lamiinae. *Zootaxa*, *1023*, 1-760.

Morellato, L. P. C. & Leitão-Filho, H. F. (1990). Estratégias fenológicas de espécies arbóreas de floresta mesófila na Serra do Japi, Jundiaí, São Paulo. *Brazilian Journal of Biology*, *50*, 163-173.

Morellato, L. P. C. & Leitão-Filho, H. F. (1992). Padrões de frutificação e dispersão na Serra do Japi. In: L. P. C. Morellato (Ed.), História natural da Serra do Japi: ecologia e preservação de uma área florestal no Sudeste do Brasil, (111-140). Campinas, São Paulo: Editora da Unicamp.

Morellato, L. P. C., Leitao-Filho, H. L., Rodrigues, R. R. & Joly, C. A. (1990). Estratégias fenológicas de espécies arbóreas em floresta de altitude na Serra do Japi, Jundiaí, SP. *Revista Brasileira de Biologia*, *50*, 149-162.

Morellato, L. P. C., Romera, E. C., Talora, D. C., Takahasi, A., Bencke, C. C. & Zipparro, V. B. (2000). Phenology of Atlantic rain forest trees: a comparative study. *Biotropica*, *32*, 811-823.

Novotny, V. & Basset, Y. (2005). Host specificity of insect herbivores in tropical forests. Procedings of the Royal Society of London, B, *Biologial Sciences*, *272*, 1083-1090.

Paro, C. M. (2008). Dinâmica populacional e padrões de utilização de plantas hospedeiras, por 12 espécies de serra-paus (Coleoptera: Cerambycidae: Onciderini) na Serra do Japi, Jundiaí-SP [PhD thesis]. Campinas-SP: Universidade Estadual de Campinas.

Paro, C. M., Arab, A. & Vasconcellos-Neto, J. (2011). The host-plant range of twig-girdling beetles (Coleoptera: Cerambycidae: Lamiinae: Onciderini) of the Atlantic rainforest in southeastern Brazil. *Journal of Natural History*, *45*, 1649-1665.

Paro, C.M., Arab, A, Vasconcellos-Neto, J. (2012). Population dynamics, seasonality and sex ratio of twig-girdling beetles (Coleoptera: Cerambycidae: Lamiinae: Onciderini) of an Atlantic rain forest in south-eastern Brazil. *Journal of Natural History*, *46*, 1249-1261

Paulino-Neto, H. F., Romero, G. Q. & Vasconcellos-Neto, J. (2005). Interactions between *Oncideres humeralis* Thomson (Coleoptera: Cerambycidae) and Melastomataceae: host-plant selection and patterns of host use in South-east Brazil. *Neotropical Entomology*, *34*, 7-14.

Paulino-Neto, H. F., Vasconcellos-Neto, J. & Carmello-Guerreiro, S. M. (2006). The biology of *Oncideres humeralis* Thorms (Coleoptera: Cerambycidae: Lamiinae) and new Cerambycidae–Melastomataceae host-plant associations. *Studies on Neotropical Fauna and Environment*, *41*, 227-233.

Pinto, H. S. (1992). Clima da Serra do Japi. In: L. P. C. Morellato (Ed.), História natural da Serra do Japi: ecologia e preservação de uma área florestal no Sudeste do Brasil (pp. 30-38). Campinas, São Paulo: Editora da Unicamp.

Prance, G. T., Beentje, H., Dransfield, J. & Johns, R. (2000). The tropical Flora remains undercollected. *Annals of the Missouri Botanical Garden*, *87*, 67-71.

Price, P. W., Denno, R. F., Eubanks, M. D., Finke, D. L. & Kaplan, I. (2011). Insect Ecology: Behavior, Populations and Communities. Cambridge, UK: Cambridge University Press.

Rao, M., Terborg, J. & Nuñez, P. (2001). Increased Herbivory in forest isolates: implications for plant community structure and composition. *Conservation Biology*, *15*, 624-633.

Rice, M. E. (1985). New host association for Cerambycidae (Coleoptera) from selected species of Leguminosae and Rutaceae. *Journal of the New York Entomological Society*, *93*, 1223-1225

Rice, M. E. (1989). Branch Girdling And Oviposition Biology Of *Oncideres pustulatus* (Coleoptera: Cerambycidae) On *Acacia Farnesiana*. *Annals of the Entomological Society of America*, *82*, 181-186.

Rice, M. E. (1995). Branch girdling by *Oncideres cingulata* (Coleoptera: Cerambycidae) and relative host quality of persimmon, hickory, and elm. *Annals of the Entomological Society of America*, *88*, 451-455.

Rodrigues, R. R. & Shepherd, G. J. (1992). Análise da variação estrutural e fisionômica da vegetação e características edáficas, num gradiente altitudinal na Serra do Japi. In: L. P. C. Morellato (Ed.), História natural da Serra do Japi: ecologia e preservação de uma área florestal no Sudeste do Brasil (64-96). Campinas, São Paulo: Editora da Unicamp.

Rodríguez-del-Bosque. L. A. & Garza-Cedillo R. D. (2008). Survival, emergence, and damage by *Oncideres pustulata* (Coleoptera: Cerambycidae) on *Huisache* and *Leucaena* (Fabaceae) in Mexico. *Southwestern Entomologist*, *33*, 209-217.

Romero, G. Q., Vasconcellos-Neto, J. & Paulino-Neto, H. F. (2005). The effects of the wood-boring *Oncideres humeralis* (Coleoptera, Cerambycidae) on the number and size structure of its host-plants in south-east Brazil. *Journal of Tropical Ecology*, *21*, 233-236.

Sullivan, J. J. (2003). Density-dependent shoot-borer herbivory increases the age of first reproduction and mortality of neotropical tree saplings. *Oecologia*, *136*, 96-106.

Tavakilian, G., Berkov, A., Meurer-Grimes, B. & Mori, S. (1997). Neotropical tree species and their faunas of xylophagous longicorns (Coleoptera: Cerambycidae) in French Guiana. *Botanical Review*, *63*, 303-355.

Uribe-Mú, C. A. & Quesada, M. (2006). Preferences, patterns and consequences of branch removal on the dioecious tropical tree *Spondias purpurea* (Anacardiaceae) by the insect borer *Oncideres albomarginata chamela* (Cerambycidae). *Oikos*, *112*, 691-697.

Viana, V. M. & Tabanez A. J. (1996). Biology and conservation of forest fragments in the Brazilian Atlantic moist forest. In: J.Schelhas , & R. Greenberg (Eds.), *Forest patches, tropical landscapes*, (151-167). Washington DC, USA: Island Press.

In: Beetles: Biodiversity, Ecology and Role in the Environment ISBN: 978-1-63463-380-2
Editor: Camilla Stack

Chapter 5

BIOLOGY, ECOLOGY AND STRATEGIES FOR CONTROL OF STORED-GRAIN BEETLES: A REVIEW

Thiago H. Napoleão, Afonso C. Agra-Neto, Bernardo R. Belmonte, Emmanuel V. Pontual and Patrícia M.G. Paiva *

Departamento de Bioquímica, Centro de Ciências Biológicas, Universidade Federal de Pernambuco, Cidade Universitária, Recife, Pernambuco, Brazil

ABSTRACT

Beetle species belonging to the coleopteran families Bruchidae, Curculionidae, Laemophoeidae, Silvanidae and Tenebrionidae, as well as beetle-like insects from the psocopteran family Liposcelidae, are responsible for serious damages to agricultural products and resources. These beetles can be primary and/or secondary pests, feeding on integral and healthy grains or attacking those already damaged. The affected grains lose weight and germination power, are decreased in nutritive value and vigor, and are impaired by hygiene and sanitary conditions. This chapter summarizes information on biological and ecological aspects of stored product pests such as life cycle, fecundity, longevity, growth rate, voracity, natural habitats and hosts, and infestation focuses of beetles. These aspects are important for the development and choice of control measures. Classic strategies include mechanical methods, biological control, and the use of insecticide formulations such as powders, emulsions, aerosols, and microcapsules. Fumigation with phosphine has been the main strategy for control but this insecticide is highly volatile and toxic and increasingly there are reports of insect resistance to it. In this sense, natural insecticides, such as plant extracts, secondary metabolites, essential oils and lectins, have been investigated for insecticidal activity on destructive beetles. In summary, this chapter provides a brief and updated view of the state of the art on beetles that act as stored grain pests.

*Correponding author: E-mail: ppaivaufpe@yahoo.com.br Tel: +55 81 21268540; fax: +55 81 21268576.

1. Damages to Agriculture Caused by Stored-Grain Pests

Grains such as sorghum, maize, rice and wheat are the main components of basic food in many countries. Since the harvest of grains occurs seasonally while the market demand is relatively unceasing, the storage step has a critical role in the global economy (FAO, 1994; Bilia et al., 2004). In this sense, the deterioration of stored products is a problem for agriculture, mainly in tropical regions, where the seeds are often subjected to high temperatures and humidity. In addition to physical and chemical factors, the stored grains can also be damaged by biological agents such as insect pests (Tavares and Vendramim, 2005; Silveira et al., 2006; Alencar et al., 2011). The insects belonging to the orders Coleoptera (beetles) and Psocoptera (booklice and barklice, also referred as beetle-like) cause most of the damage to stored grains. Figure 1 shows representative drawings of important species of the coleopteran families such as Bruchidae, Curculionidae, Laemophoeidae, Silvanidae and Tenebrionidae, and the psocopteran family, Liposcelidae.

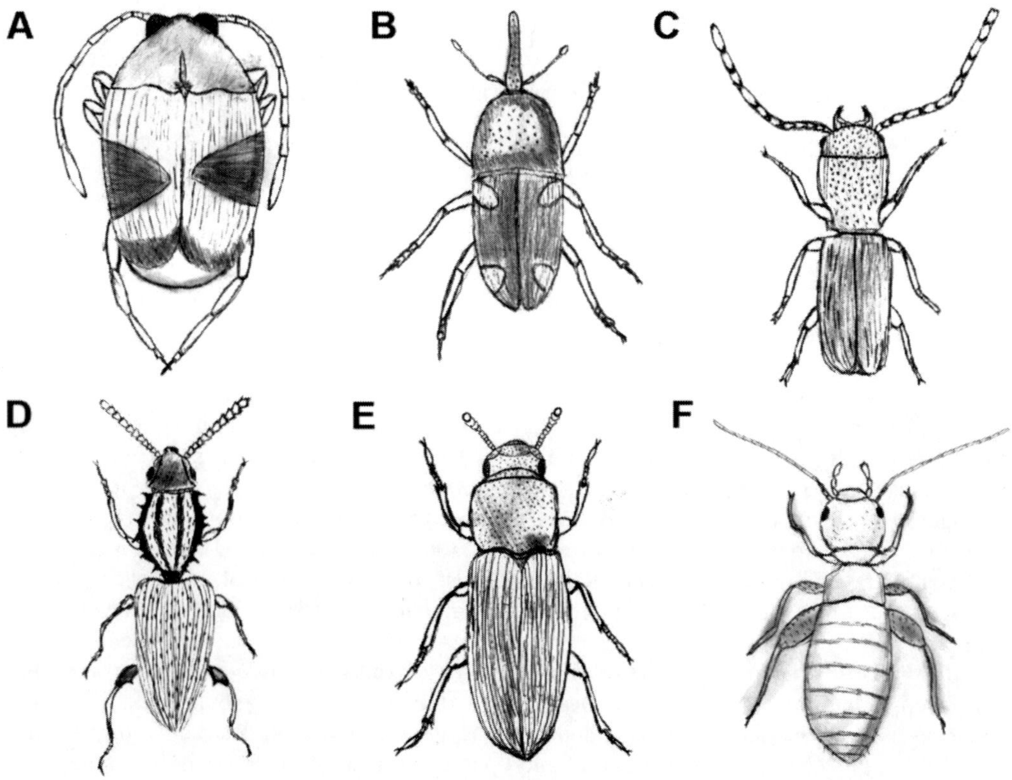

Figure 1. Representative drawings of stored-grain pests belonging to the respective families and species: (A)Bruchidae, *Callosobruchus maculatus*; (B) Curculionidae, *Sitophilus zeamais*; (C) Laemophoeidae, *Cryptolestes ferrugineus*; (D) Silvanidae, *Oryzaephilus surinamensis*; (E) Tenebrionidae, *Tribolium castaneum*; (F) Liposcelidae, *Liposceles* sp.

The insects that attack grains can be considered primary pests, which attack healthy grains throughout their development, and secondary pests, which are only able to attack grains that had been previously damaged (Silveira et al., 2006). The density of the insect population, the exposure time and the coexistence of primary and secondary pests dramatically increase the deterioration (Antunes et al., 2011; Tefera et al., 2011; Copatti et al., 2013). Alencar et al. (2011) assessed the effects of *Sitophilus zeamais* and *Tribolium castaneum* on maize and verified that the coexistence of these species during infestation promoted damage 20 times higher than that detected when these species infested the seeds separately. Similarly, Copatti et al. (2013) found more significant losses (91.48%) in rice grains concurrently attacked by *S. zeamais* and *Laemophloeus minutes*.

In general, insect pests usually feed on the endosperm because of its high nutritional content (Faroni, 1992; Nawrocka et al., 2010) and the most prevalent characteristics of damaged grains include reduction of dry matter, nutrient content and germination power. Caneppele et al. (2003) detected a loss of dry matter around 0.36% per day in maize grains infested during 150 days by *S. zeamais*. Mutungi et al. (2014) evaluated the effects of *Callosobruchus maculatus* on maize after six months of storage and found that 95% of the grains had their germination power reduced by more than 50%. This same pest also reduced the content of lipids and carbohydrates of beans by over 50% (Akintunde, 2012). Wheat grains infested by *Liposcelis bostrychophila* showed weight reduction of approximately 15% after 90 days of exposure (Kucerova, 2002).

The affected grains also have impaired hygiene and sanitary conditions. The water generated by the respiratory metabolism of insects leads to increased moisture in the storage environment, which leads to the multiplication of other deteriorating agents such as fungi (FAO, 1985; Puzzi, 1986; Jayas and White, 2003) and fungi that develop in storage conditions can produce mycotoxins that are harmful to the human health (Moreno-Martinez et al., 2011; Smith et al., 2012; Suleiman et al., 2013). Ahmedani et al. (2011) observed a strong correlation between the increase of moisture content and weight loss of wheat infested by the beetle *Trogoderma granarium*.

2. Biology and Ecology of Beetles and Beetle-Like Insects That Attack Stored Grains

2.1. Bruchidae (Coleoptera)

The insects from Bruchidae family (approximately 1,300 species) are cosmopolitan and develop in the seeds of several plants, mainly leguminous, cultivated in almost all parts of the world. Some authors classify this family as a subfamily of the Crysomelidae family called Bruchinae. The bruchids are usually classified in two groups: the species that lay the eggs in the fruits of host plants and whose larvae feed on the seeds; and the species that lay the eggs directly on the seeds and thus act as stored-grain pests (Lima, 1952; Gallo et al., 2002; Buzzi, 2010).

Adult bruchids are usually less than 10 mm in length and the body has an oval shape (Figure 1A). Some of the characteristics of Bruchidae family were described by Lima (1952), Athié and de Paula (2002), Gallo et al. (2002), Kingsolver (2004) and Buzzi (2010) are: the head is free with a short and flat rostrum as well as serrated or pectinate antennae with 11 segments; the elytra are striated and do not cover all the abdomen, leading the last tergum (called pygidium) exposed; and the posterior legs are more robust than the others. The larvae are white or yellow and have a robust body, with a tiny and curved head that can retract into the thorax. After oviposition, the newly-hatched larvae penetrate into the grains and build a chamber consuming the entire cotyledon. The larva passes through four molts and remains feeding until the end of the last instar, after which it enters the pupa stage. The pupation may occur completely inside the larval chamber, may start within and complete out of the chamber, or may occur completely out of the seed where the larva developed.

Some of the most relevant pests from Bruchidae are *Acanthoscelides obtectus*, *Bruchus pisorum*, *Callosobruchus maculatus*, *Callosobruchus phaseoli* and *Zabrotes subfasciatus* (Athié and de Paula, 2002; Gallo et al., 2002). *A. obtectus* is a primary pest and the eggs may be laid on the pods in the field or directly on the stored seeds. The optimal conditions for development of larvae are ca. 30°C and 70% relative humidity. Although the period of adult life is short, this species has great infestation ability because the adults are good fliers and the life cycle is quickly completed (approximately 23 days) (Lorini, 2010). *C. maculatus* is native from Africa but distributed along all the tropics and subtropics. The most important hosts of *C. maculatus* are the beans from the *Vigna* genus and this beetle has a high ability for cross-infestation, attacking the tillage and the storage environment (Athié and de Paula, 2002; Gallo et al., 2002). An interesting study conducted by Cope and Fox (2003) revealed that the *C. maculatus* females distributed the eggs during oviposition according to the size of the seeds, aiming to optimize the use of resources. The oviposition of this species is stimulated by the alkanes present on the surface wax that cover the attacked seeds (Parr et al., 1998; Adhikary et al., 2014).

2.2. Curculionidae (Coleoptera)

The family Curculionidae represents the more numerous in the Animal Kingdom, with approximately 50,000 species described (Gallo et al., 2002; Buzzi, 2012) and about 30 species of this family are pests of stored products around the world. The curculionids attack mainly fruits and seeds and a few subfamilies and the genus feed on dead vegetable materials. The most economically important species are *Sitophilus zeamais* (Figure 1B), *Sitophilus oryzae* and *Sitophilus granarium*, which can be found in storages of wheat, oat, rye, rice and maize and are also able to deteriorate beans, nuts, birdseeds, sunflower seeds and processed foods such as pasta.

According to Lima (1952), Athié and de Paula (2002), Gallo et al. (2002) and Buzzi (2010), the adults have a fairly elongated rostrum, straight or bent, which contains the chewing mouthparts at the end. The antennae are geniculate-capitate or geniculate-clavate with the scape inside a groove. The elytra are associated with the pro-thorax, usually cover the entire abdomen and can be glabrous, hairy and scaled. The scales may confer metallic coloration (green, blue, violet and golden). The posterior wings can be well-developed, rudimentary, obsolete or absent. The curculioniform larvae, which are apodal, robust, slightly bent and with a darker head, develop within the fruits, stems and seeds, consuming all the content present in these tissues. The pupae have a whitish color.

S. zeamais is found in all the warm and tropical regions. They are one of the major pests in stored grains in Brazil. This species possess a large number of hosts, including maize, wheat, sorghum and rice. It may also develop in processed cereals and food (Athié and de Paula, 2002; Gallo et al., 2002). *S. zeamais* have also been found attacking fruits such as apples, peaches and grapes (Botton et al., 2005). A remarkable characteristic of *S. zeamais* is the presence of reddish spots on the elytra and the curved rostrum present in the head is shorter and thicker in males. The adults are able to fly, quickly infesting the grains in the field and storage and are able to easily penetrate in the grains, and have a high potential for cross-infestation (Antunes and Dionello, 2010).

Danho et al. (2002) reported that the proportion of grains infected by *S. zeamais* is greater with decreasing in the amount of grains available. The females can live up to 140 days, 104 of these corresponding to oviposition period, and the average number of eggs per female is 282 (Botton et al., 2005). The females seal the hole made in the grain with a protein-rich secretion so that it is not possible to view the place of oviposition.

2.3. Laemophoeidae (Coleoptera)

The genus (*Cryptolestes*) is unique in the family Laemophoeidae that has importance as stored-grain pest. The *Cryptolestes* species feed on cereals, oleaginous seeds, nuts and dry fruits, and their presence is an indicator of very inadequate conditions in storage since these beetles develop in places already infested by other insects and fungi. The adults (Figure 1C) have a reddish-brown color, about 1.5–3.0 mm in length with a dorsoventral flattening. The antennae are filiform with 11 segments. The larvae penetrate grains with damaged or imperfect coats but the breaks only need to be microscopic to allow the entering of the larva. The eggs are elongated and more tapered at one of the ends. Each female is capable of laying 200 to 500 eggs and the oviposition occurs in the debris of plant material (Rillet, 1949; Athié and de Paula, 2002).

The species *Cryptolestes ferrugineus* is distributed at tropical, subtropical and temperate regions. It usually appears after an infestation by *Sitophilus* or *Rhyzopertha* and is able to develop in the grains of rye, wheat, maize, rice, oats, barley, corn, sunflower, flax, and soybeans (Rillet, 1949). *C. ferrugineus* also feeds on several types of fungi (mycophagous habit) found in storage. During cold seasons, they tend to move to the inner region of the grain mass (warmer region), and are present in the peripheral regions in warmer seasons (Athié & Cesar de Paula, 2002).

2.4. Silvanidae (Coleoptera)

The Silvanidae family includes about 500 species. The body of these insects is narrowed, brownish and densely punctured, with variable dimensions (1–15 mm in length), and the dorsoventral region is flattened (Figure 1D). Their antennae are clavate with 11 segments and their elytra entirely cover the abdomen. Several species are mycophagous and some are important pests of grain products (Athié and de Paula, 2002).

The species *Oryzaephilus surinamensis* (sawtoothed grain beetle) has a cosmopolitan distribution and is found infesting cereals, flour, spices, dry fruits, pasta, chocolate and even jerky beef. These beetles did not develop well in oleaginous seeds. It is a major pest of stored barley and is classified as a secondary pest because it is only able to cause scratches or scars on healthy and whole grains; however it quickly develops in broken grains. The adults have limited flight ability and thus the infestations are usually resulting from residual populations present in the storage or previous contamination of grains (Athié and de Paula, 2002; Gallo et al., 2002; Beckel et al., 2002, 2007). Each female lays 37 eggs on average (Beckel et al., 2007).

2.5. Tenebrionidae (Coleoptera)

The family Tenebrionidae (darkling beetles) is present in tropical and temperate regions. About 15,000 species are described, commonly xerophiles with nocturnal habits. About 80 species are reported to act as stored-grain pests and are found attacking cereals and flours. The most important species that act as pests belong to the genus *Tribolium*, *Gnatocerus*, *Alphitobius*, *Tenebrio*, and *Latheticus*; all these beetles possess thoracic and abdominal glands of defense that secrete benzoquinones and hydroquinones. In addition to storage pests, in this family other crop pests are present such as mycetophagous, coprophagous, predators, myrmecophilous and polyphagous species (Lima, 1952; Athié and de Paula, 2002; Buzzi, 2010).

The adults have a variable coloration (black, brown, reddish-brown, cinereous) and body size (3–10 mm in length, with species that can reach 18 mm). The head is very small and narrower than the prothorax (Figure 1E). The antennae are filiform, moniliform or serrated (more common) with 11 segments. The exoskeleton is remarkably thick and stiff, shiny and glabrous. The elytra cover the entire abdomen and the wings are often stunted. The legs are ambulatorial (cursorial) or less frequently fossorial (adapted for digging). The larvae are elateriform (worm-like) with sclerotic, glabrous and shiny integument and have short legs (Lima, 1952; Athié and de Paula, 2002; Gallo et al., 2002; Buzzi, 2010).

Tenebrio mollitor (mealworm) is considered a pest because the larvae are able to feed on stored grains (Siemianowska et al., 2013). The species *Tribolium castaneum* is supposed to have an Indian origin and is distributed in tropical and subtropical regions being very tolerant to arid conditions. It attacks all kinds of ground cereals and is an important secondary pest that infests cereals, coffee, cocoa, soybeans, dried fruits, nuts, cotton seed and also stored milk powder. Occasionally, *T. castaneum* may attack stored peas and beans (Athié and de Paula, 2002).

2.6. Liposcelidae (Psocoptera)

The insects belonging to this family are known as booklice or barklice. Although they comprise a taxonomically distinct group, sometimes these insects are referred as "beetle-like". They often live under dead bark, leaves and grass, in the nests of birds and mammals, on shelves, inside cracks of steps and furniture, and dusty places, among other locations. It grows easily in environments with relatively high temperature and humidity. Liposcelidae species can feed on fungi, algae, lichens, and pollen as well as on eggs and fragments of dead insects. Only a few species behave as pests of grain, for example, *Liposcelis bostrychophila*, which is able to damage rice. They are easily attracted by the presence of flour as well as moldy or wet food. Since these beetles are apterous the auto-dispersion is limited (Turner, 1998; Athié and de Paula, 2002; Buzzi, 2010; Chin et al., 2010).

The short (1–10 mm in length) body of these insects is tiny, flat and fragile (Figure 1F), with a yellowish or pale gray color and a semi-transparent appearance. The head is relatively large with a long and filiform (thread-like) antenna with 15 segments. The thorax is small and the abdomen is bigger than the rest of the body. Although the egg is approximately 1/3 of the female size, about 3-4 eggs are laid daily per female, resulting in a mean of 100 eggs in three weeks in the summer season. The metamorphosis is incomplete, being the nymphal stages (4 for females and 3 for males) and adult forms are very similar only differing in size and color (Turner, 2002; Athié and Cesar de Paula, 2002; Buzzi, 2010).

3. Current Strategies for Control of Stored-Grain Beetles

According to Gallo et al. (2002), three main techniques are usually applied to control pests in stored grains: fumigation, pulverizing and spraying. Other methods listed by Bond (1984) are sanitation, refrigeration, aeration, heating, drying, gamma radiation, microwaves, and infra-red radiation, as well as the use of insect growth regulators, predators, and pathogens. It also includes the use of insect resistant packaging.

Fumigation is the method most often used and can be applied to bulk or bagged products. It consists of the use of chemical compounds that are volatile at storage temperatures and toxic to the insect pests. However, only these characteristics are not enough to indicate the use of a substance. The compound should not be corrosive to containers or other materials in storage, should not react with the products originating irreversible residues and should not cause damage to the grains (Bond, 1984; Gallo et al., 2002). The fumigation technique is effective in eliminating insects in different stages of their life cycle.

The most used compounds in fumigation are methyl bromide and phosphine (aluminum phosphate and magnesium phosphate), which target the respiratory system of the insect. However, these substances are strongly toxic to non-target organisms, including humans (Gallo et al., 2002). Methyl bromide is acutely very toxic, mainly affecting the central nervous system, and intoxication may lead to death (Yang et al., 1995). Symptoms of methyl bromide poisoning are vomiting, headache, vertigo, imbalance while walking, slurred speech, and tremulousness of the upper limbs (Balogabal et al., 2011). Epidemiological evidence indicates that occupational exposure to methyl bromide is linked to incidence of human prostate cancer (Budnik et al., 2012). Phosphine is also toxic to non-target invertebrates and vertebrates disrupting the sympathetic nervous system, energy metabolism and the redox state of the cell (Nath et al., 2011). Exposure to high levels of phosphine causes acute respiratory problems, cough, headaches, dizziness, numbness, general fatigue and gastrointestinal disturbance. The chronic exposure results in anemia, bronchitis, gastrointestinal disorders, speech and motor disturbances, toothache, weakness, weight loss, mandible necrosis and spontaneous fractures (Takamiya, 2007).

The pulverizing method consists of mixing chemical powders with the grains and is recommended for treatment of small amounts of grains. Examples of chemicals used are bifenthrin, deltamethrin, fenitrothion, and pirimiphos (Gallo et al., 2002). The spraying method is achieved by micro pulverizations using an atomizer unit.

The excessive use of an insecticide inevitably results in a great selection pressure, which favors the proliferation of resistant individuals. Resistance of *S. zeamais* populations to the insecticides malathion, lindane, deltamethrin and phosphine has been described (Perez-Mendoza, 1999) and the development of strong resistance to phosphine by booklice has contributed to increase the importance of this species as a stored-grain pest (Collins et al., 2001). Opit et al. (2012) reported the detection in Oklahoma of a *Tribolium castaneum* population being 119-fold more resistant to phosphine in comparison with a laboratory susceptible strain, and highlighted that there is an increasing trend of phosphine resistance in the last 21 years. Figure 2 summarizes the steps involved in the emergence of a resistant insect population due to the prolonged, excessive and unplanned use of an insecticide.

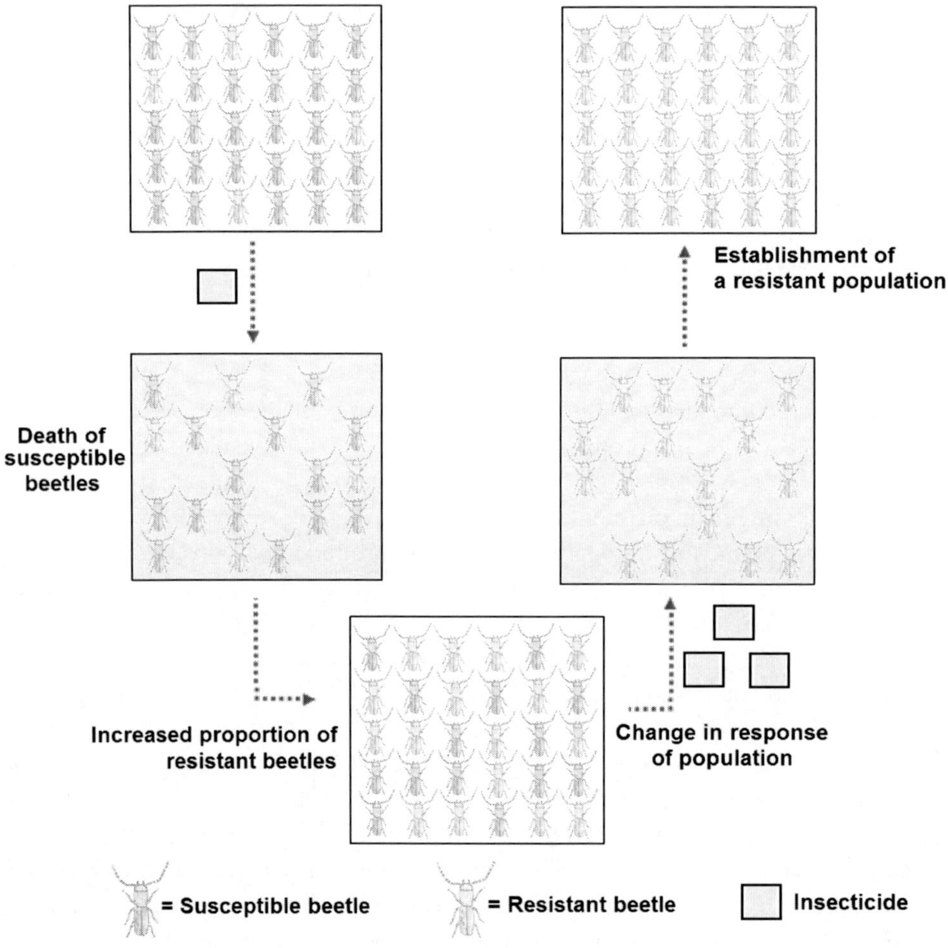

Figure 2. Establishment of an insect resistant population due to the excessive, intermittent and unplanned use of an insecticide.

4. Natural Insecticides for Control of Stored-Grain Beetles

The emergence of resistant populations and the risks that synthetic insecticides pose to human health and the environment have intensified the search for natural insecticides as an alternative in the control of agricultural pests. In general, natural products are environmentally safer because they are biodegradable and exhibit a greater selectivity than synthetic chemicals (Menezes, 2005; Duke et al., 2010; Kishore et al., 2011; Freitas et al., 2014).

Many parts of plants contain bioactive compounds that are involved in chemical defense against insect attack. A lot of naturally ant-insect substances have been extracted from plants in water and organic solvents. The entomotoxicity of plant preparations is mainly characterized by neurotoxic actions, feeding inhibition, digestion impairment, developmental delay and alterations of reproductive and behavioral aspects (Kim et al., 2003; Menezes, 2005; Correa and Salgado, 2011).

Several plant extracts have shown insecticidal activity against stored-grain beetles, affecting survival and physiology. Hexane extracts from African nutmeg (7.5 and 10 mg/100mL) was strongly toxic to *Callosobruchus maculatus* adults promoting 100% mortality after 1 hour of exposure (Ogunsina et al., 2011). Treatment of *Sitophilus oryzae* with hexane extract from the *Capparis decidua* stem by 16 hours interfered on insect physiology reducing the levels of glycogen, protein, and amino acid as well as activity of enzymes phosphatases, transaminases, dehydrogenases and acetylcholinesterase involved in the physiology processes (Upadhyay, 2013). The powdered leaves of *Azadirachta indica* containing saponins and azadirachtin promoted a mortality of 50% of larvae and adults of *Tribolium castenum*, and also reduced more than 10% amylase activity (Sami, 2014).

Essential oils are complex mixtures of volatile secondary metabolites, mainly extracted from aromatic plants. They are able to affect survival, behavior and physiology of insect pests. Wang et al. (2011) showed that the essential oil from *Illicium fargesii* promoted mortality of *S. zeamais* through contact (LC_{50}=28.95 μg/adult) and by fumigation (LC_{50}=11.36 mg/L). Mossi et al. (2014) reported that the essential oil from *Ocotea odorifera* containing camphor (43%) and safrole (42%) as major constituents killed *S. zeamais* (DL_{50}=14.1μl cm^2) by contact after 24 h and was also a repellent agent. Food deterrence (74.52%) and reduction in oviposition rate (35.66%) were detected after treatment of *S. oryzae* with oil from *Aegles quinces* leaves (Mishra et al., 2014). Gusmão et al. (2013) determined that essential oils from *Eucalyptus citriodora*, *Eucalyptus staigeriana*, *Cymbopogon winterianus* and *Foeniculum vulgare* were toxic to *Callosobruchus maculatus* by contact (LC_{50} ranging from 178.13 to 345.57 ppm) and fumigation (LC_{50} ranging from 2.58 to 7.85 μL/L of air), when used as repellent agents and reduced the oviposition rate as well as the emergence of adults in comparison with the control group.

In parallel, another progress in the search for alternative insecticides is the possible use of nanoparticles containing essential oils, since these systems are characterized by slow and persistent liberation of the oil in its active form. Gonzalez et al. (2014) determined that polyethylene glycol nanoparticles containing commercial essential oils (geranium and bergamot) have elevated residual toxicity against *Tribolium castaneum* and affected nutritional physiology reducing the relative growth rate, relative consumption rate and the efficiency of conversion of ingested food.

The deleterious effects of isolated secondary metabolites from different classes on insects are also reported. Sirinol, a compound isolated from garlic emulsion, showed a repellent effect on *T. castaneum* at a concentration of 10% while the metabolite allicin, also isolated from this plant, was toxic by fumigation to adults of *T. castaneum*, *Oryzaephilus surinamensis* and *Cryptolestes ferrugineus* with LC_{50} of 0.38, 0.51 and 0.51 mL/L in air, respectively (Jahromi et al., 2012; Lu et al., 2013). The ar-turmerone metabolite extracted from the rhizome of *Curcuma longa* affected the survival of *S. zeamais* at 1% (m/m), after six days of exposure by contact (Tavares et al., 2013).

Peptides and proteins isolated from plants also show insecticidal activity and thus have potential use in strategies for control of pests of stored products. Peptides purified from *Cicer arietinum* (chickpea) seeds were able to kill 83% and 100% of *S. oryzae* adults after 7 and 14 days of feeding (Mouhouche et al., 2009). Fields et al. (2010) demonstrated that peptide mixtures obtained from peas reduced the feeding and increased the mortality of *S. oryzae*. These authors also reported a synergistic effect between these peptides and insecticidal saponins and attributed this effect to the ability of saponins in impairing the hydrolysis of peptides by digestive enzymes at insect gut. Consequently, the increased time of contact between the peptides and insect gut probably allows the improvement of insecticidal action.

Zottich et al. (2014) revealed that the lipid transfer protein isolated from *Coffea canephora* seeds (Cc-LTP1, 0.5 %) inhibited the development of *C. maculatus* larvae, reduced the weight and the number of larvae, as well as decreased the oviposition rate. The same study showed that Cc-LTP1 inhibited the α-amylase activity of larval gut and was able to interact with the endoplasmic reticulum, mitochondria and microvilli of columnar cells from larvae.

Lectins are proteins whose structure contains carbohydrate-binding sites, which are able to interact with glycosylated molecules present in the lumen of insect gut, on the surface of epithelial cells or/and in the peritrophic matrix (Napoleão et al., 2012). The binding of lectins to glycosylated proteins at the midgut of insect larvae interferes with the nutrient uptake and the efficiency of diet utilization, resulting in a drop in mass gain. In this sense, lectins are insecticides that promote mortality or delay development of insects. The *N*-acetylglucosamine-specific lectin isolated from *Griffonia simplicifolia* leaf (GSII) at 1.0% (w/v) increased the WDST (within seed developmental time) of *C. maculatus* twice (Zhu et al., 1996) and the lectin isolated from *Myracrodruon urundeuva* leaves had a strong deterrent effect on adults of *S. zeamais*, and also reduced the activities of amylases, proteases, phosphatases, trypsin and endoglucanases (Napoleão et al., 2013). The lectin isolated from *Bauhinia monandra* leaf (BmoLL) was active on *C. maculatus* and *Z. subsfasciatus* larvae reducing the survival rate (LC_{50} of 0.5% and 0.3%, w/v, respectively) and decreasing body weight (Macedo et al., 2007). The authors demonstrated that the larvicidal effects of BmoLL probably involve resistance to proteolysis by larval enzymes and interaction with molecules at the membrane from midgut cells. *Talisia esculenta* seed lectin (TEL) also induced mortality of *C. maculatus* larvae (LC_{50}=1.0% w/v), was resistant to hydrolysis by cysteine proteases from larval midgut and bound proteins from the midgut of larvae (Macedo et al., 2004).

Products based on entomopathogenic fungi have also attracted the interest of researchers because they usually do not generate negative impacts to the environment and humans; they exhibit a residual effect that allows prolonged protection to stored grain, and the fungi applied are selective to the insect and do not develop in the grains (Alves et al., 2008; Michereff Filho et al., 2009). The fungi *Beauveria bassiana* caused mortality of *C. maculatus* (LC_{50}=3.17×10^6 conidia/mL) and *S. granarius* (LC_{50}= 6.08×10^7 conidia/mL) after 9 days (Shams et al., 2011). Nabaei et al. (2012) reported that the combination between diatomaceous earth and the entomopathogenic fungi *B. bassiana* or *Metharizium anisopliae* resulted in high mortality rates of the *C. maculatus* adult with improved median lethal time. Pimentel and Ferreira (2012) reported on the insecticidal activity against *S. zeamais* of the products Metarril® (LC_{50}=181.9 ml/L for 3 days) and Boveril® (LC_{50}=2.1 g/L for 3 days). These products are formulations based on the entomopathogenic fungi *M. anisopliae* and *B. bassiana*, respectively. The fungi *Isaria fumosorosea* was able to induce the mortality of*S. oryzae* adults and it was demonstrated that the concentration was not a critical parameter that determined the speed of insect death (Kavallieratos et al., 2014). The insecticide spinosad, produced by fermentation of the soil actinobacteria, *Saccharopolyspora spinosa*, has been indicated as a promising insecticide for control of several species that attack stored grains such as *Cryptolestes ferrugineus*, *Cryptolestes pusillus*, *Liposcelis entomophila, Sitophilus oryzae*, *Sitophilus granarius*, *Tribolium castaneum* and *Tribolium confusum* (Chintzoglou et al., 2008; Vayias et al., 2010; Hertlein et al., 2011).

Conclusion

This chapter summarizes information on all families of stored-grain pests in relation to morphology, physiology, ecology, grains attacked and current strategies for control of insects. The finding of alternative insecticides will allow their use in rotation programs, which canlead to increasing efficiency in insect control as well as minimizing the development of resistance. The potential use as an insecticide of plant compounds (such as secondary metabolites, essential oils and peptides) and entomopathogenic fungi is described.

Acknowledgments

The authors express their gratitude to the *Conselho Nacional de Desenvolvimento Científico e Tecnológico* for research grants and fellowship (P.M.G. Paiva). We are also grateful to the *Coordenação de Aperfeiçoamento de Pessoal de Nível Superior* (CAPES), the Brazilian Ministry for Science, Technology and Innovation (MCTI) and the *Fundação de Amparo à Ciência e Tecnologia do Estado de Pernambuco* for financial support. B.R. Belmonte would like to thank FACEPE for their graduate scholarship. E.V. Pontual would like to thank CAPES and FACEPE for their post-doctoral scholarship.

References

Adhikary, P., Mukherjee, A. & Barik, A. (2014). Role of surface wax alkanes from *Lathyrus sativus* L. seeds for attraction of *Callosobruchus maculatus* (F.) (Coleoptera: Bruchidae). *Journal of Stored Products Research*, *59*, 113-119

Ahmedani, M. S., Afazal, S. N., Haque, M. I., Nadeem, S., Hussain, T. & Naz, S. (2011). Quantitative losses and physical damage caused to wheat kernel (*Triticum aestivum*) by khapra beetle infestation. *Pakistan Journal of Botany*, *43*, 659-668.

Akintunde, E. M. (2012). Reduction of the nutritional values of cowpea infested with *Callosobruchus maculatus* (Coleoptera: Bruchidae). *Agriculture Science Developments*, *1*, 1-7.

Alencar, E. R., Faroni, L. R. D., Ferreira, L. G., Costa, A. R. & Pimentel, M. A. G. (2011). Qualidade de milho armazenado e infestado por *Sitophilus zeamais e Tribolium castaneum*. *Revista Engenharia na Agricultura*, *19*, 9-18.

Alves, S. B., Lopes, R. B., Vieira, S. A. & Tamai, M. A. (2008) Fungos entomopatogênicos usados no controle de pragas na América Latina. In: Alves, S.B., Lopes, R.B. (Eds.) Controle microbiano de pragas na América Latina: Avanços e Desafios. Piracicaba: FEALQ, 69-110.

Antunes, L. E. G., Viebrantz, P. C., Gottardi, R. & Dionello, R. G. (2011). Características físico-químicas de grãos de milho atacados por *Sitophilus zeamais* durante o armazenamento. *Revista Brasileira de Engenharia Agrícola e Ambiental*, *15*, 615-620.

Antunes, L. E. G. & Dionello, R. G.(2010). Bioecologia de *Sitophilus zeamais* Motschulsky 1885 (Coleoptera: Curculionidae). Hipertext article. Available in: http://www.infobibos.com/Artigos/2010_2/Sitophilus/index.htm

Assem, L. & Takamiya, M. (2007). Phosphine General information. Institute of Environment and Health, Cranfield University.

Athié, I. & de Paula, D. C. (2002). Insetos de grãos armazenados: aspectos biológicos e identificação. São Paulo: Varela.

Balagopal, K., Muthusamy, K., Alexander, M. & Mani, S. (2011). Methyl bromide poisoning presenting as acute ataxia. *Neurology India*, *59*, 768-769.

Beckel, H. (2002). Detecção da resistência de *Oryzaephilus surinamensis* (L.) (Coleoptera: Silvanidae), praga de grãos de cevada armazenada, a inseticidas químicos. Passo Fundo: Embrapa Trigo.

Beckel, H. D. S., Lorini, I. & Lazzari, S. (2007). Rearing method of *Oryzaephilus surinamensis* (L.) (Coleoptera, Silvanidae) on various wheat grain granulometry. *Revista Brasileira de Entomologia*, *51*, 501-505.

Bilia, D. A. C., Fancelli, A. L., Marcos Filho, J. E. & Machado, J. A. (1994). Comportamento de sementes de milho híbrido durante o armazenamento sob condições variáveis de temperatura e umidade relativa do ar. *Scientia Agricola*, *51*, 153-157.

Bond, E. (1984). Manual of fumigation for insect control. FAO Plant Production and Protection Paper. Food & Agriculture Organization of the United Nations. 432 p.

Botton, M., Lorini, I., Loeck, A. E. & Afonso, A. P. S. (2005). O gorgulho do milho *Sitophilus zeamais* (Coleptera: Curculionidae) como praga em frutíferas de clima temperado. Embrapa Uva e Vinho. Circular Técnica.

Budnik, L. T., Kloth, S., Velasco-Garrido, M. & Baur, X. (2012). Prostate cancer and toxicity from critical use exemptions of methyl bromide: environmental protection helps protect against human health risks. *Environmental Health*, *11*, 5.

Buzzi, Z. J. (2010). Entomologia didática**.** Curitiba: Editora da UFPR.

Caneppele, M. A. B., Caneppele, C., Lazzari, A. F. & Lazzari, S. M. N. (2003). Correlation between the infestation level of *Sitophilus zeamais* Motschulsky, 1855 (Coleoptera,

Curculionidae) and the quality factors of stored corn, *Zea mays* L. (Poaceae). *Revista Brasileira de Entomologia*, *47*, 625-630.

Chin, H. C., Jeffery, J., Ahmad, N. W., Kiang, H. S., Omar, B., Othman, H. & Lim, L. H. (2010). First Report of *Liposcelis bostrychophila* Badonnel (Psocoptera: Liposcelidae) as a Museum Insect Pest in Malaysia. *Sains Malaysiana*, *39*, 329-331.

Chintzoglou, G., Athanassiou, C. G. & Arthur, F. H. (2008). Insecticidal effect of spinosad dust, in combination with diatomaceous earth, against two stored-grain beetle species. *Journal of Stored Products Research*, *44*, 347-353.

Collins, P. J., Daglish, G. J., Nayak, M. K., Ebert, P. R., Schlipalius, D., Chen, W., Pavic, H., Lambkin, T. M., Kopittke, R. & Bridgeman,B. W. (2001). Combating resistance to phosphine in Australia. In: Donahaye, E.J, Navarro, S., Leesch J.G. (Eds.), Int. Conf. Controlled Atmosphere and Fumigation in Stored Products. Fresno: Executive Printing Services. 593-607.

Copatti, E. C., Marcon, R. K. & Machado, B. M. (2013). Avaliação de dano de *Sitophilus zeamais*, *Oryzaephilus surinamensis* e *Laemophloeus minutus* em grãos de arroz armazenados *Revista Brasileira de Engenharia Agrícola e Ambiental*, *17*, 855-860.

Cope, M. J. & Fox, C. W. (2003). Oviposition decisions in the seed beetle, *Callosobruchus maculatus* (Coleoptera: Bruchidae): effects of seed size on superparasitism. *Journal of Stored Products Research*, *39*, 355–365.

Correa, J. C. R. & Salgado, H. R. N. (2011). Atividade inseticida das plantas e aplicações: revisão. *Revista Brasileira de Plantas Medicinais*, *13*, 500-506.

Danho, M., Gaspar, C. & Haubruge, E. (2002). The impact of grain quantity on the biology of *Sitophilus zeamais* Motschulsky (Coleoptera: Curculionidae): oviposition, distribution of eggs, adult emergence, body weight and sex ratio. *Journal of Stored Products Research*, *38*, 259-266.

Duke, S. O., Cantrell, C. L., Meepagala, K. M., Wedge, D. E., Tabanca, N. & Schrader, K. K. (2010). Natural toxins for use in pest management. *Toxins*, *2*, 1943-1962.

FAO. Food and Agriculture Organization of United Nations (1985). Prevention of post-harvest food losses: a training manual**.** Pest Biology and Identification.

FAO. Food and Agriculture Organization of United Nations (1994). The role of storage in the economy. Grain storage techniques, 109.

Faroni, L. R. D. A. (1992). Manejo das pragas dos grãos armazenados e sua influência na qualidade do produto final. *Revista Brasileira de Armazenamento*, *17*, 36-43.

Fields, P. G., Woods, S. & Taylor, W. G. (2010). Triterpenoid saponins synergize insecticidal pea peptides: effect on feeding and survival of *Sitophilus oryzae* (Coleoptera: Curculionidae). *The Canadian Entomologist*, *142*, 501-512.

Freitas, R. S., Queiroz, M. E. L. R., Faroni, L. R. A., Heleno, L. L. & Moura, V. V. (2014). Development of a solid-liquid extraction method with low-temperature partitioning for the determination of insecticides in ozonized maize grain. *Química Nova*, *37*, 238-243.

Gallo, D., Nakano, O., Silveira Neto, S., Carvalho, R. P. L., Batista, G. C., Berti Filho, E., Parra, J. R. P., Zucchi, R. A., Alves, S. B., Vendramin, J. D., Marchini, L. C., Lopes, J. R. S. & Omoto, C. (2002). Entomologia agrícola. Piracicaba: FEALQ.

Gonzalez, J. O. W., Gutierrez, M. M., Ferrero, A. A. & Band, B. F. (2014). Essential oils nanoformulations for stored-product pest control – Characterization and biological properties. *Chemosphere*, *100*, 130-138.

Gusmão, N. M. S., Oliveira, J. V., Navarro, D. M. A. F., Dutra, K. A., Silva, W. A. & Wanderley, M. J. A. (2013). Contact and fumigant toxicity and repellency of *Eucalyptus citriodora* Hook., *Eucalyptus staigeriana* F., *Cymbopogon winterianus* Jowitt and *Foeniculum vulgare* Mill. essential oils in the management of *Callosobruchus maculatus* (FABR.) (Coleoptera: Chrysomelidae, Bruchinae). *Journal of Stored Products Research*, *54*, 41-47.

Hertlein, M.B., Thompson, G. D., Subramanyam, B., Christo B. & Athanassiou, G. (2011). Spinosad: A new natural product for stored grain protection. *Journal of Stored Products Research*, *47*, 131-146.

Jahromi, M. G., Pourmirza, A. A. & Safaralizadeh, M. H. (2012). Repellent effect of sirinol (garlic emulsion) against *Lasioderma serricorne* (Coleoptera: Anobiidae) and *Tribolium castaneum* (Coleoptera: Tenebrionidae) by three laboratory methods. *African Journal of Biotechnology*, *11*, 280-288.

Jayas, D. S. & White, N. D. G. (2003). Storage and drying of grain in Canada: low cost approaches. *Food Control*, *14*, 255-261.

Kavallieratos, N. G., Athanassiou, C. G., Aountala, M. M. & Kontodimas, D. C. (2014). Evaluation of the entomopathogenic fungi *Beauveria bassiana*, *Metarhizium anisopliae*, and *Isaria fumosorosea* for control of *Sitophilus oryzae*. *Journal of Food Protection*, *77*, 87-93.

Kim, S. I., Roh, J. I., Kim, D. H., Lee, H. S. & Ahn, Y. J. (2003). Insecticidal activities of aromatic plant extracts and essential oils against *Sitophilus oryzae* and *Callosobruchus chinensis*. *Journal of Stored Products Research*, *39*, 293-303.

Kingsolver, J. M. (2004). Handbook of the Bruchidae of the United States and Canada (Insecta, Coleoptera). United States Department of Agriculture. Agricultural Research Service Technical Bulletin Number 1912.

Kishore, N., Mishra, B. B., Tiwari, V. K. & Tripathi, V. (2011). A review on natural products with mosquitosidal potentials. In: Tiwari, V.K., Mishra, B.B. (Eds.). Opportunity, Challenge and Scope of Natural Products in Medicinal Chemistry. Kerala: Research Signpost, 335-365.

Kucerova, Z. (2002). Weight losses of wheat grains caused by psocid infestation (*Liposcelis bostrychophila*: Liposcelididae: Psocoptera). *Plant Protection Science*, *38*, 103–107.

Lima, A. C. (1952). Coleópteros. In: Insetos do Brasil, 7° Tomo, Capítulo XXIX. Série Didática N° 9. Escola Nacional de Agronomia.

Lorini, I., Krzyzanowski, F. C., Franca-Neto, J. B. & Henning, A. A. (2010). Principais pragas e métodos de controle em sementes durante o armazenamento - Série Sementes. Londrina: Embrapa Soja. Circular Técnica.

Lu, Y., Zhong, J., Wang, Z., Liu, Z. & Wan, Z. (2013). Fumigation toxicity of allicin against three stored product pests. *Journal of Stored Products Research*, *55*, 48-54.

Macedo, M. L. R., Castro, M. M. & Freire, M. G. M. (2004). Mechanisms of the insecticidal action of TEL (*Talisia esculenta* lectin) against *Callosobruchus maculatus* (Coleoptera: Bruchidae). *Archives of Insect Biochemistry and Physiology*, *56*, 84–96.

Macedo, M. L. R., Freire, M. G. M., Silva, M. B. R. & Coelho, L. C. B. B. (2007). Insecticidal action of *Bauhinia monandra* leaf lectin (BmoLL) against *Anagasta kuehniella* (Lepidoptera: Pyralidae), *Zabrotes subfasciatus* and *Callosobruchus maculatus* (Coleoptera: Bruchidae). *Comparative Biochemistry and Physiology, Part A*, *146*, 486-498.

Menezes, E. L. A. (2005). Inseticidas botânicos: seus princípios ativos, modo de ação e uso agrícola. Rio de Janeiro: Seropédica - Embrapa Agrobiologia.

Michereff Filho, M., Faria, M., Wraight, S. P. & Silva, K. F. A. S. (2009). Micoinseticidas e micoacaricidas no Brasil: como estamos após quatro décadas? *Arquivos do Instituto Biológico*, *76*, 769-779.

Mishra, B. B., Tripathi, S. P. & Tripathi, P. M. (2014). Chronic activity of plant volatiles essential oils in management of rice weevil *Sitophilus oryzae* (Coleoptera: Curculionidae). *Journal of Entomology*, *11*, 78-86.

Moreno-Martinez, E., Chapa-Oliver, A. M., Mejía-Teniente, L., Torres-Pacheco, I., Guevara-González, R. G., Vazquez-Cruz, M. A., Cervantes-Landaverde, J. J. & Preciado-Ortiz, R. E. (2011). Genetic resistance to drought in maize and its relationship in aflatoxins production. In: Guevara-González, R.G. (Ed.). *Aflatoxins - Biochemistry and Molecular Biology*. Rijeka: InTech, 151-161.

Mossi, A. J., Zanella, C. A., Kubiak, G., Lerin, L. A., Cansian, R. L., Frandoloso, F. S., Prá, V. D., Mazutti, M. A., Costa, J. A. V. & Treichel, H. (2014). Essential oil of *Ocotea odorifera*: An alternative against *Sitophilus zeamais*. *Renewable Agriculture and Food Systems*, *29*, 161-166.

Mouhouche, F., Fleurat-Lessard, F. & Bouznad, Z. (2009). Laboratory assessment of toxic activity of purified peptides extracted from chickpea seeds to two strains of the rice weevil *Sitophilus oryzae* (L.) (Coleoptera: Curculionidae). *Journal of Stored Products Research*, *45*, 261-266.

Mutungi, C. M., Affognon, H., Njoroge, A. W., Baributsa, D. & Murdock, L. L. (2014). Storage of mung bean (*Vigna radiata* [L.] Wilczek) and pigeon pea grains (*Cajanus cajan* [L.] Millsp) in hermetic triple-layer bags stops losses caused by *Callosobruchus maculatus* (F.) (Coleoptera: Bruchidae). *Journal of Stored Products Research*, doi: 10.1016/j.jspr.2014.03.004

Nabaei, N., Mehrvar, A., Saber, M. & Bagheri, M. (2012). Efficacy of entomopathogenic fungi in combination with diatomaceous earth against *Callosobruchus maculatus* (Coleoptera: Bruchidae). *Acta Entomologica Sinica*, *55*, 1282-1288.

Napoleão, T. H., Belmonte, B. R., Pontual, E. V., Albuquerque, L. P., Sá, R. A., Paiva, L. M., Coelho, L. C. B. B. & Paiva, P. M. G. (2013). Deleterious effects of *Myracrodruon urundeuva* leaf extract and lectin on the maize weevil, *Sitophilus zeamais* (Coleoptera, Curculionidae). *Journal of Stored Products Research*, *54*, 26–33.

Napoleão, T. H., Pontual, E. V., Lima, T. A., Santos, N. D. L., Sá, R. A., Coelho, L. C. B. B., Navarro, D. M. A. F. & Paiva, P. M. G. (2012). Effect of *Myracrodruon urundeuva* leaf lectin on survival and digestive enzymes of *Aedes aegypti* larvae. *Parasitology Research*, *110*, 609-616.

Nawrocka, A., Grundas, S. & Grodek, J. (2010). Losses caused by granary weevil larva in wheat grain using digital analysis of X-ray images. *International Agrophysics*, *24*, 63-68.

Nath, N. S., Bhattacharya, I., Tuck, A. G., Schlipalius, D. I. & Ebert, P. R. (2011). Mechanisms of phosphine toxicity. *Journal of Toxicology*, ID 494168.

Ongusina, O. O., Oladimeji, M. O. & Lajide, L. (2011). Insecticidal action of hexane extracts of three plants against bean weevil, *Callosobruchus maculatus* (F.) and maize weevil, *Sitophilus zeamais* Motsch. *Journal of Ecology and the Natural Environment*, *3*, 23-28.

Opit, G. P., Phillips, T. W., Aikins, M. J. & Hasan, M. M. (2012). Phosphine resistance in *Tribolium castaneum* and *Rhyzopertha dominica* from stored wheat in Oklahoma. *Journal of Economic Entomology*, *105*, 1107-1114.

Parr, M. J., Tran, B. M. D., Simmonds, M. S. J., Kite, G. C. & Credland, P. F. (1998). Influence of some fatty acids on oviposition by the bruchid beetle, *Callosobruchus maculatus*. *Journal of Chemical Ecology*, *24*, 1577-1593.

Perez-Mendonza, J. (1999). Survey of insecticide resistance in Mexican populations of maize weevil, *Sitophilus zeamais* Motschulsky (Coleoptera: Curculionidae). *Journal of Stored Products Research*, *35*, 107-115.

Pimentel, M. A. G. & Ferreira, E. G. (2012). Toxicidade de produtos formulados à base de fungos entomopatogênicos para o caruncho-do-milho. *Revista Brasileira de Milho e Sorgo*, *11*, 209-215.

Puzzi, D. (1986). Abastecimento e Armazenamento de grãos. Campinas: Instituto Campineiro de Ensino Agrícola.

Rilett, R. O. (1949). The biology of *Laemophloeus ferrugineus* (Steph.). *Canadian Journal of Research*, *27*, 112-148.

Sami, A. J. (2014). *Azadirachta indica* derived compounds as inhibitors of digestive alpha-amylase in insect pests: Potential bio-pesticides in insect pest management. European *Journal of Experimental Biology*, *4*, 259-264.

Shams, G., Safaralizadeh, M. H., Imani, S., Shojai, M. & Aramideh, S. (2011). A laboratory assessment of the potential of the entomopathogenic fungi *Beauveria bassiana* (Beauvarin) to control *Callosobruchus maculatus* (F.) (Coleoptera: Bruchidae) and *Sitophilus granarius* (L.) (Coleoptera: Curculionidae). *African Journal of Microbiology Research*, *5*, 1192-1196.

Siemianowska, E., Kosewska, A., Aljewicz, M., Skibniewska, K. A., Polak-Juszczak, L., Jarocki, A. & Jędras, M. (2013). Larvae of mealworm (*Tenebrio molitor* L.) as European novel food. *Agricultural Sciences*, *4*, 287-291.

Silveira, R. D., Faroni, L. R. D. A., Pimentel, M. A. G. & Zocolo, G. J. (2006). Influência da temperatura do grão de milho, no momento da pulverização, e do período de armazenamento, na mortalidade de *Sitophilus zeamais* e *Tribolium castaneum*, pela mistura bifenthrin e pirimifós-metil. *Revista Brasileira de Armazenamento*, *31*, 120-124.

Smith, L. E., Stoltzfus, R. J. & Prendergast, A. (2012). Food chain mycotoxin exposure, gut health, and impaired growth: A conceptual framework. *Advances in Nutrition*, *3*, 526-531.

Suleiman, R. A., Rosentrater, K. A. & Bern, C. J. (2013). Effects of deterioration parameters on storage of maize: A review. *Journal of Natural Sciences Research*, *3*, 147-165.

Tavares, M. A. G. C. & Vendramim, J. D. (2005). Bioatividade da Erva-de-Santa-Maria, *Chenopodium ambrosioides* L., sobre *Sitophilus zeamais* Mots. (Coleoptera: Curculionidae). *Neotropical Entomology*, *34*, 319-323.

Tavares, W. S., Freitas, S. S., Grazziottib, G. H., Parentec, L. M. L., Lião, L. M. & Zanuncio, J. C. (2013). Ar-turmerone from *Curcuma longa* (Zingiberaceae) rhizomes and effects on *Sitophilus zeamais* (Coleoptera: Curculionidae) and *Spodoptera frugiperda* (Lepidoptera: Noctuidae). *Industrial Crops and Products*, *46*, 158–164.

Tefera, T., Mugo, S. & Likhayo, P. (2011). Effects of insect population density and storage time on grain damage and weight loss in maize due to the maize weevil *Sitophilus zeamais* and the larger grain borer *Prostephanus truncates*. *African Journal of Agricultural Research*, *6*, 2249-2254.

Turner, B. D. (1998). The pest status of psocids in the UK. *Pesticide Outlook*, *9*, 27-30.

Turner, B.D. (2002) Current trends in stored product research. *International Pest Control*, *44*, 110-113.

Upadhyay, R.K. (2013). *Capparis decidua* solvent and aqueous extracts affect biochemical and enzymatic parameters in *Sitophilus oryzae* Linn. (Coleoptera: Curculionidae). *World Journal of Zoology*, *8*, 245-255.

Vayias, B.J., Athanassiou, C.G. & Buchelos, C.T.H. (2009). Effectiveness of spinosad combined with diatomaceous earth against different European strains of *Tribolium confusum* Jacquelin du Val (Coleoptera: Tenebrionidae): influence of commodity and temperature. *Journal of Stored Products Research*, *45*, 165-176.

Wang, C.F., Liu, P., Yang, K., Zeng, Y., Liu, Z.L., Du, S.S. & Dend, Z.W. (2011). Chemical composition and toxicities of essential oil of *Illicium fragesii* fruits against *Sitophilus zeamais*. *African Journal of Biotechnology*, *10*, 18179-18184.

Yang R.S., Witt K.L., Alden C.J. & Cockerham L.G. (1995). Toxicology of methyl bromide. *Reviews of Environmental Contamination and Toxicology*, *142*, 65-85.

Zhu K., Huesing, J.E., Shade, R.E., Bressan, R.A., Hasegawa, P.M. & Murdock, L.L. (1996). An insecticidal N-acetylglucosamine-specific lectin gene from *Griffonia simplicifolia* (Leguminosae). *Plant Physiology*, *110*, 195-202.

Zottich, U., Da Cunha, M., Dias, G.B., Rabelo, G.R., Oliveira, A.E.A., Carvalho, A.O., Fernandes, K.V., Nascimento, V.V. & Gomes, V.M. (2014). The toxicity of a lipid transfer protein (Cc-LTP1) from *Coffea canephora* seeds on the larval development of *Callosobruchus maculatus* (Coleoptera: Bruchidae). *The Protein Journal*, doi: 10.1007/s10930-014-9575-9.

In: Beetles: Biodiversity, Ecology and Role in the Environment ISBN: 978-1-63463-380-2
Editor: Camilla Stack

Chapter 6

DECAYING MATTERS: COLEOPTERA INVOLVED IN HETEROTROPHIC SYSTEMS

Paryse Nadeau[1], Monic Thibault[1], Finbarr G. Horgan[2], Jean-Philippe Michaud[1,3], Franck Gandiaga[1], Charles Comeau[1] and Gaétan Moreau[1]

[1]Département de biologie, Université de Moncton, Moncton, New Brunswick, Canada

[2]Crop and Environmental Sciences Division, International Rice Research Institute, DAPO Box 7777, Metro Manila, The Philippines

[3]Royal Canadian Mounted Police, Calgary General Investigations Section, Serious Crimes Branch, Calgary, Alberta, Canada

ABSTRACT

In terrestrial ecosystems, a rich assemblage of Coleoptera species contributes to the breakdown of plant and animal organic materials such as dead wood, fruits, leaf litter, feces, and carcasses, including human cadavers. This taxon represents a key component of energy flow in many ecosystems both because heterotrophic Coleoptera species contribute to nutrient cycling and because they are important trophic links in food webs. However, the taxon as a whole has not received much consideration in the literature and similarities between terrestrial heterotrophic systems have yet to be investigated. This chapter examines the diversity and ecological roles of Coleoptera fauna involved in terrestrial heterotrophic communities and reviews the main mechanisms and models of heterotrophic succession. The impacts on heterotrophic Coleoptera of resource patch characteristics (e.g., abundance, distribution, and resource quality) and of anthropogenic alterations of habitats supporting resource patches (e.g., agriculture, forest management, urbanization, global changes) are described and discussed.

1. INTRODUCTION

In his book "*The Diversity of Life*", the influential and widely cited Edward O. Wilson argued that without insects, mankind would survive only a few months and the environment would collapse into chaos (Wilson, 1999). To understand the implications of this message, imagine a world without decomposition: ecosystems would be buried under plant and animal organic material such as dead wood, fruits, dung, and animal carcasses. Without insects, the situation would not be much better because the direct action of insects in the decomposition process facilitates the fragmentation of organic material, which allows the return of elements to the ground during nutrient cycling (Poole, 1974).

Among the main macroscopic actors involved in decomposition are many Coleoptera families and subfamilies such as the Aphodiinae, Cerambycidae, Dermestidae, Scolytinae, and Silphidae, to name a few. Other than contributing to the decomposition of organic matter, these beetles fulfill different roles in their ecological communities. For example, they will attract beetles and other organisms that act as their natural enemies, therefore permitting the establishment of higher-order trophic levels. Some beetle species, notably from the Staphylinidae, will consume both the organic matter and other organisms that contribute to the decomposition process. By feeding on the resource, some beetle species will also facilitate colonization of the resource by other species. Considering the different niches they occupy and the relative specificity of several beetle species to their given substrates, beetles involved in decomposition processes will often form a significant part of the biodiversity of their respective systems (Ehnström et al., 1993; Michaud et al., 2010).

In this chapter, we examine the diversity and ecological roles of the Coleoptera fauna involved in terrestrial heterotrophic communities. We define here these beetles and communities as 'heterotrophic' because they are involved in heterotrophic succession, as opposed to autotrophic succession (see Section 3.1). Up until now, the different heterotrophic systems have been described and studied by different branches of biology (e.g., community ecology, conservation, forensic entomology). As such, the similarities between these systems have not been subject to a thorough review, thus limiting the development of unifying principles or concepts that may apply to these systems. As a result, current models of succession in heterotrophic systems were transposed from autotrophic systems but, as will be demonstrated below, have not been empirically assessed. This chapter describes the different terrestrial heterotrophic communities of beetles, compares the main mechanisms and models of heterotrophic succession, and examines the impacts of anthropogenic disturbances of these systems on the Coleoptera fauna.

2. COLEOPTERA INVOLVED IN HETEROTROPHIC SYSTEMS

Heterotrophic beetles represent a large and diversified taxon because of the contrasted substrates these insects colonize. In this section, we examine the substrates used by heterotrophic beetles and describe the specifics of these substrates. This section begins with animal carcasses and will then describe heterotrophic communities on fecal and vegetal heterotrophic resources.

2.1. Animal Carcasses

After death, animal remains that are not accessible to large vertebrate scavengers (see Coe, 1978) undergo a gradual loss of muscle and adipose tissue until the totality of the organic material is recycled into the soil. This process is attributable in part to autolysis, the *post-mortem* self-digestion of cells, but mainly to the action of the complex assemblages of heterotrophic organisms such as bacteria and insects that inhabit carcasses (Putman, 1978). Heterotrophic insects associated with animal carcasses include some families of Diptera (e.g., Calliphoridae, Muscidae, Sarcophagidae), Hymenoptera (see Payne & Mason, 1971), and Lepidoptera (see Payne & King 1968), but especially a large number of Coleoptera families (e.g., Cleridae, Dermestidae, Geotrupidae, Histeridae, Hydrophilidae, Leodidae, Nitidulidae, Scarabaeidae, Silphidae, Staphylinidae, Trogidae; Anderson & VanLaerhoven, 1996; Byrd & Castner, 2001; Benecke, 2001; Tabor et al., 2005a; Michaud et al., 2010). Beetles are attracted to carcasses by odors and gases released during decomposition (e.g., Campobasso et al., 2001; Kalinova et al., 2009; von Hoermann et al., 2011, 2012) and/or by other arthropods that use the carcass as a resource or refuge (Goff, 1993). Depending on the ecological conditions, it is not uncommon to see dozens of beetle species throughout the decomposition process and several individuals per species occupying the carcass at the same time.

Succession, the predictable arrival pattern of organisms on a disturbed or newly exposed substrate, is a classic concept of animal and plant ecology that dates back to the late 1800's (Mégnin 1883, 1887, 1894). The visitation and colonization patterns of flies and beetles on carcasses are arguably the most obvious examples of ecological succession. Flies of the family Calliphoridae are usually the earliest species to visit carcasses, arriving minutes after death (Greenberg, 1991). Female flies lay their eggs in natural body cavities such as the mouth, ears, nostrils, and anus, as well as in open wounds. The larvae emerging from these eggs can, in some cases, reduce the biomass of the carcass by up to 80% (Putman, 1978). Beetles colonize at various moments throughout the decomposition process and many species contribute to the direct reduction of biomass by feeding on the remains. However, many other species do not feed on the remains but instead prey on insects already on the carcass and/or their offspring, whereas others only use the carcass as an extension of their natural habitat.

Beetles involved in the decomposition of animal carcasses have been studied from a variety of angles. Indeed, beetles have evolved interesting defensive, competitive, and even in some cases social strategies that make them interesting to practitioners from multiple disciplines. This section provides an overview of how animal carcasses decompose and discusses the role beetles have in the process.

2.1.1. Carcass Decomposition

Carcass decomposition is a complex process that involves internal physicochemical changes best documented in medical textbooks. Carcass decomposition also involves time-dependent changes in insect community structure, also known as succession. Studies on carrion-insect succession date back to the late 1800's when researchers realized that insects could potentially be used as a tool for estimating the time of death of the deceased in criminal investigations (Mégnin, 1883, 1887, 1894). In his studies on buried and exhumed corpses, Mégnin identified eight "squads" or groups of carrion insects that colonize following a predictable pattern. In addition to identifying several specimens of the Silphidae, Histeridae, Nitidulidae, and Dermestidae families, Mégnin also included drawings for identification purposes and some notes on the families' natural histories (Mégnin, 1894). Mégnin's work was then replicated in North America (Johnston & Villeneuve, 1897; Motter, 1898), giving birth to the discipline of carrion ecology and its applied cousin, forensic entomology. Although many ecological studies have examined the insect fauna of carcasses of various animal taxa (e.g., Jaques, 1915; Illingworth, 1926; Fuller, 1934; Bornemissza, 1957; Payne, 1965; Wasti, 1972; Coe, 1978; Abell et al., 1982; Braack, 1987), to this day most studies of carrion-insect succession continue to be conducted in a forensic entomology context, where beetles can play an important role in estimating the time of death.

From the very first studies on carrion-insects (e.g., Chapman & Sankey, 1955; Bornemissza, 1957; Reed, 1958), researchers have utilized decay stages, which are essentially descriptors that take into account both the physical appearance of the carcass and the insect fauna. Decay stages have been defined and used differently by various researchers. For example, Chapman & Sankey (1955) identified three distinct stages, Reed (1958) recognized four, Bornemissza (1957) described five whereas Payne (1965) documented six stages. For years researchers believed, based on Mégnin's pioneer notion of "squads" (Mégnin, 1894), that to each decay stage was associated a specific group of carrion insects, to which belonged many beetle species (e.g., Fuller, 1934; Bornemissza, 1957; Payne, 1965; Wasti, 1972; Coe, 1978; Abell et al., 1982). This belief was so prevalent that it became the first paradigm of carrion ecology and forensic entomology (Michaud et al., unpublished). However, the notion that decay stage boundaries are associated with marked changes in carrion-insect assemblages has since been rebutted and more emphasis has been placed on the notion that succession is a continuous process (Schoenly & Reid, 1987; Boulton & Lake; 1988; Moura et al., 2005). Although decay stages have been shown to be predictable by using an accumulated degree-day model (Michaud & Moreau, 2011), beetle occurrence patterns usually either follow a Gaussian (or truncated Gaussian) or a random distribution, neither of which are closely associated to decay stages (Michaud & Moreau, 2009). However, decay stages, and particularly the description of five stages (i.e., fresh stage, bloated stage, active decay stage, advanced decay stage, and dry stage) proposed by Anderson & VanLaerhoven (1996) and adapted from Payne (1965) (Table 1), continue to be used today mainly because they are convenient descriptors of post-mortem changes but also because they are useful reference points in court for forensic practitioners (Goff, 1993; Anderson & VanLaerhoven, 1996; Tibbett & Carter, 2009).

2.1.2. Main Beetle Families Involved in Carcass Decomposition

This section presents a non-exhaustive list of the main beetle families one can expect to find on animal carcasses in North America. Beetles involved in carcass decomposition have received little attention compared to flies and only a few families have been extensively studied, such as the Silphidae, Staphylinidae, Histeridae, and Dermestidae.

Silphidae

Silphidae, commonly known as carrion beetles, are found on animal carcasses all over the world and as such are well accounted for in carrion ecology and forensic entomology studies. Carrion beetles are also popular with amateur entomologists and the general populace due to their impressive size and in some cases their flamboyant color patterns. Larvae uniquely feed on decomposing tissues, whereas adults feed on decomposing tissues and sometimes fly larvae. In North America, the family is comprised of 30 species divided into two subfamilies,

Silphinae and Nicrophorinae, which have different developmental and resource-use strategies (Anderson & Peck, 1985).

The genus *Nicrophorus* (Nicrophorinae subfamily) has been by far the most studied carrion beetle, due to its interesting social behaviors and competitive strategies. One species, *Nicrophorus americanus* Oliver, is on the US endangered species list since the late 1980's (Lomolino et al., 1995). *Nicrophorus* got its common name of "burying beetle" due to its tendency to bury small carcasses as a way to gain a competitive advantage for its offspring against microbes and fly larvae (Pukowski, 1933; Milne & Milne, 1976). The genus has also been examined in a social context due to its bi-parental care and communal breeding (Scott & Traniello, 1990; Trumbo & Fiore, 1994; Scott, 1998; Eggert & Sakaluk, 2000; Scott et al., 2007), traits that are rather unusual in insects, making it worthy of being featured in E.O. Wilson's seminal book "*The insect societies*" (Wilson, 1971). Moreover, *Nicrophorus* specimens are believed to be involved in a symbiotic relationship with mites that feed on fly eggs. By transporting mites, the beetle gains a competitive advantage over fly larvae, whereas the mites gain access to a previously unattainable food source (Springett, 1968; Wilson & Knollenberg, 1987; Schwarz & Koulianos, 1998). However, the validity of this alleged symbiotic relationship has been challenged (see Blackman & Evans, 1994; Wilson & Knollenburg, 1987). Finally, *Nicrophorus* beetles are also known for their use of antimicrobial anal secretions as a defense mechanism against microbial competitors (e.g., Rozen et al., 2008; Cotter & Kilner, 2010; Arce et al., 2012). Species of the Silphinae are essentially the opposite of Nicrophorinae: they are much less studied, they do not bury their food source, they do not demonstrate parental care or communal breeding, and they avoid competition with flies by waiting for them to wander away from the carcass (Anderson & Peck, 1985).

Table 1. Description of decay stages for carcass and cadaver decomposition

Stage	Carcass description (from Anderson & VanLaerhoven, 1996)	Beetle fauna (from Payne, 1965)
Fresh	Starts upon death; decomposition is initiated by anaerobic gut bacteria and aerobic bacteria from the environment; cell autolysis occurs due to internal chemical degradation, but tissue autolysis also takes place under the effects of bacterial enzymes (Campobasso et al., 2001); initial fly oviposition	
Bloated	Inflation of carcass abdomen caused by the accumulation of gases released by bacteria	Arrival of Scarabaeidae
Active decay	Deflation of the carcass due to maggots boring through the abdomen	Arrival of Silphidae
Advanced decay	Most of the carcass has been consumed; maggots start to migrate away from the carcass	Silphidae and Scarabaeidae adults start to migrate away from the carcass; arrival of Dermestidae
Dry	Only bones, cartilage, skin, and hair remain	Numerous Staphylinidae and Silphidae larvae occur

Studies that examine carrion beetles from an ecological standpoint remain few. Anderson (1982) examined seasonal and temporal activity along with the habitat association and preference as it relates to carcass size in both subfamilies of carrion beetles. More recently, the survival of *Nicrophorus* beetle populations was found to be, at least in one instance, associated with resource pulses comprised of salmon carcasses in British Columbia (Hocking et al., 2006). Also, a decrease in decomposition rates at lower ambient temperatures could be due to the absence of some of the larger carrion beetle species from the beetle assemblage (Farwig et al., 2014). From a forensic perspective, carrion beetles are staples of baseline succession studies and some species have predictable temperature-dependent life cycles (e.g., Velásquez & Viloria, 2009; Ridgeway et al., 2014) and occurrence patterns (Michaud & Moreau, 2009; Matuszewski, 2011; Matuszewski & Szafałowicz, 2013), making them useful tools in time of death estimation.

Staphylinidae

Staphylinidae, also known as rove beetles, are common on animal carcasses throughout decomposition. In North America, the family is comprised of over 2000 species, most of which are predatory both as larvae and adults and found in all types of decomposing organic matter, including animal carcasses (Downie & Arnett, 1996). Being predatory in nature and skilled fliers, rove beetles can be observed feeding on fly larvae and adult flies alike. Some rove beetles also have abdominal glands that secrete a variety of chemicals used as a defense mechanism (Huth & Dettner, 1990; Dettner, 1993). Although some knowledge is available regarding their habitat preferences and seasonality (Hanski & Hammond, 1986; Márquez, 2003), overall very little is known about the ecology of the Staphylinidae.

In a forensic setting, the temperature-dependent development of rove beetles has not been examined in detail. However, *Creophilus maxillosus* L. has a predictable occurrence pattern on animal carcasses, making it potentially useful for time of death estimation (Matuszewski, 2012; Perez et al., 2014). Also, habitat and seasonal preferences noted in some rove beetle species could assist investigators in determining the season in which a death occurred or whether a body has been moved after death (Mądra et al., 2014). Although some species are forensically useful, in general terms rove beetles tend to be, except for the most common and readily identifiable species, underreported in carrion and forensic studies largely because they are difficult to identify. As such, their diversity is underestimated and we suspect that new species records await documentation (e.g., Majka et al., 2009a, 2009b, 2009c).

Histeridae

Histeridae, also known as clown beetles, are common on all sorts of decomposing substrates, including carcasses (Downie & Arnett, 1996). The Histeridae is comprised of 135 species in Canada alone but is one of the most understudied families of all beetles as it relates to biology and ecology (Bousquet & Laplante, 2006). Members of this family are usually predacious both as adults and larvae. Their abundance is generally underestimated because they are dark colored, they are active at night, and usually stay concealed underneath carcasses during the day (Byrd & Castner, 2001). From a forensic perspective, some Histeridae species have a predictable occurrence pattern on carcasses, making this family a potential tool for time of death estimations (Matuszewski & Szafałowicz,, 2013).

Dermestidae

Dermestidae, commonly called skin beetles, are better known for being pests of stored products (e.g., Green, 1967; Wright et al., 2002) and museum specimens (Su & Scheffrahn,

1990) than as scavengers of carcasses. For example, the species *Dermestes maculatus* De Geer is a common pest of dry fish where it competes against *Necrobia rufipes* De Geer (Family Cleridae), another beetle also commonly found on carcasses (Odeyemi, 1997). Dermestidae play an important role in the decomposition process as both larvae and adults feed directly on the remains (Downie & Arnett, 1996). Although adults can visit the carcass as early as 3 to 5 days after death, larvae are usually observed only when most of the soft tissues have been devoured by other insects and all that remains is cartilage, dry skin, and hair (Early & Goff, 1987). As such, the occurrence of this beetle's larvae is a good indication that a significant amount of time has elapsed since death (Anderson & VanLaerhoven, 1996; Kulshrestha & Satpathy, 2001). However, most carrion or forensic studies do not last long enough for the occurrence of dermestids to be recorded and thus this genus is often missing from successional species lists. The temperature-dependent development of some species has been examined (Richardson & Goff, 2001) and at least one species is involved in cannibalism of pupae by larvae (Archer & Elgar, 1998). Furthermore, some species can colonize remains indoors (Schroeder et al., 2002) and are useful for toxicological analysis to detect the presence of drugs (Miller et al., 1994).

Other Families

Hide beetles (Family Trogidae) are necrophagous keratin feeding beetles that scavenge hair, hooves, and feathers during the advanced stages of decay. As such, they are not attracted in great numbers to hairless carcasses such as pigs and human corpses, by comparison with birds and other mammal carcasses (but see Strümpher et al., 2014). In addition, they are easy to overlook because their dorsal surface is dull brownish, roughly textured, and often covered with mud (Borror & White, 1970). As such, we suspect that their occurrence and diversity is underestimated in carrion and forensic studies. While most Cleridae are predaceous, a few clerids from the genus *Necrobia*, often referred to as bone beetles, are primarily scavengers in this system. One species, the cosmopolitan *N. rufipes* De Geer, can feed on decayed or dry meat. It has been found on Egyptian mummies and erroneously described as a new species under the name *Necrobia mummiarum* (Hope, 1836). The *Necrobia* genus is generally observed on carcasses from the active decay stage to the dry stage. Similarly, the genera of sap beetles (Family Nitidulidae) *Nitidula* and *Omosita* are primarily scavengers of bones and dry carrion. However, little work has been done to determine the forensic value of this family (Byrd & Castner, 2001). Other frequently observed beetles on carcasses include the Scarabaeinae (Family Scarabaeidae), earth-boring dung beetles (Family Geotrupidae), some Sphaeridiinae (Family Hydrophilidae), round fungus beetles (Family Leiodidae), and carabids (Family Carabidae).

2.1.3. Impact on Carrion-Related Communities of Resource Patch Characteristics

The decomposition of animal carcasses is influenced by a range of biotic and abiotic factors that may have variable influences on key players in the system (e.g., biochemical processes, microorganisms, arthropods). Therefore, the impact of a given factor on the system will depend on the effect of this factor on each of the key players. Currently, carrion and forensic studies mostly examine biotic and abiotic factors because of their impact on the determination of postmortem intervals in forensic investigations. Table 2 lists the factors that have been studied and briefly describes their general effects on the system and, if available, on beetle fauna. However, insufficient research attention is still allocated to issues dealing with improper replication in forensic work (Michaud et al., 2012; Moreau et al. 2015) and to a number of factors that could affect decomposition, particularly the potential effect of carcass interdependence due to improper spacing in field experiments.

Table 2. Factors affecting beetle community structure and decomposition processes on carcasses

Category	Specific perturbation	Effects	References
Access to the carcass	Effect of carcass clothing	Little effect on beetles	Kelly et al., 2009; Voss et al., 2011
	Major access limitation (wrapping, burial, carcasses placed inside a building)	Delayed decomposition and can delay or stop colonization; sometimes greater beetle diversity indoor	Rodriguez & Bass, 1983; Goff, 1992; Ahmad et al., 2011
Carcass manipulation	Maggot removal/sampling	Decelerated decomposition; change in the complex of insects; increased abundance of some beetle species	Michaud & Moreau, 2013
	Carcass freezing and thawing	Increased aerobic decomposition and accelerated rates of carcass disarticulation; no effect on Staphylinid beetle colonization	Micozzi, 1986; Bugajski et al., 2011
	Insecticides (Malathion, Parathion)	Delayed decomposition and insect colonization (Malathion) or no effect (Parathion)	Gunatilake & Goff, 1989; Wolff et al., 2004
	Antemortem ethanol and drug (oxycodone) intake	No effect on beetles	Tabor et al., 2005b; Monthei, 2009
Carcass abundance	Frequent exposure of carcasses in a given site	Insignificant effect of carcass enrichment on fauna but small sample size	Shahid et al., 2003; Schoenly et al., 2005
Carcass quality	Carcass species	Faster decay in smaller animals; no difference between 23-Kg pigs and a human	Payne, 1965; Schoenly et al., 2007
	Carcass size	Little effect on the pattern of succession	Hewadikaram & Goff, 1991

2.2. Feces and Dung

Dung is a waste product from the digestive tract of animals. The main components of dung are undigested food residues, products of the gut flora and fauna, including micro-organisms, and secretions and cellular debris from the gut mucosa. Fresh dung has a high nutritional value for the organisms that can utilize it, particularly as regards to its content in nitrogen (about 2–5% in herbivore dung). However, dung quality is highly variable and largely dependent on the quality of food ingested and the nature of digestion. For example, among herbivorous animals, non-ruminants tend to have lower fecal nitrogen levels than ruminants (deer, cattle). To date, the ecology of dung beetles in omnivore and carnivore dung has received little attention (Hanski, 1987a; Hanski & Cambefort, 1991a).

The water content of dung varies considerably between species of animal, their age and condition, as well as the availability of water in their food and environment (Hanski, 1987a; Horgan, 2002). For example, dung produced by desert mammals is generally dry (as low as 40% water) while mammals of wet temperate environments can produce dung with a water content of over 90% (Hanski, 1987a; Holter & Scholtz, 2007). The water and nutrient contents of dung are key factors in determining the suitability of dung for beetle colonizers and largely drive succession among the heterotrophic insect communities that utilize dung (Gittings & Giller, 1998). The dung of herbivorous mammals also contains large amounts of cellulose, hemicellulose, and lignin that are difficult for beetles to break down (Hanski, 1987a). Depending on the region, these compounds may ultimately be consumed by termites, earthworms or microorganisms (Putman, 1983; Freymann et al., 2008).

The beetle fauna associated with dung decomposition has been generally well studied. This is in part due to their recognized role in ecosystem functioning (Nichols et al., 2008), their success in the biological control of exotic dung (Doube et al., 1991), and their application as indicators of anthropogenic disturbances and global change (Nichols et al., 2007). Several dung beetles have also become convenient focal species for evolutionary and behavioral research (Halffter & Edmonds, 1982) – particularly since many true dung beetles (Scarabaeinae subfamily) demonstrate marked sexual dimorphism.

2.3.1. Dung Decomposition

It is difficult to categorize decomposing dung on the basis of changes in appearance over time or to highlight recognizable and sequential decomposition stages. This is because of the tremendous variability in the composition and form of dung produced by different animal species (Hanski & Cambefort, 1991; Chame, 2003), and because dung decomposition is heavily influenced by environmental conditions, particularly ambient weather (Hanski, 1980; Horgan, 2002). Furthermore, the rates and appearance of decomposition are largely determined by the abundance, total biomass, and composition of the heterotrophic invertebrate community that utilizes the dung (Horgan, 2005a; Nichols et al., 2008). For example, in tropical forests, competition for dung may be so intense that the dung is entirely removed or consumed within hours, whereas dung in tropical pastures may last several months until it eventually erodes due to weathering (Horgan, 2002; Horgan, 2005a, 2005b).

It is usual to describe dung decomposition according to changes in dung microbial communities or by the succession of organisms arriving at the dung (Mohr, 1943; Koskela & Hanski, 1977; Hanski, 1980; Holter, 1982; Gittings & Giller, 1998; Horgan, 2002; Marchiori et al., 2003). Immediately after defecation, the dung begins to change, partly through weathering, including exposure to light, and partly due to the activities of microorganisms already present in the dung. Flies and beetles often arrive to dung in predictable sequences, with flies predominantly arriving first; however, the average time of successional occurrence, and the sequence of arrival of different species may vary according to dung type and dung interactions with macrohabitat or weather (Hanski, 1980; Gittings & Giller, 1998).

Flies and beetles that arrive late in succession may depend on the activities of early colonizers to condition the dung. For example, dung beetle activity at early stages of succession can substantially increase bacterial numbers in dung (Lussenhop et al., 1980) – these bacteria are often the food for later colonizers (Landin, 1961). Insects arriving to cattle dung in tropical regions also demonstrate predictable sequences in colonization albeit with most activity occurring in the first one or two days after defecation (Horgan, 2002; Marchiori et al., 2003).

Earthworms tend to occur in dung at the later stages of succession. In colder climates particularly, ingestion and mixing of dung by earthworms is essential for efficient decomposition (Holter, 1977; Putman, 1983). For example, Denholm-Young (1978) estimated that earthworms were responsible for between 30 and 40% of dung dry weight consumption in winter, and between 50 and 60% in summer in temperate regions. During prolonged dry seasons in tropical regions, dehydrated dung often remains on the soil surface until the rains arrive. Termites contribute to the decomposition of dry dung in dry tropical pastures (Freymann et al., 2008). The mechanical break-up of dung pats by birds and mammals may also enhance dung decomposition and erosion (Horgan & Berrow, 2004).

Mohr (1943) presented five outlines of succession in dung that have been corroborated several times since: (*i*) most species at the beginning of succession are specialists (mainly dipterans in the study of Mohr), (*ii*) fewer species are observed as droppings get old, (*iii*) the earliest species in the succession have the shortest life histories and shortest occurrence in dung (again mainly flies), (*iv*) succession (physical and biochemical changes in the dung) is faster in the early stages of decomposition, and (*v*) the environment of the dropping has a profound influence on patterns of succession.

2.2.2. Beetles Associated with Dung Decomposition

There are five main beetle families associated with dung decomposition. These are the Hydrophilidae, Histeridae, Staphylinidae, and the Scarabaeidae and Geotrupidae (Super family Scarabaeiodea). In tropical regions, Hybosoridae and Trogidae also frequently occur at dung, but these are more generally associated with the decomposition of carrion and cadavers. Two other families of minor significance are the Ptillidae (Staphylinoidea) and the Cryptophagidae (Cucujoidea), which are mainly fungivorous (Mohr, 1943; Richards & Davies, 1977; Gill, 1991). The histerids and staphylinids are mainly predatory, although some coprophagous species do exist (notably the Oxytelinae; Koskela & Hanski, 1977).

Scarabaeidae

The Scarabaeidae are solid tough beetles with short stout forelegs adapted for digging. In some species, the hindlegs are also highly adapted for preparing and rolling balls of dung. Coprophagy is associated with good olfactory perception and a characteristic feature of the Scarabaeidae is their lamellate club antennae. In the Geotrupidae, the lamellae occur in a spherical bulb at the antennal apex. Dung beetles mainly feed on proteins contained in the dung liquid. Therefore, dung beetle mouth parts show extensive morphological changes over their plant-feeding relatives. In adults, the labium and labrum are heavily bristled. The galea and lacinia of the maxillae are also heavily bristled with regular rows of long bristles on the stipes; however, the main adaptations occur in the mandibles: only the basal part of the mandibles remains strongly scleritized, the apical area being broad and membranous.

This contains a few strong bristles and scleritized longitudinal channels that form a strainer to which the salivary glands are connected (Landin, 1961; Halffter & Edmonds, 1982).

Hydrophilidae

Hydrophilids, as the name suggests, are generally associated with aquatic habitats, but a number of species are noted dung dwellers. Their diets are poorly understood, but members of the genus *Sphaeridium* are well adapted to utilize dung and superficially resemble true dung beetles (Scarabaeinae). *Sphaeridium* species have long antennae that terminate in a large three-segmented club indicative of good olfaction. The forelegs are relatively short and stout with large well-developed coxae adapted for digging. The mouthparts show some similarities with the Scarabaeidae by having membranous and heavily bristled lacinia, galea, and apical stripites. However, their mandibles are heavily scleritized and are more similar to plant-feeding beetles (Richards & Davies, 1977). Other hydrophilids (i.e., *Cercyon* spp.) show little adaptation to coprophagy and are not good diggers. These beetles often occur on the surface of the dung and their larvae are mainly predatory (Mohr, 1943).

2.2.3. Competition for Dung Resources and the Efficiency of Decomposition

Because the occurrence (availability and volume) of fresh dung is often unpredictable (although some animals demonstrate spatial and temporal regularity in defecation patterns), dung as a resource for insect feeding and reproduction is generally highly limited, particularly in natural areas such as forests and meadows (Hanski, 1987a). However, even in areas where dung is apparently abundant (such as in African savannas or cattle ranches), dung as a resource to sustain insect populations may be normally scarce due to rapid decomposition rates. These factors – low availability and rapid change – have resulted in the evolution of dung insect communities that are highly adapted for the efficient pre-emption and utilization of dung under conditions of high intra- and interspecific competition. Intense competition and a high degree of niche specialization have resulted in an extraordinarily high diversity of dung inhabiting beetles. For example, there are over 3000 described Scarabaeinae from Africa, with over 1600 occurring in the Ethiopean region alone (Waterhouse, 1974; Cambefort & Walter, 1991). Dung beetle communities in north temperate regions tend to be dominated by the Aphodiinae (Scarabaeidae; Mohr, 1943; Holter, 1977; Gittings & Giller, 1998), which are often inconspicuous in tropical regions where they tend to occur late in succession (Horgan, 2002). Hydrophilids and staphylinids are also poorly represented in tropical regions (Legner & Olton, 1970).

The Scarabaeidae can be separated into four main functional groups based on the manner by which they pre-empt dung. Ball rollers (telocoprids) form balls of dung, which they roll some distance from the source pat (Figure 1). The dung is then buried in a superficial tunnel or under leaf litter and a nest is constructed. Tunnellers (paracoprids) form tunnels directly beneath the source dung pat (Figure 1). The tunnels are then provisioned with dung and eggs are laid. Dwellers (endocoprids) do not relocate the dung but instead construct simple nests within or beneath the dung pat. Some species oviposit directly on the dung pat. A number of species use dung that has been buried by others. These have been termed 'kleptocoprids' (Halffter & Edmonds, 1982; Horgan, 2008). Within each group, nesting behaviors can be further divided based on reoccurring behavioral patterns, many of which are of taxonomic significance (Halffter & Edmonds, 1982). Dung relocation is regarded as an evolved strategy to avoid intense competition.

Hanski (1987b) outlined a number of other competitive strategies observed among coprophagous insects. These include: (*i*) attaining a rapid start in resource utilization, including ovoviviparity as observed among some flies, (*ii*) having fast growth rates (i.e., many Aphodiinae), including adaptations for endothermy and group feeding, (*iii*) ability to switch to frugivory or predation during unfavorable conditions – for example, in tropical regions, some dung beetles will also utilize fallen fruits, carrion, bird eggs or decomposing vegetation (Gill, 1991), and (*iv*) having good post-feeding survival including the ability to pupate at a small size.

Some regions of the world are naturally depauperate in native dung beetles and as a consequence, competition and resource utilization are low. This is particularly noticeable on oceanic islands that lack large indigenous herbivores. North America also has a relatively low number of native dung beetle species and consequently low rates of dung decomposition. Incompletely decomposed dung coupled with the accidental introduction of nuisance dung breeding flies (i.e., *Haematobia irritans* L. and *Musca autumnalis* De Geer; Mohr, 1943) prompted the deliberate introduction to the United States of dung beetles, mainly from Europe, for biological control purposes (Waterhouse, 1974). This has resulted in heterotrophic dung beetle communities that are a mix of native and introduced species.

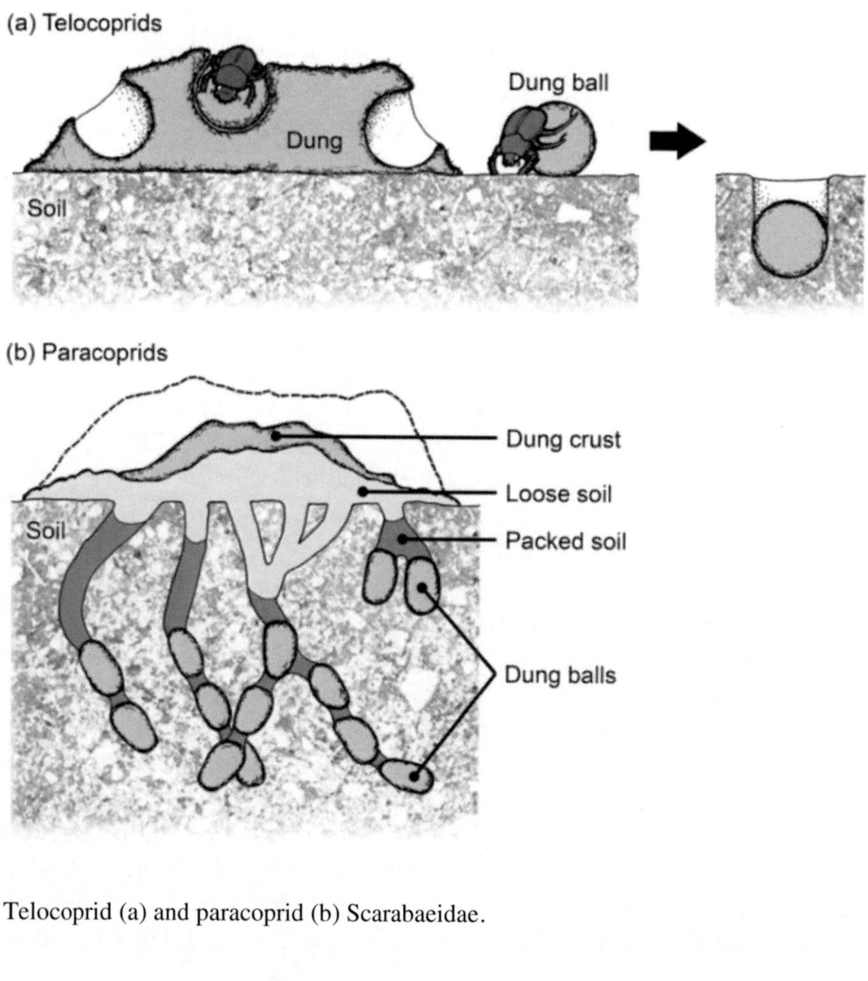

Figure 1. Telocoprid (a) and paracoprid (b) Scarabaeidae.

Table 3. Factors affecting dung beetle community structure and decomposition processes

Factor	Effects on beetles and decomposition	References
Region	Dung beetle communities vary between regions with often intense competition and fast decomposition rates in the tropics and apparently low levels of competition and slow decomposition in temperate regions	Legner & Olton, 1970; Hanski & Cambefort, 1991a
Altitude	Species diversity decreases with altitude, often with congeneric or conspecific shifts; lower temperatures at higher altitudes may also contribute to lower rates of decomposition	Hanski, 1983; Horgan, 2005b
Macrohabitat	Dung beetles demonstrate marked habitat specificity; seasonal or anthropogenic changes in habitat structure clearly alter dung beetle community structure and function	Nichols et al., 2009
Soil types	Soil types, soil moisture content, and soil hardness affect the ability of dung beetles to bury and utilize dung	Horgan, 2002
Sunlight and temperature	Dung under direct sunlight is subject to drying and the formation of crusts; crusts may decrease colonization and reduce rates of dung decomposition; beetle species that tolerante dry dung often occur late in succession	Gittings & Giller, 1998; Horgan, 2002
Presence and abundance of predators	Few studies have examined the predators of dung beetles and their effects on decomposition; however, lower than expected abundances of competitively dominant species suggest that predators can alter dung beetle community structure	Horgan & Berrow, 2004; Horgan & Fuentes, 2005
Availability of dung	The temporal and spatial availability of dung on the soil surface is a determinant of the intensity of competition between dung beetles; the position of individual dung pats relative to new or old droppings, and variations in pat conspicuousness lead to aggregated distributions of dung beetles	Giller & Doube, 1994; Horgan, 2005a, 2005b, 2006
Dung type and quality	Dung beetles show marked preferences for certain types of dung and perform better on high quality or preferred dung types	Cambefort & Walter, 1991; Gittings & Giller, 1998; Finn & Giller, 2002; Whipple & Hoback, 2012
Animal diet and food quality	Dung beetle assemblages vary according to the habitat source of the dung, i.e., assemblages respond differently to pats produced from rough pastures compared to well-managed pastures	Finn & Giller, 2002; Hutton & Giller, 2003
Dung water content and rainfall	Decomposition is higher where dung pats remain moist due to shade, irrigation or rainfall; beetle colonization of dung and reproductive output are greater during wet months where dung remains moist	Dickinson et al., 1981; Edwards, 1991
Time of dropping	Dung dropped during the day is colonized by fewer nocturnal species compared to dung dropped at night and vice-versa	Montes de Oca & Halffter, 1995
Size of pat	Large dung pats attract greater numbers of beetles, but often result in lower densities of insects, reduced competition, and lower decomposition rates	Finn & Giller, 2000; Horgan, 2005b
Age of dung	Dung age is among the principal determinants of colonization and resource use in dung beetles, with different species showing clear adaptations to utilize dung of specific ages	Gittings & Giller, 1998; Menéndez & Gutiérrez, 1999; Horgan, 2002; Sabu et al., 2006

Table 3. (Continued)

Factor	Effects on beetles and decomposition	References
Herbicides	Herbicides can result in reduced emergence of dung beetles from treated soils	Martínez et al., 2000
Insecticides	Systemic insecticides applied to agricultural livestock can reduce colonization and survival of dung beetles in contaminated dung, altering community structure and likely affecting decomposition processes	O'Hea et al., 2010
Invasive species	The establishment of exotic dung beetles alters native beetle assemblages and can improve decomposition efficiency; however, in some cases there may be competitive exclusion of indigenous beetles from certain habitats and changes in niche dimensions	Montes de Oca & Halffter, 1995

Dung beetles have also been introduced to Australia, but for a different reason. Australia has a well-established indigenous dung beetle fauna with over 300 species. These are mainly adapted to utilize marsupial dung in dry environments (Waterhouse, 1974; Doube et al., 1991). Australian dung beetles were largely incapable of utilizing the moist dung associated with introduced cattle, leading to problems of decreasing pasture productivity, an increased incidence of nuisance flies, and potential for the spread of diseases. By the early 1970s about 3 million hectares of pasture land were destroyed each year in Australia by non-decomposed cattle dung, rejection of fouled herbage by cattle, and the smothering of pasture (Waterhouse, 1974).

Beginning in the 1960s, several dung beetles were introduced to Australia, mainly from South Africa, but also from Europe and Asia (Doube et al., 1991). These introductions have lowered the wastage of herbage, decreased nutrient loss from dung, increased soil porosity, and reduced the survival of disease-causing helminths. However, it remains unclear whether exotic dung beetles have actually reduced the abundance of dung breeding flies in Australia, and as a result, attention has shifted toward the role in fly mortality played by several species of phoretic mites that are transported by dung beetles between dung pats (Beaulieu & Weeks, 2007).

2.2.4. Impact on Heterotrophic Dung Communities of Resource Patch Characteristics

Dung decomposition varies greatly between regions and habitats, and over seasons. Several factors have been shown to influence the colonization of dung by beetles. In some cases these factors have also been shown to affect dung decomposition rates. Table 3 lists some factors known to affect the structure of dung beetle communities, the reproductive success of the beetles, and the potential or observed effects on dung decomposition processes. The table indicates that dung beetle community structure, including changes in that structure over the course of succession, and decomposition rates are governed by several resource patch characteristics, all of which are directly or indirectly affected by the nature and habitat of the dung producer and by ambient climatic conditions. However, to date, a number of factors, particularly the role of predators and the influence of chemical contaminants in the dung or macrohabitat, continue to receive little research attention.

2.3. Dead Wood

Dead wood is one of the most readily available substrates in forest ecosystems. Senescence, competition, fire, windthrow, snow, pathogens, insects, and logging are natural or anthropogenic disturbances that can cause tree mortality (Harmon et al., 1986), which ends up creating resources for organisms feeding and/or depending on dead wood. Tree mortality is random, both in its spatial and temporal occurrence and the factors causing tree mortality differ substantially between different forest types (Fridman & Ståhl, 2001). Woody debris can be found in many different forms throughout forests, namely logs, crowns, snags, stumps, and branches. They are usually classified using a two-size category, either fine or coarse woody debris, which is determined by debris diameter (Harmon & Sexton, 1996).

Dead wood performs different roles in forest ecosystems. It can allow or facilitate the germination of some plant species, affect geomorphic processes in streams and rivers, and be used as a nesting site or habitat by salamanders, birds, fungi, bacteria, and arthropods (Harmon et al., 1986). Dead wood also contributes fuel to forest fires (Tinker & Knight, 2000) and account for a significant proportion of forest biomass and stored carbon.

2.3.1. Dead Wood Decomposition

Compared to the other systems described above and below where decomposition processes are relatively fast, decaying dead wood can take several decades to fully decompose. Several characteristics of dead wood influence decay rates, such as temperature, moisture, oxygen and carbon dioxide, substrate quality, size, and decomposing organisms (Harmon et al., 1986). Dead wood decay stages have been classified using different methods and with varying numbers of decay classes, ranging from three to six classes for downed woody debris, and up to eight classes for standing dead wood (Maser et al., 1979; Esseen et al., 1992; Friedman & Walheim, 2000; Martikainen et al., 2000; Siitonen & Saaristo, 2000; Imbeau & Desrochers, 2002; Similä et al., 2003; Hammond et al., 2004; Jacobs et al., 2007a). Dead wood decay classifications are usually based on visual characteristics of woody debris.

2.3.2. Beetles Associated with Dead Wood Decomposition

The term 'saproxylic' is derived from the Greek words 'sapros', meaning decayed, and 'xylon', meaning wood. Saproxylic beetles can be defined as insects that are dependent, during some part of their life cycle, upon dead or dying wood (lying or standing), wood inhabiting fungi, or other saproxylic organisms (Speight, 1989). This being said, saproxylic beetles are furthermore divided into different guilds, depending on their feeding habits (Vanderwel et al., 2006). These guilds include phloeophagous species that feed on phloem tissue and cambium, xylophagous species that feed deeper in wood, specializing on xylem tissue, fungivorous species that feed on fungi, spores, and mycelium, detritivorous species that feed on decaying plant and animal materials, and predators that feed on other saproxylic beetles (Key, 1993; Grove, 2002). Some authors will even go one step further and divide the fungivorous guild into distinct feeding guilds that relates to the fungus these species feed upon. These guilds include bolitophagous species that feed on the fruiting bodies or mushrooms, mycetophagous species that feed on fungal hyphae, and myxomycophagous species that feed on slime molds (Dollins et al., 2008; Wood, 2012).

Table 4. Recent studies on saproxylic beetle assemblages that have made use of different classification methods for dead wood decay

Author(s)	Dead wood classification	Beetle collection and association with decay
Hammond et al., 2004	Described three decay classes for boles and snags	Beetles collected using emergence traps and window traps are associated with decay classes
Lachat et al., 2006	Used four decay classes adapted from Bretz-Guby & Dobbertin (1996)	Beetles reared from dead wood and collected with emergence traps are associated with decay classes
Vanderwel et al., 2006	Used four decay classes adapted from Maser et al. (1979)	Beetles collected using emergence traps are associated with decay classes
Jacobs et al., 2007b	Described three decay classes for logs and seven decay classes for snags	Beetles collected using window traps are associated with the succession patterns defined by Esseen (1992)
Saint-Germain et al., 2007	Used four decay classes adapted from Maser et al. (1979)	Beetles collected by bole dissection are associated with decay classes
Grove et al., 2008	Used time since tree felling in months	Tracks beetle assemblages collected using emergence traps along dead wood decay progression
Ulyshen & Hanula, 2010	Used time since tree death in months for young, middle-aged, and old logs	Beetles collected using emergence traps and bole dissection are associated with the three overlapping phases (phloem, subcortical phase, and rot phase) described by Savely (1939)
Wood, 2012	Decribed six decay classes	Beetles collected using window traps, emergence traps, and bole dissections are associated with three decay categories (early, moderate, and advanced)
Lee et al., 2014	Used six decay classes adapted from Hofgaard (1993)	Beetles collected by bole dissection are associated with three decay categories; early (decay classes 1-2), intermediate (decay classes 3-4), and advanced (decay classes 5-6)

Saproxylic beetles are often among the first animals to colonize dead wood and as such play a vital role in forest ecosystems. This is attributed to the decomposing actions of these organisms, which return nutritious elements to the ground and favors carbon cycling (Swift, 1977; Esseen et al., 1997). Saproxylic beetles are part of many different food webs and also represent a significant component of forest diversity. Even though some beetle species tend to be specific in their feeding and colonization requirements, other species are generalists and colonize a wide range of different dead wood types (Müller & Bütler, 2010). In addition, some beetle species are opportunistic and use dead wood if it is readily available, without necessarily depending on it (Speight, 1989). These beetles are harder to classify, some authors considering them as facultative saproxylic species while other authors exclude them from the saproxylic taxon (Grove & Forster, 2011).

Because dead wood decay is slow, several studies have focused on specific parts of the decay gradient or on specific beetle groups and species that often have economic implications. As a consequence, only a handful of the recent studies listed in Table 4 have investigated the associations between dead wood decay and assemblages of saproxylic beetles along the full spectrum of wood decay, although such research is on the rise. Here, we raise the issue that a lack of standardization in the description of decay stages may have limited the detection of unifying trends. Granted that the visual classification of dead wood may be the simplest and more cost-efficient method for determining the state of wood decay (Newberry et al., 2004), one fundamental question still remains: is this classification method relevant to saproxylic beetle biology? Can the beetles distinguish different decay classes based on dead wood appearance? Generally, saproxylic beetles are faithful to their feeding habits (Jonsell et al., 1998), both in terms of larval and adult requirements, and are only present on dead wood that meets their requirements. For example, depending on the diameter and length category of the woody debris, the assemblage of organisms can vary, some host-species requirements being only met on either fine or coarse woody debris (Jonsell et al., 2007; Brin et al., 2011). Host-species associations also depend on the position of woody debris (standing vs. lying down), as well as on tree essence (softwood vs. hardwood) and species (Grove, 2002; Franc, 2007). We suppose that current methodologies used to describe dead wood decomposition and associated beetle fauna could be standardized in a way that allows for comparisons between studies. As such, we separate dead wood decay into three categories: early, intermediate, and advanced decay.

The early decay stage, the shortest of the decay stages, occurs while the woody debris are fresh. A peak in beetle abundance and richness characterizes this stage. This peak occurs in part because of the attractant volatiles released by recently felled or stressed trees (Lee et al., 2014). During this stage, saproxylic beetles mainly attack the sapwood (Esseen et al., 1997) and thus generally exhibit a thick exoskeleton and cylindrical shape to bore into still firm wood. Among the families and subfamilies present during this stage are bark beetles (Scolytinae), wood borers (Cerambycidae, Buprestidae, Bostrichidae, Brentidae), phloem feeding weevils (Curculionidae), and their associated predators (Staphylinidae, Cleridae, Nitidulidae, Histeridae, Rhizophagidae, Monotomidae) (Esseen et al., 1992; Vanderwel et al., 2006; Jacobs et al., 2007b; Wood, 2012; Lee et al., 2014). Together, these beetles are mainly responsible for the mechanical breakdown of woody debris (Harmon et al., 1986). By creating a network of galleries in the wood, cerambycids and bark beetles increase the surface area available to bacteria and fungi (Siitonen, 2001). Species associated with the early decay stage are not found during subsequent stages because once the quick-decaying sapwood and conductive tissues are deteriorated, the beetles will disperse to find new fresh substrates.

The intermediate decay stage begins when woody debris exhibit a decrease in bark coverage and conductive tissue (cambium, phloem, and xylem), as well as an increase in colonizing fungi. This stage is characterized by a decline in the abundance and richness of wood borers and bark beetles, as well as by a shift in the beetle assemblage toward fungivores (Ciidae, Trogossitidae, Cryptophagidae, Leiodidae, Erotylidae) and their predators (Staphylinidae, Cleridae, Cucujidae) (Esseen et al., 1992; Lachat et al., 2006; Grove et al., 2008; Wood, 2012; Lee et al., 2014). The intermediate decay stage is normally the stage with the lowest abundance and species richness of both larval and adult beetles (Lee et al., 2014), and is marked by a transition between the beetle assemblages of the early and advanced decay stages. The activity of fungi and bacteria creates habitats rich in organic matter and other dead wood products within individual pieces of wood, allowing for the feeding of saprophagous species (Scarabaeidae, Lucanidae) (Araya, 1993; Vanderwel et al., 2006). The beetles that are active during this stage generally exhibit a dorsoventrally flattened body (Wood, 2012).

The length of the intermediate decay stage depends on different attributes of the wood and species that colonize it, and as such can be quite variable.

The advanced decay stage begins when woody debris exhibit fully decomposed sapwood, decomposing heartwood, a thick moss covering, and high moisture content. During this stage, the availability of essential minerals and nutrients such as calcium, zinc, iron, phosphorous, and nitrogen increases (Brais et al., 2006). Transition from the intermediate to the advanced stage of decay is characterized by a shift toward fungivorous (Melandryidae, Anobiidae), detrivorous (Dermestidae), and predacious (Staphylinidae, Carabidae, Elateridae) species, but the abundance and species richness remain similar to that of the intermediate stage (Esseen et al., 1992; Hammond et al., 2004; Vanderwel et al., 2006; Wood, 2012; Lee et al., 2014). There is an ongoing debate as to whether carabid beetles qualify as saproxylic or not. Some carabids will use dead wood to complete part of their life cycle but most litter-dwelling carabids are opportunistic and only utilize dead wood as a refuge (Work et al., 2010). The high moisture content and moss covering of wood in this decay stage makes it possible for species that are susceptible to desiccation to establish and flourish (Esseen et al., 1997; Lee et al., 2014). This stage ends when dead wood pieces are completely transformed into humus.

2.3.3. Impact on Saproxylic Beetles of Resource Patch Characteristics

Dead wood decomposition is a continuous process influenced by several factors. Some of these factors depend on the wood debris itself, such as orientation, diameter, length, and species. Other factors may be external, such as temperature, precipitation, and proximity to the forest edge (Dajoz, 1998). Table 5 lists some of the factors influencing dead wood decomposition and saproxylic beetle communities. This being said, the saproxylic beetle species composition changes continuously along the dead wood decay gradient, and different assemblages of saproxylic beetles can overlap along different decay stages. In the absence of colonization by saproxylic beetles, dead wood decay rates would be greatly reduced. The fungi and microorganisms that usually make their way into dead wood via wood borer galleries or by being carried on beetle elytras would no longer colonize, affecting late succession fungal-feeding insects (Zhong & Schowalter, 1989). The type and quality of dead wood is linked to the cause of wood mortality and affects the assemblage of early successional species on dead wood. Windthrow and logging tend to kill healthy trees, and generate nutritious debris (Bouget & Duelli, 2004), but debris produced this way might still produce anti-herbivore defenses. Nevertheless, specialized species such as members of the Scolytinae subfamily can overcome these defenses and then facilitate the colonization of other species (Stockland et al., 2012). Desiccation and pathogens kill trees slowly, generating poorer, more easily accessible debris. Alternatively, senescing trees die progressively and will often exhibit more than one decay stages, thus providing a wider range of habitats. Therefore, depending on how the woody debris was generated, slightly different succession of saproxylic beetles could be documented (Stockland et al., 2012).

2.4. Other Heterotrophic Resources

Beetle assemblages can be observed on sources of accumulated vegetal matter other than dead wood, such as fallen fruits and leaf litter (Hammond, 1990). Fruits and leaf litter fall on the ground when they reach maturity and thus contribute to increased soil nutrients (Barlow, 1950; Miller, 1984).

2.4.1. Fruits

Some beetle species actively contribute to crop pollination and/or seed dispersal and thus are beneficial for fruit production (Bernhardt, 1995; Wang & Smith, 2002; Ghazoul, 2005). However, fruit producers often see beetles as a nuisance because several species feed on unharvested and/or harvested fruits (e.g., Potter & Held, 2002; Galvan et al., 2008; Koch & Galvan, 2008; Hammons et al., 2009; Keszthelyi, 2012; Goguen & Moreau, 2013). Consequently, past research on frugivorous beetles has mainly focused on early detection of crop and seed pest species (e.g., Sallabanks & Courtney, 1993; Johnson et al., 2009; Bartelt & Hossain, 2010; Rondon et al., 2011). Nevertheless, recent studies in berry fields have increasingly used frugivorous beetles as indicators of ecosystem health (e.g., Comeau et al., 2012, 2013; Cutler et al., 2012).

Working with fleshy fruits, Padayachi et al. (2014) described three decay classes for dead fruits in his study of beetles. Successive classes were determined based on fruit external appearance, firmness, and surface color. The first decay class begins when the fruit falls to the ground; the fruit is still intact, has a smooth surface, and exhibits the same color as when attached to the plant. The second decay class starts when the fruit begins to ferment; at that point, half of the fruit surface has changed color and is moderately wrinkled. The third and last decay class occurs when the fruit is dry and substantially decomposed; the entire surface of the fruit has changed color and is deeply wrinkled. Padayachi et al. (2014) then examined family-level associations between beetles and decay stages but their inferences were hampered by their small sample size. To our knowledge, no other study has examined the succession of frugivorous beetles in relation to fruit decay, even though many beetle families are known to feed on fruits. Some species from the families Carabidae, Coccinellidae, Nitidulidae, Scarabaeidae, and the subfamily Curculioninae (Curculionidae) can feed respectively on figs, pumpkins, strawberries, granadilla fruits, and palm fruits, just to name a few (Hernández, 2011; Koch et al., 2004; Paarmann et al., 2001; Padayachi et al., 2014; Myers, 2004). In addition, frugivorous beetles that damage fruit pulp may increase the susceptibility of fruits to pathogens (Herrera, 1982). In these instances, fruit decay after the attack of beetles will be driven by the action of fungal communities (Eckert & Sommer, 1967).

Soft ripe fruits can be colonized by beetles in pre-harvest and post-harvest conditions while hard shell fruits need to be predigested by large vertebrates such as mammals before being colonized by beetles (Janzen, 1977). Frugivorous females usually lay their eggs on the soil, proximate to accumulated organic matter (Brandhorst-Hubbard et al., 2001). When the larvae hatch, they are attracted to voltaic organic compounds emitted by ripe fruits (Domek & Johnson, 1990). Because of the damage they inflict on fruits, feeding larvae can facilitate pathogen infections (Tang et al., 2005) that may cause infected fruits to ferment and release additional voltaic organic compounds that attract more beetles (Domek & Johnson, 1990). Alternatively, some beetle larvae can emit aggregation pheromones that attract other beetles, usually of the same species (Domek & Johnson, 1990). Depending on their size, exocarp thickness, and resistance to pathogen infection (Herrera, 1982; Tang et al., 2005), ripe fruits colonized by beetles will last from a few days to several weeks, or even longer (Swinburne, 1973; Erwin, 1979; Tang et al., 2005).

Table 5. Factors influencing dead wood decay and saproxylic beetle communities

Factor	Effect on beetles and decomposition	References
Quality	Different decay stages will attract different beetle assemblages and communities	Esseen et al., 1992; Similä et al., 2003; Jonsson et al., 2005
Quantity	Greater amounts of dead wood are beneficial for saproxylic beetles and crucial for conservation, while a decrease in dead wood quantity has a negative effect	Siitonen, 2001; Grove, 2002; Gibb et al., 2006
Size	Beetle communities differ with the size of woody debris (e.g., fine vs. coarse woody debris)	Brin et al., 2011; Wood, 2012; Lassauce et al., 2012
Position	Beetle species composition differs between standing dead wood and downed woody debris	Janssen et al., 2011; Bouget et al., 2012
Tree mortality type	Debris from senescing trees, burnt and windthrown trees, and logged trees host different beetle assemblages due to difference in quality and anti-herbivore defenses	Stockland et al., 2012
Tree species	Beetle communities differ between softwood and hardwood substrates; species are generally faithful to one type	Irmler et al., 1996; Jonsell, 2007
Distance from forest edge	Beetles assemblages differ between debris in forest interior and debris located close to a forest edge	Murcia, 1995; Wermelinger et al., 2007; Nadeau et al., *in press*
Sun exposure	Sun exposure affects the decay rates of debris and beetle communities; beetle abundance is higher on sun exposed debris	Lindhe et al., 2005; Brunet & Isacsson, 2009
Weather	Warmer temperature changes the decay rate, beetle colonization, and beetle emergence from woody debris.	Ulyshen & Hanula, 2009; Jacobs & Work, 2012

2.4.2. Leaf Litter

Leaf litter accumulates throughout the ecological succession of forest ecosystems, and as forests age, the litter gets thicker (Miller, 1984). While studies usually describe leaf litter decomposition in terms of time since the onset of decay (e.g., Harmon et al., 1990; Coûteaux et al., 1991), Berg & Staaf (1980) described leaf litter decomposition using a model including two phases – the early stage and the late stage. Climate, turnover in concentration of the major nutrients, and water-soluble substances are responsible for the rapid loss in litter biomass in the early stage. Conversely, slower decomposition rates of leaf litter in the late stage is driven by the lignin content of the litter. When decomposition rates approach a limit value, leaf litter is considered humus.

To our knowledge, no study has examined the succession of beetles according to the different phases of leaf litter decomposition. Schaefer (1996) divided soil fauna in their respective feeding guilds (i.e., saprophagous, microphytophagous, zoophagous) depending on the body width of invertebrates (i.e., micro, meso, macro). The soil fauna, but especially the microfauna, is known to be responsible for 15 and 30% of carbon and nitrogen turnover, respectively (Anderson, 1995). The microfauna can improve soil fertility by increasing microbial activity and mineralization of the leaf litter (Kautz & Topp, 2000). Litter dwelling beetles in the Carabidae and Staphylinidae families are typically included in the zoophagous macrofauna because they feed on other invertebrates and use the leaf litter as a refuge against predation and desiccation (Magura et al., 2004a). By feeding on other leaf litter invertebrates, they only indirectly affect litter decomposition (Magura et al., 2005). Zoophagous litter dwelling beetles are more abundant and speciose when woody debris and leaf litter are abundant, mainly due to temperature and moisture gradients influencing the richness and abundance of their prey (Topp et al., 2006).

3. Mechanisms Affecting Heterotrophic Community Development

Because most decomposing substrates consists of organic material of vegetal or fecal origin (Swift et al., 1979), it is not surprising that the vast majority of ecological studies on heterotrophic community development have focused on dung, fruits, fungi, or dead wood. Indeed, more than half a century ago, Allee et al. (1949) reported that decomposing carcasses, another substrate hosting heterotrophic communities, had long been a neglected ecological system. Surprisingly, Allee's statement remains valid today. Conversely, dung pats have been, by far, the most studied system hosting heterotrophic communities. However, it should be noted that vegetal and fecal organic material is generally used to study aggregation and resource partitioning and seldom used to directly study heterotrophic succession, while the opposite is generally true for carcasses and cadavers.

3.1. Heterotrophic Succession

Ecological succession is probably the best known ecological concept. The vast majority of literature on the subject deals with plant succession, also known as autotrophic succession. Autotrophic succession has been described following deforestation (Guariguata & Ostertag, 2001), glacier melts (e.g., Chapin et al., 1994; Jones & del Moral, 2005), forest fires (Posamentier et al., 1981), volcanic eruptions (e.g., del Moral & Wood, 1993; Grishin et al., 1996), as well as following the onset of a new environments such as a sand dunes (Olff et al., 1993). Succession has also been described in some algae (Murray & Littler, 1978; Sousa, 1979; Ohno et al., 1990), in animal communities (e.g., Adams, 1908; Shelford, 1912; Farrell, 1991), and in bacteria (e.g., Kuramae et al., 2010; Gillilland et al., 2012).

Another type of succession – heterotrophic succession – defined as succession occurring on dead organic matter, has also been described for organisms colonizing decomposing material such as wood (e.g., Boulanger & Sirois, 2007; Weslien et al., 2011), feces (e.g., Hanski, 1980; Menéndez & Gutiérrez, 1999; Sabu et al., 2006), carcasses (e.g., Bornemissza, 1957; Anderson & VanLaerhoven, 1996), and corpses (Rodriguez & Bass, 1983; Schoenly et al., 2007). However, heterotrophic succession differs from autotrophic succession by its relatively short duration, lasting from days to decades, by comparison to the centuries or millennia required for autotrophic primary succession. In addition, biomass, species diversity, and substrate (i.e., soil) increase over time in autotrophic succession, while the opposite is observed in heterotrophic systems, which lack primary production and thus act as non-equilibrium islands (Beaver, 1977). As such, the successional endpoint of heterotrophic systems is complete recycling of the substrate into soil and not the achievement of an equilibrium point that resembles the 'climax' or 'polyclimax' of autotrophic systems (Poole, 1974; Braack, 1987; Begon et al., 1996; Carter et al., 2007).

3.1.1. Mechanisms of Heterotrophic Succession

The literature on heterotrophic succession is, for the most part, scarce with regards to empirical tests that could potentially explain how organisms succeed each other. Two hypotheses, however, have been proposed: the facilitation model and the gas-emanation hypothesis.

The facilitation model was formulated by Connell & Slatyer (1977) in an attempt to develop a testable hypothesis of the role of facilitation in ecological succession. Connell & Slatyer (1977) also formulated two additional models of succession. The three models can be stated as follows:

Facilitation. Only some pioneer species from the pool of species that arrive to a site following a disturbance can colonize the site. The environment then becomes less favorable for hosting these pioneers but more favorable for other species. Therefore, changes in the environment caused by pioneer species facilitate subsequent colonization by other species. Pioneer species are then eliminated. This sequence continues as long as the presence of the resident species facilitates colonization by other species.

Tolerance. From the pool of species that arrive at a site following a disturbance, only those that can survive as adults will settle. The environment then becomes less favorable for these species, but has little or no effect on colonization by other species. Therefore, other species can colonize the site, despite the presence of pioneer species. Thus, pioneer species tolerate the presence of new species, but will eventually be eliminated. This sequence continues as long as new species can coexist with the species occupying the site.

Inhibition. From the pool of species arriving at a site following a disturbance, only those species that can survive as adults will settle. The environment then becomes less favorable for pioneer species, but also for species that could potentially colonize the site. As pioneer species persist, they inhibit colonization of the site by other species.

In all three cases, the long-term result is the same: a new occupancy can only occur if the community is damaged or removed. Changes in the community thus depend on the environment and the specific characteristics of the species that have the potential to colonize.

Without empirical validation, and despite major differences between autotrophic and heterotrophic systems, Connell & Slatyer (1977) stated that the facilitation model of succession was probably applicable to heterotrophic systems such as dung, litter, logs, and carcasses. In the same vein, Schoenly & Reid (1987) advanced that succession patterns on carcasses were consistent with the facilitation model of Connell & Slatyer (1977) and proposed an experimental design to directly test the model in a future study, whereas Smith & Baco (2003) stated that facilitation had a role to play in community assembly on decomposing whales on the deep-sea floor. Facilitation is also routinely cited as the main driver of heterotrophic succession on dead wood (e.g., Schroeder, 1997; Økland et al., 2009; Victorsson, 2012). To our knowledge, only three studies have empirically supported the facilitation model of succession in heterotrophic systems: Renvall (1995) and Weslien et al.

(2011) on dead wood, and Slade et al. (2007) on feces. More empirical research and direct testing of the facilitation model will be required to further our understanding of temporal dynamics of communities on decomposing substrates. We also suggest that the inhibition model and the tolerance model be empirically tested as well. For example, the inhibition model appears to better describe situations where *Nicrophorus* spp. (Silphidae) adults bury small carcasses to inhibit fly colonization, hereby gaining a competitive advantage. Competitive behavior (i.e., inhibition) also dominates the literature on dung beetles (e.g., Hanski, 1991; Horgan & Fuentes, 2005; Simmons & Ridsdill-Smith, 2011). Similarly, in a study that examined turnover rates for both plant and carrion-insect successions, Anderson (2007) concluded that the main drivers of plant succession were competition, dispersion limitation, and abiotic limitation, but she could not conclude about which factors drove carrion-insect succession. Recently, Bastow (2012) argued that none of Connell & Slatyer's (1977) models fits the initial stages of succession on plant litter.

The gas-emanation hypothesis precedes the facilitation model by almost a century. The hypothesis is actually rooted in Mégnin's (1894) concept of insect "squads" and how those insect "squads" are attracted to specific gases released sequentially during decomposition. This hypothesis has never been subjected to direct empirical validation, although some indirect evidence appears to support it. For example, many chemical compounds emanating from carcasses have been identified (see Vass, 2012). In addition, many insect species are known to respond to the presence of decomposition-related olfactory stimuli (e.g., von Hoermann et al., 2012; Johansen et al., 2014). To properly test Mégnin's hypothesis, it will be necessary to examine the specific order in which gases are released from the carcass and to evaluate whether the said order is consistent with the succession of carrion-insects. The applicability of Mégnin's hypothesis to other systems, however, is still a matter of debate. Feces are already partly decomposed when exposed to the environment and do not generate a succession of gases in the same way that bloating carcasses do, and neither does dead wood or litter. Nonetheless, both feces and dead wood will emit odors that act as attractants for beetles (e.g., Brattli et al., 1998; Jonsell et al., 2003; Dormont et al., 2004, 2010).

3.1.2. Decomposition or Decay Stages

The use of decomposition or decay stages to describe the physical appearance of a decomposing substrate is common to all fields of heterotrophic succession. As indicated in Section 2, the number and characteristics of the different stages can vary greatly according to author. Figure 2 presents some of the most cited stage descriptions for three different substrates.

At first glance, a few striking similarities between succession in different substrates is apparent (Figure 2). First, the three systems are characterized by a series of initial stages of relatively short duration. Second, the three systems present a long to very long final stage of decay, by comparison with the initial stages. Often, the last and penultimate stages are combined because they are indistinguishable (Mohr, 1943; Payne, 1965). Third, the three systems present a stage of intermediate duration between the initial and final stages (i.e., stage IV in Figs. 2a and b; stage III in Figure 2c). Leaf litter and fruit decay are not presented in the figure due to an absence in the ecological literature of quantitative data on decomposition of these substrates. Nevertheless, stage descriptions presented in Section 2.4.1. and 2.4.2. seem to fit the current analysis.

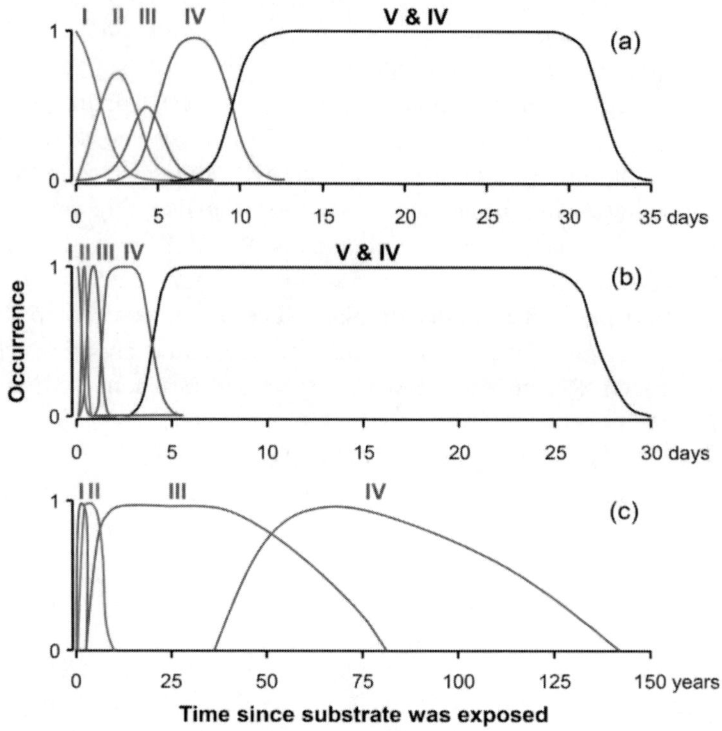

Figure 2. Decay stages (Roman numerals) of (a) animal carcasses, (b) dung pats, and (c) dead wood, as a function of the time since the substrate was exposed. Figure (a) was adapted from Payne (1965) and Michaud & Moreau (2011), (b) was adapted from Mohr (1943), and (c) was adapted from Esseen et al. (1992). In (a) and (b), an occurrence of one means that the substrates are all in a given stage at that time while an occurrence of one in (c) indicates the peak in abundance of species associated with that decay stage.

Is there any specific link between insect assemblages and decay stages? Historically, the general belief in the scientific community was that such a link existed, with assemblages being sometimes identified using Clementian terminology such as ‘microsere’ and ‘microseral stages’ (e.g., Mohr, 1943; Reed, 1958; Johnson, 1975). However, this belief was challenged by Schoenly & Reid (1987) who demonstrated that insect succession on carcasses generally followed a continuum of gradual changes, which was later corroborated by Boulton & Lake (1988) and Moura et al. (2005). Similarly, the microseral approach has largely been abandoned in dung beetle work (but see Lee & Wall, 2006). By contrast, insect succession in dead wood is still largely examined in terms of discrete assemblages associated with decay stages (Esseen, 1992; Stokland et al., 2012).

Other characteristics shared by these systems are that Diptera and Coleoptera are the major players in the breakdown of organic material, except maybe for leaf litter where earthworms and woodlice play a significant role, and cattle dung in cool temperate regions that is largely consumed by earthworms (Denholm-Young, 1978; Zimmer et al., 2005). Furthermore, the Diptera and Coleoptera occur sequentially on heterotrophic substrates. Thus, it is well known that Diptera will first colonize dung pats and animal carcasses, sometimes within minutes, and Coleoptera will colonize later (Mégnin, 1894; Mohr, 1943). Less is known about saproxylic Diptera as succession in this taxon is poorly understood in comparison with saproxylic Coleoptera (Stubbs, 1982; but see Hövemeyer & Schauermann, 2003).

3.2. Species Aggregation

A feature of heterotrophic insect communities is their often high diversity of species and exposure to generally intense interspecific competition. In many cases this has led to anatomical, physiological, and behavioral adaptations toward rapid resource utilization, high reproductive rates or an enhanced competitive ability. The intense activity around decaying substrates has prompted the question of how so many species manage to co-exist on such rare, temporally unstable, and patchily distributed resources (i.e., Atkinson & Shorrocks, 1981; Holter, 1982; Shorrocks & Rosewell, 1987; Sevenster & van Alphen, 1993, 1996; Giller & Doube, 1994; Horgan, 2006).

One mechanism that has received considerable attention is the "aggregation model of coexistence" (Hartley & Shorrocks, 2002). This model suggests that the individuals of a single species, when clumped together, inhibit their own population growth to a greater extent than that of heterospecific populations. Therefore, even when species use the same type of resource, their coexistence is facilitated where the distribution of individuals differs across patches. The aggregation model is supported by empirical data from the insect communities of rotting carcasses (Hanski, 1987c; Ives 1991), rotting fruits (Shorrocks & Rosewell, 1987; Sevenster & van Alphen, 1993, 1996), fungi (Wertheim et al., 2000), dung (Giller & Doube, 1994; Hutton & Giller, 2004), and stream detritus (Murphy et al., 1998) that indicate higher levels of intraspecific compared to interspecific aggregation and therefore reduced potential for interspecific competition among community members. However, to what extent the observed aggregation patterns are governed by unperceived or undetected resource partitioning is still unknown.

Furthermore, a defining concept of the model is that relative strengths of intra- and interspecific aggregation should be directly related to decomposer species diversity: evidence for this is still weak and the research approaches subject to issues of cause and effect such that there is still no clear indication of whether aggregation determines species richness or vice-versa (Krijger & Sevenster, 2001; Woodcock et al., 2002; Horgan, 2006). Nevertheless, methodological issues aside, the model – together with classical resource partitioning – offers a clear and probable explanation for the high diversity in heterotrophic communities.

4. Effects on Coleoptera of Anthropogenic Alterations of Habitats Supporting Resource Patches

Biodiversity loss, in all its forms, in response to anthropogenic environmental changes is generally considered to reduce ecosystem function and resilience. While conservation efforts have been accomplished in recent decades, the situation has hardly improved in terms of rates of species loss (Butchart et al., 2010), although some local successes have been achieved. Thus, to understand and mitigate our impacts on ecosystems, it is still relevant to study the effect of anthropogenic environmental changes on different taxa, especially sensitive ones. As such, heterotrophic beetles represent valuable indicators of environmental changes because their communities reflect human impacts both on heterotrophic substrates and surrounding habitats.

4.1. Agriculture and Pasture Management

The expansion and intensification of agriculture has brought about rapid and dramatic changes in the quality of habitat for decomposer beetles (Nichols et al., 2007). Reductions in the numbers of birds and small mammals on modern farms, particularly on large monocultures of cereals or annual crops under intensive management, will reduce available resources (dung and carcasses) for decomposers; however, shady crops, such as coffee and fruit trees are less inhospitable (Nichols et al., 2007; Horgan, 2009); the poisoning of animals with toxic pesticides creates contaminated carcasses with consequences for decomposers. For example, decomposers and scavengers often avoid molluscicide-killed snails and slugs or have altered behaviors during and after feeding on the cadavers of poisoned snails compared to behaviors on snails killed by freezing. In some cases decomposer beetles may be directly killed by the toxic pesticides in their food (Langan et al., 2004; Horgan, unpublished), thus reducing decomposition rates (Horgan, unpublished).

Dung beetles are an important component of pasture biodiversity with a recognized role in the clearing of dung produced by domestic livestock, particularly in the tropics and subtropics (Nichols et al., 2009). The effects of pasture management and livestock husbandry on dung beetle communities will have direct consequences for ecosystem functioning. The quality of pastures provided for animals as well as the extent of shade provided by field boundaries or trees has been shown to affect dung beetle assemblages and potentially affects decomposition processes (Finn & Giller, 2002; Horgan, 2002; Hutton & Giller, 2003). Furthermore, the increasing use of agrochemicals, either for weed control or as anti-parasitic drenches can affect beetle behavior, physiology, and function (Martínez et al., 2000; O'Hea et al., 2010). Efforts are needed to better design pasture management and ranching methods to conserve dung beetles and optimize their benefits.

4.2. Forest Management

4.2.1. Deforestation

The high diversity of dung and carrion beetles and their importance in ecosystem functioning has made these a popular indicator group for monitoring habitat quality and anthropogenic disturbances. A high degree of micro- and macrohabitat specificity, particularly in the tropics, means that dung beetle communities are often drastically altered by changing agricultural and forest management practices (Nichols et al., 2007). Extensive deforestation in tropical regions has led to noted declines in dung beetle diversity and the invasion of newly opened agricultural areas by a limited number of specialized open habitat species and some robust generalists (Nichols et al., 2007; Horgan, 2007). The predominance of cattle ranching in the neotropics particularly has resulted in a small group of dung beetles, including two or three species of European and African origin, becoming the most abundant and widespread synantropic insect species (Horgan, 2007). The conversion of forests to small holding farms is also associated with a change in dung beetle fauna. Shade from coffee and banana trees have been noted to promote dung beetle diversity, but the extent of such plantations and their proximity to intact forests appear to greatly influence species richness (Horgan, 2009). Even where good forest cover is maintained by selective logging and other forest conservation strategies, dung beetle diversity has been noted to decline (Nichols et al., 2007). This may be directly related to a reduction in forest mammals, which often avoid forest edges, secondary forests or are hunted to a greater extent because of easier access by human hunters.

Little has been done on deforestation in other heterotrophic systems. One study that compared carcasses in forests and adjacent pastures has shown that both had similar Diptera fauna but different assemblages of Coleoptera because one abundant scarab species always occurred in open fields while species of the family Leodidae were generally only documented in forests (Michaud et al., 2010). In a study of carrion beetles, Dekeirsschieter et al. (2011) reported that one of the seven species of Silphidae found on carcasses in forests – *Oiceoptoma thoracica* L. – was not recovered on carcasses in agricultural fields.

4.2.2. Plantation Establishment

A recent intensification in the establishment of plantation forests has occurred because companies aim to rapidly increase timber production while reducing operational costs (Stephens & Wagner, 2007). Depending on plant species used and regional climate, plantation rotations vary in length from half a decade to a century, and even longer (Evans, 1992; Lambert & Turner, 2000; Lindenmayer & Hobbs, 2008; Chauhan et al., 2006; Jarvis & Linder, 2007). Clearfelling drastically alter the structure of habitats, which in turn considerably modifies the composition of faunal communities (Pawson et al., 2008).

While beetles can be grouped according to their feeding preferences (see Section 2.3), they can also be grouped according to their preferential habitat, e.g., open habitat specialists, forest generalists, and mature forest specialists (Niemelä et al., 1993). Open habitat specialists are favored by the occurrence of natural and anthropogenic disturbances such as clearcuts and fires (Elek et al., 2001; Saint-Germain et al., 2005). Forest generalists are common species in forest habitats (Jonsell et al., 2005) while mature forest specialists occur in old forests that generally exhibit closed canopies (Pedley et al., 2014). As stands develop, saproxylic beetle assemblages change, and leaf litter beetles appear (Jacobs et al., 2007b; Stenbacka et al., 2010). Therefore, a mosaic of forest landscapes is needed to maintain abundant and diverse beetle assemblages (Koivula, 2002; Riffell et al., 2011; Sverdrup-Thygeson et al., 2014). However, when dead wood is collected for biofuel production and when ground is scarified after clear-felling operations, saproxylic and leaf litter beetles have no habitat and resources to rely on (Gustafsson et al., 2010; Work et al., 2010; Thibault & Moreau, unpublished). Moreover, dead wood is normally scarce at the beginning of plantation establishment and by the time plantations start to naturally produce dead wood, they are usually harvested (Lassauce et al., 2011; Brin et al., 2008). Thus, it comes to no surprise that several saproxylic beetles are considered threatened or near threatened in areas that have historically relied on tree plantation for forestry operations (Nieto & Alexander, 2010).

4.2.3. Thinning

As forests age, trees mature and occupy more space. Eventually the canopy closes and tree growth rates start to decline because of limited growth space, light, soil moisture, and soil nutrients (Canham et al., 2004). To avoid further decline of growth rates, forests will often be treated using an intermediate partial harvesting strategy called thinning, which reduces the density of trees in a stand by a fixed proportion or volume (Mayor & Rodà, 1993). Thinning operations also add dead wood to the ground and create openings in the canopy that favor the growth of understory vegetation (Zeide, 2001). Thus, this treatment diversifies microhabitats in forests, which can favor different beetle assemblages and alter ongoing succession (Magura et al., 2000; Nadeau et al., *in press*). Such operation, however, can also modify food webs (Moreau et al., 2006), deter mature forest specialists and, in the long term, results in a decline of woody debris in forest ecosystems because self-thinning (see Westoby, 1984) is reduced. In addition, thinned forests are largely devoid of large diameter dead wood, which has a direct effect on saproxylic beetles and typical succession patterns (Siitonen, 2001; Müller et al., 2007). Because saproxylic beetles play a crucial role in forest ecosystems as decomposers and facilitaters, further research on these insects and heterotrophic succession on dead wood is required to make better conservation decisions concerning dead wood and forest management.

4.3. Urbanization

If converting a natural forest into a plantation forest is a dramatic disturbance for beetles (see section 4.2), imagine how dramatic it is for a beetle assemblage to see its forest transformed into a shopping center (Alberti et al., 2003; McKinney, 2008; Martinson & Raupp, 2013). Urbanization impacts beetles according to perturbation severity, highly urbanized areas exhibiting more profound shifts in beetle assemblages than moderately urbanized areas (McDonnell et al., 1997). The urbanization-rural gradient also affects the magnitude of physical changes in those habitats. For example, beetle habitats can be modified or lost, and human disturbances can be mild or severe (Magura et al., 2013). Remnant habitats are not only more isolated, but they are also missing connectivity structures (Collins et al., 2000).

Forest stands reduced in quantity and size due to urbanization are affected by greater edge effects. These edge effects can be of abiotic nature, such as warmer temperatures and drier soils, and they can also be of biotic nature, such as a reduction in dead wood and leaf litter quantity (Karjalainen, 1991). Forest beetle assemblages in urban areas will accordingly change toward open habitat species and generalist beetle species because of highly homogenous habitats unsuitable for specialist species (Gray, 1989; Desender et al., 1999; Magura et al., 2004b).

Dung beetles usually avoid feces with low cellulose content, and as such favor herbivorous animal feces, which are usually scarce in urban areas. A dog's diet is not totally herbivorous, but omnivorous, as they consume table food which is high in cellulose and/or carbohydrate content (Carpaneto et al., 2005). Human and dog dung might not be the ideal resource for a highly abundant and diverse assemblage of dung beetles but sewage sites and dog parks are habitats were dung beetles can find substitute resources in urban areas (Faithfull, 1992; Huang et al., 2001).

Only a handful of studies have addressed beetle succession on decaying cadavers in rural areas (e.g., Grassberger & Frank, 2004; Perez et al., 2005; Segura et al., 2009). These studies mainly focused on fly succession in rural to urban gradients (e.g., Baumgartner, 1988; Fischer, 2000; Leblanc & Strongman, 2002; Hwang & Turner, 2005). The only study, to our knowledge, that compared the succession patterns of beetles along a rural to urban gradient documented a 3-days delay in beetle colonization in urban compared to rural areas, as well as a lower abundance and species richness in urban areas (Hans, 2011). Dekeirsschieter et al. (2011) also reported that no Silphidae visited two piglet carcasses exposed in urban sites.

4.4. Pollution and Global Changes

Pollution and global changes are not only linked together and affecting beetles, but also the whole food web (Bytnerowicz et al., 2007; Lagisz & Laskowski, 2008). Pollution and global changes can be positive or negative in ecosystems (Allen et al., 2010), and the negative effects are more severe in highly urbanized areas due to increased human activities (VijayaVenkataRaman et al., 2012). However, their linked effect on heterotrophic beetle succession has not directly been tested.

For example, heavy metals emitted by smelting industries may be acquired by litter dwelling beetles through dermal absorption when feeding on leaf litter or contaminated prey (Stone et al., 2001). Carabidae species have developed detoxification and excretion processes through protein synthesis and catabolism activity, but these processes are not without metabolism costs (Calow, 1991; Hawkins, 1991). Stone et al. (2001) found that when carabid beetles are contaminated with high doses of heavy metals, their survival time decreases. Moreover, Carabidae species in polluted sites have smaller body sizes (Braun et al., 2004) and increased egg production but the quality of said eggs is lower than in unpolluted sites (Lagisz & Laskowski, 2008). Global warming increases nitrogen deposition, which increases the rate of leaf litter accumulation but reduces its decomposition rate (Schulze, 2000), which might be beneficial to litter dwelling beetles and their prey.

Most studies on beetles and climate change focus on range extension of industrial pest species, such as bark beetles in forestry settings (Jönsson et al., 2009; Seidl et al., 2008; Bentz et al., 2010; Cudmore et al., 2010), and phytophagous beetles in agricultural settings (Bale et al., 2002; Andrew & Hughes, 2004). Outbreaks of bark beetles weaken trees, and may lead to increases in dead wood quantity (Schelhaas et al., 2003), an important factor in saproxylic beetle succession. Climate change affects dead wood in different ways. The decay process can be altered by changes in humidity and radiation intensity, which would in turn affect beetle succession (Milad et al., 2011). Climate change also produces periods of intensified drought, which can increase the risk of forest fire (Dale et al., 2001) and subsequent beetle succession.

Because substrate quality is such an important aspect of beetle succession, whether it is dead wood, dung, fruits, leaf litter or carcasses, studies on the effects of climate change and pollution on heterotrophic beetle succession have hardly scratched the surface and should be an object of future research.

CONCLUSION

Any biological system, when considered by itself, appears unique and distinct from any other system. However, the appreciation of the fundamental mechanisms underlying a given system requires the identification of the mechanisms that govern similar systems. In this chapter, we have aimed to identify what is different, but especially what is common to the different terrestrial heterotrophic systems in which beetles are involved. In doing so, we hope to stimulate future crossover between the different heterotrophic fields that could lead to the development of analytical tools, models, and unifying principles that will benefit all fields of decomposition ecology.

REFERENCES

Abell, D. H., Wasti, S. S. & Hartmann, G. C. (1982). Saprophagous arthropod fauna associated with turtle carrion. *Applied Entomology and Zoology*, *17*, 301-307.

Adams, C. C. (1908). The ecological succession of birds. *The Auk*, *25*, 109-153.

Ahmad, N. W., Lim, L. H., Dhang, C. C., Chin, H. C., Abdullah, A. G., Mustaffa, W. N. W., Kian, C. W., Jeffery, J., Hashim, R. & Azirun, S. M. (2011). Comparative insect fauna succession on indoor and outdoor monkey carrions in a semi-forested area in Malaysia. *Asian Pacific Journal of Tropical Biomedicine*, *1*, S232-S238.

Alberti, M., Marzluff, J. M., Shulenberger, E., Bradley, G., Ryan, C. & Zumbrunnen, C. (2003). Integrating humans into ecology: Opportunities and challenges for studying urban ecosystems. *BioScience*, *53*, 1169-1179.

Allee, W. C., Emerson, A. E., Park, O., Park, T. & Schmidt, K. P. (1949). *Principles of Animal Ecology*. Saunders W.B. Co. Ltd., Philadelphia, Pennsylvania.

Allen, C. D., Macalady, A. K., Chenchouni, H., Bachelet, D., McDowell, N., Vennetier, M., Kitzberger, T., Rigling, A., Breshears, D. D., Hogg, E. H., Gonzalez, P., Fensham, R., Zhang, Z., Castro, J., Dmidova, N., Lim, J. H., Allard, G., Running, S. W., Semerci, A. & Cobb, N. (2010). A global overview of drought and heat-induced tree mortality reveals emerging climate change risks for forests. *Forest Ecology and Management*, *259*, 660-684.

Anderson, G. S. & VanLaerhoven, S. L. (1996). Initial studies on insect succession on carrion in south western British Columbia. *Journal of Forensic Science*, *41*, 617-625.

Anderson, J. M. (1995). Soil organisms as engineers: Microsite modulation of macroscale processes. In: Jones, C. G., Lawton, J. H. (Eds.) *Linking Species and Ecosystems*. 94-106. Chapman and Hall, New York, New York.

Anderson, K. J. (2007). Temporal patterns in rates of community change during succession. *The American Naturalist*, *169*, 780-793.

Anderson, R. S. (1982). Resource partitioning in the carrion beetle (Coleoptera: Silphidae) fauna of southern Ontario: Ecological and evolutionary considerations. *Canadian Journal of Zoology*, *60*, 1314-1325.

Anderson, R. S. & Peck, S. B. (1985). *The Carrion Beetles of Canada and Alaska (Coleoptera: Silphidae and Agyrtidae), The Insects and Arachnids of Canada, Part 13.* Agriculture Canada publication 1778, Ottawa, Canada.

Andrew, N. R. & Hughes, L. (2004). Species diversity and structure of phytophagous beetle assemblages along a latitudinal gradient: Predicting the potential impacts of climate change. *Ecological Entomology*, *29*, 527-542.

Araya, K. (1993). Relationship between the decay types of dead wood and occurrence of lucanid beetles (Coleoptera: Lucanidae). *Applied Entomology and Zoology*, *28*, 27-33.

Arce, A. N., Johnston, P. R., Smiseth, P. T. & Rozen, D. E. (2012). Mechanisms and fitness effects of antibacterial defences in a carrion beetle. *Journal of evolutionary biology*, *25*, 930-937.

Archer, M. S. & Elgar, M. A. (1998). Cannibalism and delayed pupation in hide beetles, *Dermestes maculatus* DeGeer (Coleoptera: Dermestidae). *Australian Journal of Entomology*, *37*, 158-161.

Atkinson, W. D. & Shorrocks, B. (1981). Competition on a divided and ephemeral resource: A simulation model. *Journal of Animal Ecology*, *50*, 461-471.

Bale, J. S., Masters, G. J., Hodkinson, I. D., Awmack, C., Bezemer, T. M., Brown, V. K., Butterfield, J., Buse, A., Coulson, J. C., Farrar, J., Good, J. E. G., Harrington, R., Hartley, S., Jones, T. H., Lindroth, R. L., Press, M. C., Symrnioudis, I., Watt, A. D. & Whittaker, J. B. (2002). Herbivory in global climate change research: Direct effects of rising temperature on insect herbivores. *Global Change Biology*, *8*, 1-16.

Barlow, H. W. B. (1950). Studies in abscission. I. Factors affecting the inhibition of abscission by synthetic growth-substances. *Journal of Experimental Botany*, *1*, 264-81.

Bartelt, R. J. & Hossain, M. S. (2010). Chemical ecology of Carpophilus sap beetles (Coleoptera: Nitidulidae) and development of an environmentally friendly method of crop protection. *Terrestrial Arthropod Reviews*, *3*, 29-61.

Bastow, J. (2012). Succession, resource processing, and diversity in detrital food webs. In: Wall, D. H., Bardgett, R. D., Behan-Pelletier, V., Herrick, J. E., Jones, T. H., Ritz, K., Six, J., Strong, D. R., van der Putten, W. H. (Eds.) *Soil Ecology and Ecosystem services*., 117-135. Oxford University Press, Oxford, UK.

Baumgartner, D. L. (1988). Spring season survey of the urban blowflies (Diptera: Calliphoridae) of Chicago, Illinois. *Great Lakes Entomologist*, *21*, 119-121.

Beaulieu, F. & Weeks, A. R. (2007). Free-living mesostigmatic mites in Australia: Their roles in biological control and bioindication. *Australian Journal of Experimental Agriculture*, *47*, 460-478.

Beaver, R. A. (1977). Non-equilibrium 'island' communities: Diptera breeding in dead snails. *The Journal of Animal Ecology*, *46*, 783-798.

Begon, M., Harper, J. L. & Townsend, C. R. (1996). *Ecology: Individuals, Populations and Communities*. Third edition. Blackwell Science, Oxford, UK.

Benecke, M. (2001). A brief history of forensic entomology. *Forensic Science International*, *120*, 2-14.

Bentz, B. J., Régnière, J., Fettig, C. J., Hansen, E. M., Hayes, J. L., Hicke, J. A., Kelsey, R. G., Negron, J. F. & Seybold, S. J. (2010). Climate change and bark beetles of the western United States and Canada: Direct and indirect effects. *BioScience*, *60*, 602-613.

Berg, B. & Staaf, H. (1980). Decomposition rate and chemical changes of Scots pine needle litter. II. Influence of chemical composition. *Ecological Bulletins*, *32*, 373-390.

Bernhardt, K. G. (1995). Seed burial by soil burrowing beetles. *Nordic Journal of Botany*, *15*, 257-260.

Blackman, S. W. & Evans, G. O. (1994). Observations on a mite (*Poecilochirus davydovae*) predatory on the eggs of burying beetles (*Nicrophorus vespilloides*) with a review of its taxonomic status. *Journal of Zoology*, *234*, 217-227.

Bornemissza, G. F. (1957). An analysis of arthropod succession in carrion and the effect of its decomposition on the soil fauna. *Australian Journal of Zoology*, *5*, 1-12.

Borror, D. J. & White, R. E. (1970). *A Field Guide to the Insects of North America and Mexico*. Houghton Mifflin Compagny, Boston, Massachusetts.

Bouget, C. & Duelli, P. (2004). The effects of windthrow on forest insect communities: A literature review. *Biological Conservation*, *118*, 281-299.

Bouget, C., Nusillard, B., Pineau, X. & Ricou, C. (2012). Effect of deadwood position on saproxylic beetles in temperate forests and conservation interest of oak snags. *Insect Conservation and Diversity*, *5*, 264-278.

Boulanger, Y. & Sirois, L. (2007). Postfire succession of saproxylic arthropods, with emphasis on Coleoptera, in the north boreal forest of Quebec. *Environmental Entomology*, *36*, 128-141.

Boulton, A. J. & Lake, P. S. (1988). Dynamics of heterotrophic succession in carrion arthropod assemblages: A comment on Schoenly and Reid (1987). *Oecologia*, *76*, 477-480.

Bousquet, Y. & Laplante, S. (2006). *Coleoptera Histeridae: The Insects and Arachnids of Canada, part 24*. NRC Research Press, Ottawa, Canada.

Braack, L. E. O. (1987). Community dynamics of carrion-attendant arthropods in tropical African woodland. *Oecologia*, *72*, 402-409.

Brais, S., Paré, D. & Lierman, C. (2006). Tree bole mineralization rates of four species of the Canadian eastern boreal forest: Implications for nutrient dynamics following stand-replacing disturbances. *Canadian Journal of Forest Research*, *36*, 2331-2340.

Brandhorst-Hubbard, J. L., Flanders, K. L. & Appel, A. G. (2001). Oviposition site and food preference of the green June beetle (Coleoptera: Scarabaeidae). *Journal of Economic Entomology*, *94*, 628-633.

Brattli, J. G., Andersen, J. & Nilssen, A. C. (1998). Primary attraction and host tree selection in deciduous and conifer living Coleoptera: Scolytidae, Curculionidae, Cerambycidae and Lymexylidae. *Journal of Applied Entomology*, *122*, 345-352.

Braun, S. D., Jones, T. H. & Perner, J. (2004). Shifting average body size during regeneration after pollution–a case study using ground beetle assemblages. *Ecological Entomology*, *29*, 543-554.

Bretz-Guby, N. A. & Dobbertin, M. (1996). Quantitative estimates of coarse woody debris and standing trees in selected Swiss forests. *Global Ecology and Biogeography Letters*, *5*, 327-341.

Brin, A., Bouget, C., Brustel, H. & Jactel, H. (2011). Diameter of downed woody debris does matter for saproxylic beetle assemblages in temperate oak and pine forests. *Journal of Insect Conservation*, *15*, 653-669.

Brin, A., Meredieu, C., Piou, D., Brustel, H. & Jactel, H. (2008). Changes in quantitative patterns of dead wood in maritime pine plantations over time. *Forest Ecology and Management*, *256*, 913-921.

Brunet, J. & Isacsson, G. (2009). Influence of snag characteristics on saproxylic beetle assemblages in a south Swedish beech forest. *Journal of insect conservation*, *13*, 515-528.

Bugajski, K. N., Seddon, C. C. & Williams, R. E. (2011). A comparison of blow fly (Diptera: Calliphoridae) and beetle (Coleoptera) activity on refrigerated only versus frozen-thawed pig carcasses in Indiana. *Journal of Medical Entomology*, *48*, 1231-1235.

Butchart, S. H., Walpole, M., Collen, B., van Strien, A., Scharlemann, J. P., Almond, R. E. A., Baillie, J. E. M., Bomhard, B., Brown, C., Bruno, J., Carpenter, K. E., Carr, G. M., Chanson, J., Chenery, A. M, Csirke, J., Davidson, N. C., Dentener, F., Foster, M., Galli, A., Galloway, J. N., Genovesi, P., Gregory, R. D., Hockings, M., Kapos, V., Lamarque, J. F., Leverington, F., Loh, J., McGeoch, M. A., McRae, L., Minasyan, A., Morcillo, M. H., Oldfield, T. E. E., Pauly, D., Quader, S., Revenga, C., Sauer, J. R., Skolnik, B., Spear, D., Stanwell-Smith, D., Stuart, S. N., Symes, A., Tierney, M., Tyrrell, T. D., Vié, J. C. & Watson, R. (2010). Global biodiversity: Indicators of recent declines. *Science*, *328*, 1164-1168.

Byrd, J. H. & Castner J. L. (2001). Insects of forensic importance. In: Byrd, J. H., Castner, J. L. (Eds.) *Forensic Entomology, the Utility of Arthropods in Legal Investigations*., 43-79. CRC Press, Boca Raton, Florida.

Bytnerowicz, A., Omasa, K. & Paoletti, E. (2007). Integrated effects of air pollution and climate change on forests: A northern hemisphere perspective. *Environmental Pollution*, *147*, 438-445.

Calow, P. (1991). Physiological costs of combating chemical toxicants: Ecological implications. Comparative Biochemistry and Physiology Part C: *Comparative Pharmacology*, *100*, 3-6.

Cambefort, Y. & Walter, P. (1991). Dung beetles in tropical forests in Africa. In: Hanski, I., Cambefort, Y. (Eds.) *Dung beetle ecology*., 198-209. Princeton University Press, Princeton, New Jersey.

Campobasso, C. P., Di Vell, G. & Introna, F. (2001). Factors affecting decomposition and Diptera colonization. *Forensic Science International*, *120*, 18-27.

Canham, C. D., LePage, P. T. & Coates, K. D. (2004). A neighborhood analysis of canopy tree competition: Effects of shading versus crowding. *Canadian Journal of Forest Research*, *34*, 778-787.

Carpaneto, G. M., Mazziotta, A. & Piattella, E. (2005). Changes in food resources and conservation of scarab beetles: From sheep to dog dung in a green urban area of Rome (Coleoptera, Scarabaeoidea). *Biological Conservation*, *123*, 547-556.

Carter, D. O., Yellowlees, D. & Tibbett, M. (2007). Cadaver decomposition in terrestrial ecosystems. *Naturwissenschaften*, *94*, 12-24.

Chame, M. (2003). Terrestrial mammal faeces: A morphometric summary and description. *Memórias do Instituto Oswaldo Cruz*, *98*, 71-94.

Chapin, F. S., Walker, L. R., Fastie, C. L. & Sharman, L. C. (1994). Mechanisms of primary succession following deglaciation at Glacier Bay, Alaska. *Ecological Monographs*, *64*, 149-175.

Chapman, R. F. & Sankey, J. H. P. (1955). The larger invertebrate fauna of three rabbit carcasses. *Journal of Animal Ecology*, *24*, 395-402.

Chauhan, S., Donnelly, R., Huang, C. L., Nakada, R., Yafang, Y. & Walker, J. C. (2006). Wood quality: In context. In: Walker, J. F. C. *Primary Wood Processing, Principles and Practice*., 121-158. Springer, Netherlands.

Coe, M. (1978). The decomposition of elephant carcasses in the Tsavo (East) National Park, Kenya. *Journal of Arid Environments*, *1*, 71-86.

Collins, J. P., Kinzig, A., Grimm, N. B., Fagan, W. F., Hope, D., Wu, J. & Borer, E. T. (2000). A new urban ecology. *American Scientist*, *88*, 416-425.

Comeau, C., Privé, J. P. & Moreau, G. (2012). Beneficial impacts of the combined use of rain shelters and reflective groundcovers in an organic raspberry cropping system. *Agriculture, Ecosystems and Environment*, *155*, 117-123.

Comeau, C., Privé, J. P. & Moreau, G. (2013). Effects of reflective groundcovers on ground beetles (Coleoptera: Carabidae) in red raspberry (*Rubus idaeus*) cropping systems. *Journal of Applied Entomology*, *137*, 264-274.

Connell, J. H. & Slatyer, R. O. (1977). Mechanisms of succession in natural communities and their role in community stability and organization. *American Naturalist*, *111*, 1119-1144.

Cotter, S. C. & Kilner, R. M. (2010). Sexual division of antibacterial resource defence in breeding burying beetles, *Nicrophorus vespilloides*. *Journal of animal ecology*, *79*, 35-43.

Coûteaux, M. M., Mousseau, M., Célérier, M. L. & Bottner, P. (1991). Increased atmospheric CO_2 and litter quality: Decomposition of sweet chestnut leaf litter with animal food webs of different complexities. *Oikos*, *61*, 54-64.

Cudmore, T. J., Björklund, N., Carroll, A. L. & Staffan Lindgren, B. (2010). Climate change and range expansion of an aggressive bark beetle: Evidence of higher beetle reproduction in naive host tree populations. *Journal of Applied Ecology*, *47*, 1036-1043.

Cutler, G. C., Renkema, J. M., Majka, C. G. & Sproule, J. M. (2012). Carabidae (Coleoptera) in Nova Scotia, Canada wild blueberry fields: Prospects for biological control. *The Canadian Entomologist*, *144*, 808-820.

Dajoz, R. (1998). *Écologie des Insectes Forestiers*. Gauthier-Villars, Paris, France.

Dale, V. H., Joyce, L. A., McNulty, S., Neilson, R. P., Ayres, M. P., Flannigan, M. D., Hanson, P. J., Irland, L. C., Lugo, A. E. & Peterson, C. J. (2001). Climate Change and Forest Disturbances. *BioScience*, *51*, 723-734.

Dekeirsschieter, J., Verheggen, F. J., Haubruge, E. & Brostaux, Y. (2011). Carrion beetles visiting pig carcasses during early spring in urban, forest and agricultural biotopes of Western Europe. *Journal of Insect Science*, *11*, 73.

del Moral, R. & Wood, D. M. (1993). Early primary succession on a barren volcanic plain at Mount St. Helens, Washington. *American Journal of Botany*, *80*, 981-991.

Denholm-Young, P. A. (1978). Studies of decomposing cattle dung and its associated fauna. Doctoral dissertation, Oxford University, UK.

Desender, K., Ervynck, A. & Tack, G. (1999). Beetle diversity and historical ecology of woodlands in Flanders. *Belgian Journal of Zoology*, *129*, 139-155.

Dettner, K. (1993). Defensive secretions and exocrine glands in free-living staphylinid beetles - their bearing on phylogeny (Coleoptera: Staphylinidae). *Biochemical Systematics and Ecology*, *21*, 143-162.

Dickinson, C. H., Underhay, V. S. H. & Ross, V. (1981). Effect of season, soil fauna and water content on the decomposition of cattle dung pats. *New Phytologist*, *88*, 129-141.

Dollin, P. E., Majka, C. G. & Duinker, P. N. (2008). Saproxylic beetle (Coleoptera) communities and forest management practices in coniferous stands in southwest Nova Scotia, Canada. *ZooKeys*, *2*, 291-336.

Domek, J. M. & Johnson, D. T. (1990). Inhibition of aggregation behavior in the green June beetle (Coleoptera: Scarabaeidae) by antibiotic treatment of food substrate. *Environmental Entomology*, *19*, 995-1000.

Dormont, L., Epinat, G. & Lumaret, J. P. (2004). Trophic preferences mediated by olfactory cues in dung beetles colonizing cattle and horse dung. *Environmental Entomology*, *33*, 370-377.

Dormont, L., Jay-Robert, P., Bessière, J. M., Rapior, S. & Lumaret, J. P. (2010). Innate olfactory preferences in dung beetles. *The Journal of Experimental Biology*, *213*, 3177-3186.

Doube, B. M., Macqueen, A., Ridsdill-Smith, T. J. & Weir, T. A. (1991). Native and introduced dung beetles in Australia. In: Hanski, I., Cambefort, Y. (Eds.) *Dung Beetle Ecology*., 255-278. Princeton University Press, Princeton, New Jersey.

Downie, N. M. & Arnett, Jr., R. H. (1996). *The Beetles of Northeastern North America*. Sandhill Crane Press, Gainesville, Florida.

Early, M. & Goff, M. L. (1987). Anthropod succession patterns in exposed carrion on the island of O'ahu Hawaii. *Journal of Medical Entomology*, *23*, 520-531.

Eckert, J. W. & Sommer, A. F. (1967). Control of diseases of fruits and vegetables by postharvest treatment. *Annual Review of Phytopathology*, *5*, 391-428.

Edwards, P. B. (1991). Seasonal variation in the dung of African grazing mammals and its consequences for coprophagous insects. *Functional Ecology*, *5*, 617-628.

Eggert, A. K. & Sakaluk, S. K. (2000). Benefits of communal breeding in burying beetles: A field experiment. *Ecological Entomology*, *25*, 262-266.

Ehnström B., Gärdenfors U. & Lindelöw Å. (1993). Rödlistade evertebrater i Sverige 1993. Data-banken för hotade arter, Uppsala, Sweden.

Elek, Z., Magura, T. & Tóthmérész, B. (2001). Impacts of non-native Norway spruce plantation on abundance and species richness of ground beetles (Coleoptera: Carabidae). *Web Ecology*, *2*, 32-37.

Erwin, T. L. (1979). Thoughts on the evolutionary history of ground beetles: Hypotheses generated from comparative faunal analyses of lowland forest sites in temperate and tropical regions. In: Erwin, T. L., Ball, G. E., Whitehead, D. R., Halper, A. L. (Eds.) *Carabid Beetles: Their Evolution, Natural History and Classification*., 539-592. W. Junk, The Hague, Netherlands.

Esseen, P. A., Ehnström, B., Ericson, L. & Sjöberg, K. (1992). Boreal forests – the focal habitats of Fennoscandia. In: Hansson, L. (Ed.) *Ecological principles of nature conservation*., 252-325. Elsevier, Amsterdam, Netherlands.

Esseen, P. A., Ehnström, B., Ericson, L. & Sjöberg, K. (1997). Boreal forests. *Ecological Bulletins*, *46*, 16-47.

Evans, J. (1992). *Plantation Forestry in the Tropics: Tree Planting for Industrial, Social, Environmental, and Agroforestry Purposes*., 252-266. Oxford University Press, Oxford, UK.

Faithfull, I. (1992). Records of native dung beetles *Onthophagus pexatus* Harold, *O. auritus* Erichson and *O. granulatus* Boheman (Coleoptera: Scarabaeidae) at dog scats and their potential for biocontrol of dog dung. *Victorian Entomologist*, *22*, 105-108.

Farrell, T. M. (1991). Models and mechanisms of succession: An example from a rocky intertidal community. *Ecological Monographs*, *61*, 95-113.

Farwig, N., Brandl, R., Siemann, S., Wiener, F. & Müller, J. (2014). Decomposition rate of carrion is dependent on composition not abundance of the assemblages of insect scavengers. *Oecologia*, *175*, 1291-1300.

Finn, J. A. & Giller, P. S. (2000). Patch size and colonization patterns: An experimental analysis using north temperate coprophagous dung beetles. *Ecography*, *23*, 315-327.

Finn, J. A. & Giller, P. S. (2002). Experimental investigations of colonization by north temperate dung beetles of different types of domestic herbivore dung. *Applied Soil Ecology*, *20*, 1-13.

Fischer, O. A. (2000). Blowflies of the genera *Calliphora*, *Lucilia* and *Protophormia* (Diptera, Calliphoridae) in South-Moravian urban and rural areas with respect to *Lucilia bufonivora* Moniez, 1876. *Acta Veterinaria Brno*, *69*, 225-231.

Franc, N. (2007). Standing or downed dead trees—does it matter for saproxylic beetles in temperate oak-rich forest? *Canadian Journal of Forest Research*, *37*, 2494-2507.

Freymann, B. P., Buitenwerf, R., Desouza, O. & Olff, H. (2008). The importance of termites (Isoptera) for the recycling of herbivore dung in tropical ecosystems; a review. *European Journal of Entomology*, *105*, 165-173.

Fridman, J. & Ståhl, G. (2001). A three-step approach for modelling tree mortality in Swedish forests. *Scandinavian Journal of Forest Research*, *16*, 455-466.

Fridman, J. & Walheim, M. (2000). Amount, structure, and dynamics of dead wood on managed forestland in Sweden. *Forest Ecology and Management*, *131*, 23-36.

Fuller, M. E. (1934). The insect inhabitants of carrion: A study of animal ecology. *Council for Scientific and Industrial Research (Australia)*, *82*, 1-62.

Galvan, T. L., Koch, R. L. & Hutchison, W. D. (2008). Impact of fruit feeding on overwintering survival of the multicolored Asian lady beetle, and the ability of this insect and paper wasps to injure wine grape berries. *Entomologia Experimentalis et Applicata*, *128*, 429-436.

Ghazoul, J. (2005). Pollen and seed dispersal among dispersed plants. *Biological Reviews*, *80*, 413-443.

Gibb, H., Pettersson, R. B., Hjältén, J., Hilszczański, J., Ball, J. P., Johansson, T., Atlegrim, O. & Danell, K. (2006). Conservation-oriented forestry and early successional saproxylic beetles: Responses of functional groups to manipulated dead wood substrates. *Biological Conservation*, *129*, 437-450.

Gill, B. D. (1991). Dung beetles in tropical American forests. In: Hanski, I. Cambefort, Y. (Eds.) *Dung Beetle Ecology.*, 211-229. Princeton University Press, Princeton, New Jersey.

Giller, P. S. & Doube, B. M. (1994). Spatial and temporal co-occurrence of competitors in South African dung beetle communities. *Journal of Animal Ecology*, *63*, 629-643.

Gillilland, M. G., Erb-Downward, J. R., Bassis, C. M., Shen, M. C., Toews, G. B., Young, V. B. & Huffnagle, G. B. (2012). Ecological succession of bacterial communities during conventionalization of germ-free mice. *Applied and environmental microbiology*, *78*, 2359-2366.

Gittings, T. & Giller, P. S. (1998). Resource quality and the colonisation and succession of coprophagous dung beetles. *Ecography*, *21*, 581-592.

Goff, M. L. (1992). Problems in estimation of postmortem interval resulting from wrapping of the corpse: A case study from Hawaii. *Journal of Agricultural Entomology*, *9*, 237-243.

Goff, M. L. (1993). Estimation of postmortem interval using arthropod development and successional patterns. *Forensic Science Review*, *5*, 81-94.

Goguen J. & Moreau G. (2013). Factors affecting head capsule development in field populations of *Altica sylvia* (Coleoptera: Chrysomelidae). *The Canadian Entomologist, 145*, 40-47.

Grassberger, M. & Frank, C. (2004). Initial study of arthropod succession on pig carrion in a central European urban habitat. *Journal of Medical Entomology, 41*, 511-523.

Gray, J. S. (1989). Effects of environmental stress on species rich assemblages. *Biological Journal of the Linnean Society, 37*, 19-32.

Green, A. A. (1967). The protection of dried sea-fish in south Arabia from infestation by *Dermestes frischii* Kug. (Coleoptera, Dermestidae). *Journal of Stored Products Research, 2*, 331-350.

Greenberg, B. (1991). Flies as forensic indicators. *Journal of Medical Entomology, 28*, 565-577.

Grishin, S. Y., del Moral, R., Krestov, P. V. & Verkholat, V. P. (1996). Succession following the catastrophic eruption of Ksudach volcano (Kamchatka, 1907). *Vegetatio, 127*, 129-153.

Grove, S. J. (2002). Saproxylic Insect Ecology and the Sustainable Management of Forests. *Annual Review of Ecology and Systematics, 33*, 1-23.

Grove, S. J., Bashford, D. & Yee, M. (2008). A long-term experimental study of saproxylic beetle (Coleoptera) succession in Tasmanian Eucalyptus obliqua logs: Findings from the first five years. In: Fattorini, S. (Ed.) *Insect Ecology and Conservation.*, 71-114. Research Signpost, Kerala, India.

Grove, S. J. & Forster, L. (2011). A decade of change in the saproxylic beetle fauna of eucalypt logs in the Warra long-term log-decay experiment, Tasmania. 1. Description of the fauna and seasonality patterns. *Biodiversity and Conservation, 20*, 2149-2165.

Guariguata, M. R. & Ostertag, R. (2001). Neotropical secondary forest succession: Changes in structural and functional characteristics. *Forest ecology and management, 148*, 185-206.

Gunatilake, K. & Goff, M. L. (1989). Detection of organophosphate poisoning in a putrefying body by analyzing arthropod larvae. *Journal of Forensic Sciences, 34*, 714-716.

Gustafsson, L., Kouki, J. & Sverdrup-Thygeson, A. (2010). Tree retention as a conservation measure in clear-cut forests of northern Europe: A review of ecological consequences. *Scandinavian Journal of Forest Research, 25*, 295-308.

Halffter, G. & Edmonds, W. D. (1982). *The nesting behavior of dung beetles (Scarabaeinae).* Institute of Ecology, Mexico D.F., Mexico.

Hammond, H. J., Langor, D. W. & Spence, J. R. (2004). Saproxylic beetles (Coleoptera) using Populus in boreal aspen stands of western Canada: Spatiotemporal variation and conservation of assemblages. *Canadian Journal of Forest Research, 34*, 1-19.

Hammond, P. M. (1990). Insect abundance and diversity in the Dumoga-Bone National Park, North Sulawesi, with special reference to the beetle fauna of lowland rainforest in the Toraust region. In: Knight, W. J., Holloway, J. D. (Eds.) *Insects and the Rain Forests of Southeast Asia (Wallacea).*, 197-254. Royal Entomological Society of London, London.

Hammons, D. L., Kurtural, S. K., Newman, M. C. & Potter, D. A. (2009). Invasive Japanese beetles facilitate aggregation and injury by a native scarab pest of ripening fruits. *Proceedings of the National Academy of Sciences, 106*, 3686-3691.

Hans, K. R. (2011). Insect signature indicating corpse movement from urban to rural areas of northeast Ohio. Doctoral dissertation, Cleveland State University, Ohio.

Hanski, I. (1980). Patterns of beetle succession in droppings. *Annales Zoologici Fennici*, *17*, 17-25.

Hanski, I. (1983). Distributional ecology and abundance of dung and carrion-feeding beetles (Scarabaeidae) in tropical rain forests in Sarawak, Borneo. *Acta Zoologica Fennica*, *167*, 1-45.

Hanski, I. (1987a). Nutritional ecology of dung and carrion feeding insects. In: Slansky, F., Rodriguez, J.G. (Eds.) *Nutritional Ecology of Insects, Mites, Spiders and Related Invertebrates*., 837-884. John Wiley & Sons, New York.

Hanski, I. (1987b). Colonisation of ephemeral habitats. In: Crawley, M., Edwards, P.. Gray, A.J. (Eds.) *Colonization, Succession and Stability*. 155-186. Blackwell Scientific, Oxford, UK.

Hanski, I. (1987c). Carrion fly community dynamics: Patchiness, seasonality and coexistence. *Ecological Entomology*, *12*, 257-266.

Hanski, I. (1991). Competition in dung beetles. In: Hanski, I., Cambefort, Y. (Eds.) *Dung Beetle Ecology*., 305-329. Princeton University Press, Princeton, New Jersey.

Hanski, I. & Cambefort, Y. (1991a). *Dung Beetle Ecology*. Princeton University Press. Princeton, New Jersey.

Hanski, I. & Cambefort, Y. (1991b). Resource partitioning. In: Hanski, I., Cambefort, Y. (Eds.) *Dung Beetle Ecology*., 330-349. Princeton University Press, Princeton, New Jersey.

Hanski, I. & Hammond, P. M. (1986). Assemblages of carrion and dung Staphylinidae in tropical rainforests in Sarawak, Borneo. *Annales Entomologicae Fennicae*, *52*, 1-19.

Harmon, M. E., Baker, G. A., Spycher, G. & Greene, S. E. (1990). Leaf-litter decomposition in the *Picea/tsuga* forests of Olympic National Park, Washington, USA. *Forest Ecology and Management*, *31*, 55-66.

Harmon, M. E., Franklin, J. F., Swanson, F. J., Sollins, P., Gregory, S. V., Lattin, J. D., Anderson, N. H., Cline, S. P., Aumen, N. G., Sedell, J. R., Lienkaemper, G. W., Cromack, K. & Cummins, Jr., K. W. (1986). Ecology of coarse woody debris in temperate ecosystems. *Advances in Ecological Research*, *15*, 133-302.

Harmon, M. E. & Sexton, J. (1996). *Guidelines for Measurements of Woody Detritus in Forest Ecosystems*. U.S. Long Term Ecological Research Network Office, Publication no. 20. University of Washington, Seattle, Washington.

Hartley, S. & Shorrocks, B. (2002). A general framework for the aggregation model of coexistence. *Journal of Animal Ecology*, *71*, 651-662.

Hawkins, A. J. S. (1991). Protein turnover: A functional appraisal. *Functional Ecology*, *5*, 222-233.

Hernández, A. (2011). Internal dispersal of seed-inhabiting insects by vertebrate frugivores: A review and prospects. *Integrative Zoology*, *6*, 213-221.

Herrera, C. M. (1982). Defense of ripe fruit from pests: Its significance in relation to plant-disperser interactions. *American Naturalist*, *120*, 218-241.

Hewadikaram, K. A. & Goff, M. L. (1991). Effect of carcass size on rate of decomposition and arthropod succession patterns. *The American Journal of Forensic Medicine and Pathology*, *12*, 235-240.

Hocking, M. D., Ring, R. A. & Reimchen, T. E. (2006). Burying beetle *Nicrophorus investigator* reproduction on Pacific salmon carcasses. *Ecological Entomology*, *31*, 5-12.

Hofgaard, A. (1993). Structure and regeneration patterns in a virgin *Picea abies* forest in northern Sweden. *Journal of Vegetation Science*, *4*, 601-608.

Holter, P. (1977). An experiment on dung removed by *Aphodius* larvae (Scarabaeidae) and earthworms. *Oikos*, *28*, 130-136.

Holter, P. (1982). Resource utilization and local coexistence in a guild of scarabaeid dung beetles (*Aphodius* spp.). *Oikos*, *39*, 213-227.

Holter, P. & Scholtz, C. H. (2007). What do dung beetles eat? *Ecological Entomology*, *32*, 690-697.

Hope, F. W. (1836). Notice of several species of insects found in the heads of egyptian mummies. *Transactions of the Royal Entomological Society of London*, *1*, 6-13.

Horgan, F. G. (2002). Shady field boundaries and the colonization of dung by coprophagous beetles in Central American pasture. *Agriculture, Ecosystems and Environment*, *91*, 25-36.

Horgan, F. G. (2005a). Effects of deforestation on the diversity, biomass and function of dung beetles on the eastern slopes of the Peruvian Andes. *Forest Ecology and Management*, *216*, 117-133.

Horgan, F. G. (2005b). Aggregated distribution of resources creates competition refuges for rainforest dung beetles. *Ecography*, *28*, 603-618.

Horgan, F. G. (2006). Aggregation and coexistence of dung beetles in montane rain forest and deforested sites in central Peru. *Journal of Tropical Ecology*, *22*, 359-370.

Horgan, F. G. (2007). Dung beetles in pasture landscapes of Central America: Proliferation of synanthropogenic species and decline of forest specialists. *Biodiversity and Conservation*, *16*, 2149-2165.

Horgan, F. G. (2008). Dung beetle assemblages in forests and pastures of El Salvador: A functional comparison. *Biodiversity and Conservation*, *17*, 2961-2978.

Horgan, F. G. (2009). Invasion and retreat: Shifting assemblages of dung beetles amidst changing agricultural landscapes in central Peru. *Biodiversity and Conservation*, *18*, 3519-3541.

Horgan, F. G. & Berrow, S. D. (2004). Hooded crow foraging from dung pats: Implications for the structure of dung beetle assemblages. *Biology and Environment: Proceedings of the Royal Irish Academy*, *104*, 119-124.

Horgan, F. G. & Fuentes, R. C. (2005). Asymmetrical competition between Neotropical dung beetles and its consequences for assemblage structure. *Ecological Entomology*, *30*, 182-193.

Hövemeyer, K. & Schauermann, J. (2003). Succession of Diptera on dead beech wood: A 10-year study. *Pedobiologia*, *47*, 61-75.

Huang, M. S., Shieh, B. S. & Liang, S. H. (2001). Food selection and reproductive success of *Paragymnopleurus sinuatus* on three kinds of dung. *Formosan Entomology*, *21*, 329-337.

Huth, A., Dettner, K. (1990). Defense chemicals from abdominal glands of 13 rove beetle species of subtribe Staphylinia (Coleoptera: Staphylinidae, Staphylininae). *Journal of Chemical Ecology*, *16*, 2691-2711.

Hutton, S. A. & Giller, P. S. (2003). The effects of the intensification of agriculture on northern temperate dung beetle communities. *Journal of Applied Ecology*, *40*, 994-1007.

Hutton, S. A. & Giller, P. S. (2004). Intra- and interspecific aggregation of north temperate dung beetles on standardised and natural dung pads: The influence of spatial scale. *Ecological Entomology*, *29*, 594-605.

Hwang, C. & Turner, B. D. (2005). Spatial and temporal variability of necrophagous Diptera from urban to rural areas. *Medical and Veterinary Entomology*, *19*, 379-391.

Illingworth, J. F. (1926). Insects attracted to carrion in Southern California. *Proceedings of the Hawaiian Entomological Society*, *6*, 397-401.

Imbeau, L. & Desrochers, A. (2002). Foraging ecology and use of drumming trees by three-toed woodpeckers. *The Journal of Wildlife Management*, *66*, 222-231.

Irmler, U., Heller, K. & Warning, J. (1996). Age and tree species as factors influencing the populations of insects living in dead wood (Coleoptera, Diptera: Sciaridae, Mycetophilidae). *Pedobiologia*, *40*, 134-148.

Ives, A. R. (1991). Aggregation and coexistence in a carrion fly community. *Ecological Monographs*, *61*, 75-94.

Jacobs, J. M., Spence, J. R. & Langor, D. W. (2007a). Variable retention harvest of white spruce stands and saproxylic beetle assemblages. *Canadian journal of forest research*, *37*, 1631-1642.

Jacobs, J. M., Spence, J. R. & Langor, D. W. (2007b). Influence of boreal forest succession and dead wood qualities on saproxylic beetles. *Agricultural and Forest Entomology*, *9*, 3-16.

Jacobs, J. M. & Work, T. T. (2012). Linking deadwood-associated beetles and fungi with wood decomposition rates in managed black spruce forests. *Canadian Journal of Forest Research*, *42*, 1477-1490.

Janssen, P., Hebert, C. & Fortin, D. (2011). Biodiversity conservation in old-growth boreal forest: Black spruce and balsam fir snags harbour distinct assemblages of saproxylic beetles. *Biodiversity and Conservation*, *20*, 2917-2932.

Jaques, H. E. (1915). The fish-breeding coleoptera of Cedar Point. *The Ohio Naturalist*, *15*, 525-528.

Janzen, D. H. (1977). Why fruits rot, seeds mold, and meat spoils. *American Naturalist*, *111*, 691-713.

Jarvis, P. G. & Linder, S. (2007). Forests remove carbon dioxide from the atmosphere: Spruce forest tales! In: Freer-Smith, P. H., Broadmeadow, M. S. J., Lynch, J. M. (Eds.) *Forestry and Climate Change*., 60-72. CAB International, Wallingford, UK.

Johansen, H., Solum, M., Knudsen, G. K., Hagvar, E. B., Norli, H. R. & Aak, A. (2014). Blow fly responses to sociochemicals produced by decaying carcasses. *Medical and Veterinary Entomology*, *28*, 26-34.

Johnson, D. T., Lewis, B. A., Bryant, R. J., Liyanage, R., Lay, J. O. & Pszczolkowski, M. A. (2009). Attractants for the green June beetle (Coleoptera: Scarabaeidae). *Journal of Economic Entomology*, *102*, 2224-2232.

Johnson, M. D. (1975). Seasonal and microseral variations in the insect populations on carrion. *American Midland Naturalist*, *93*, 79-90.

Johnston, W. & Villeneuve, G. (1897). On the medico-legal application of entomology. *Montreal Medicine Journal*, *26*, 81-90.

Jones, C. C. & del Moral, R. (2005). Patterns of primary succession on the foreland of Coleman Glacier, Washington, USA. *Plant Ecology*, *180*, 105-116.

Jonsell, M., Hansson, J. & Wedmo, L. (2007). Diversity of saproxylic beetle species in logging residues in Sweden – Comparisons between tree species and diameters. *Biological Conservation*, *138*, 89-99.

Jonsell, M., Schroeder, M. & Larsson, T. (2003). The saproxylic beetle *Bolitophagus reticulatus*: Its frequency in managed forests, attraction to volatiles and flight period. *Ecography*, *26*, 421-428.

Jonsell, M., Schroeder, M. & Weslien, J. (2005). Saproxylic beetles in high stumps of spruce: Fungal flora important for determining the species composition. *Scandinavian Journal of Forest Research*, *20*, 54-62.

Jonsell, M., Weslien, J. & Ehnström, B. (1998). Substrate requirements of red-listed saproxylic invertebrates in Sweden. *Biodiversity and Conservation*, *7*, 749-764.

Jönsson, A. M., Appelberg, G., Harding, S. & Bärring, L. (2009). Spatio-temporal impact of climate change on the activity and voltinism of the spruce bark beetle, *Ips typographus*. *Global Change Biology*, *15*, 486-499.

Jonsson, B. G., Kruys, N. & Ranius, T. (2005). Ecology of species living on dead wood – lessons for dead wood management. *Silva Fennica*, *39*, 289-309.

Kalinova, B., Podskalska, H., Růžička, J. & Hoskovec, M. (2009). Irresistible bouquet of death—how are burying beetles (Coleoptera: Silphidae: Nicrophorus) attracted by carcasses. *Naturwissenschaften*, *96*, 889-899.

Karjalainen, H. (1991). Elävä Metsä. [*Living forest. Threatened Nature and Forest management*] Forssan kijapaino Oy.

Kautz, G. & Topp, W. (2000). Acquisition of microbial communities and enhanced availability of soil nutrients by the isopod *Porcellio scaber* (Latr.)(Isopoda: Oniscidea). *Biology and Fertility of Soils*, *31*, 102-107.

Kelly, J. A., Van Der Linde, T. C. & Anderson, G. S. (2009). The influence of clothing and wrapping on carcass decomposition and arthropod succession during the warmer seasons in Central South Africa. *Journal of Forensic Sciences*, *54*, 1105-1112.

Keszthelyi, S. (2012). Evaluation of flight phenology and number of generations of the four-spotted sap beetle, *Glischrochilus quadrisignatus* in Europe. *Bulletin of Insectology*, *65*, 9-16.

Key, R. (1993). What are saproxylic invertebrates? In: Kirby, P., Drake, C. M. (Eds.) *Dead Wood Matters: The Ecology and Conservation of Saproyxlic Insects*., 5. English Nature Science, No. 7, Peterborough, UK.

Koch, R. L., Burkness, E. C., Burkness, S. J. W. & Hutchison, W. D. (2004). Phytophagous preferences of the multicolored Asian lady beetle (Coleoptera: Coccinellidae) for autumn-ripening fruit. *Journal of Economic Entomology*, *97*, 539-544.

Koch, R. L. & Galvan, T. L. (2008). Bad side of a good beetle: the North American experience with *Harmonia axyridis*. In: Roy, H. E., Wajnberg, E. (Eds.) *From Biological Control to Invasion: The Ladybird Harmonia axyridis as a Model Species*., 23-35. Springer, Netherlands.

Koivula, M. (2002). Boreal carabid-beetle (Coleoptera, Carabidae) assemblages in thinned uneven-aged and clear-cut spruce stands. *Annales Zoologici Fennici*, *39*, 131-149.

Koskela, H. & Hanski, I. (1977). Structure and succession in a beetle community inhabiting cow dung. *Annales Zoologici Fennici*, *14*, 204-223.

Krijger, C. L. & Sevenster, J. G. (2001). Higher species diversity explained by stronger spatial aggregation across six neotropical *Drosophila* communities. *Ecology Letters*, *4*, 106-115.

Kulshrestha, P. & Satpathy, D. K. (2001). Use of beetles in forensic entomology. *Forensic Science International*, *120*, 15-17.

Kuramae, E. E., Gamper, H. A., Yergeau, E., Piceno, Y. M., Brodie, E. L., DeSantis, T. Z., Andersen, G. L., van Veen, J. A. & Kowalchuk, G. A. (2010). Microbial secondary succession in a chronosequence of chalk grasslands. *The ISME Journal*, *4*, 711-715.

Lachat, T., Nagel, P., Cakpo, Y., Attignon, S., Goergen, G., Sinsin, B. & Peveling, R. (2006). Dead wood and saproxylic beetle assemblages in a semi-deciduous forest in Southern Benin. *Forest Ecology and Management*, *225*, 27-38.

Lagisz, M. & Laskowski, R. (2008). Evidence for between-generation effects in carabids exposed to heavy metals pollution. *Ecotoxicology*, *17*, 59-66.

Lambert, M. & Turner, J. (2000). Species selection and productivity. In: Lambert, M., Turner, J. (Eds.) *Commercial Forest Plantations on Saline Lands*., 68-89. Csiro Publishing, Collingwood, Australia.

Landin, B. O. (1961). Ecological studies on dung beetles. *Opuscula Entomologica, Supplementum*, *19*, 1-228.

Langan, A. M., Taylor, A. & Wheater, C. P. (2004). Effects of metaldehyde and methiocarb on feeding preferences and survival of a slug predator (*Pterostichus melanarius* (F.): Carabidae, Pterostichini). *Journal of Applied Entomology*, *128*, 51-55.

Lassauce, A., Lieutier, F. & Bouget, C. (2012). Woodfuel harvesting and biodiversity conservation in temperate forests: Effects of logging residue characteristics on saproxylic beetle assemblages. *Biological Conservation*, *147*, 204-212.

Lassauce, A., Paillet, Y., Jactel, H. & Bouget, C. (2011). Deadwood as a surrogate for forest biodiversity: Meta-analysis of correlations between deadwood volume and species richness of saproxylic organisms. *Ecological Indicators*, *11*, 1027-1039.

Leblanc, H. N. & Strongman, D. B. (2002). Carrion insects associated with small pig carcasses during fall in Nova Scotia. *Canadian Society of Forensic Science Journal*, *35*, 145-152.

Lee, C. & Wall, R. (2006). Distribution and abundance of insects colonizing cattle dung in South West England. *Journal of Natural History*, *40*, 1167-1177.

Lee, S. I., Spence, J. R. & Langor, D. W. (2014). Succession of saproxylic beetles associated with decomposition of boreal white spruce logs. *Agricultural and Forest Entomology*. doi: 10.1111/afe.12069.

Legner, E. F. & Olton, G. S. (1970). Worldwide survey and comparison of adult predator and scavenger insect populations associated with domestic animal manure where livestock is artificially congregated. *Hilgardia*, *40*, 225-266.

Lindenmayer, D. & Hobbs, R. J. (2008). *Managing and Designing Landscapes for Conservation: Moving from Perspectives to Principles*., 409-422. Blackwell Publishing Ltd., Oxford, UK.

Lindhe, A., Lindelöw, Å. & Åsenblad, N. (2005). Saproxylic beetles in standing dead wood density in relation to substrate sun-exposure and diameter. *Biodiversity & Conservation*, *14*, 3033-3053.

Lomolino, M. V., Creighton, J. C., Schnell, G. D. & Certain, D. L. (1995). Ecology and conservation of the endangered American burying beetle (*Nicrophorus americanus*). *Conservation Biology*, *9*, 605-614.

Lussenhop, J., Kumer, R., Wicklow, D. T. & Lloyd, J. E. (1980). Insect effects on bacteria and fungi in cattle dung. *Oikos*, *34*, 54-58.

Mądra, A., Konwerski, S. & Matuszewski, S. (2014). Necrophilous Staphylininae (Coleoptera: Staphylinidae) as indicators of season of death and corpse relocation. *Forensic Science International*, *242*, 32-37.

Magura, T., Nagy, D. & Tóthmérész, B. (2013). Rove beetles respond heterogeneously to urbanization. *Journal of Insect Conservation*, *17*, 715-724.

Magura, T., Tóthmérész, B. & Bordán, Z. (2000). Effects of nature management practice on carabid assemblages (Coleoptera: Carabidae) in a non-native plantation. *Biological Conservation*, *93*, 95-102.

Magura, T., Tóthmérész, B. & Elek, Z. (2004a). Effects of leaf-litter addition on carabid beetles in a non-native Norway spruce plantation. *Acta Zoologica Academiae Scientiarum Hungaricae*, *50*, 9-23.

Magura, T., Tóthmérész, B. & Elek, Z. (2005). Impacts of leaf-litter addition on carabids in a conifer plantation. *Biodiversity and Conservation*, *14*, 475-491.

Magura, T., Tóthmérész, B. & Molnár, T. (2004b). Changes in carabid beetle assemblages along an urbanisation gradient in the city of Debrecen, Hungary. *Landscape Ecology*, *19*, 747-759.

Majka C. G., Klimaszewski J., Michaud J. P. & Moreau G. (2009a). *Aleochara sekanai*: A significant southern and eastern range extension (Coleoptera: Staphylinidae). *The Coleopterists Bulletin*, *63*, 454-455.

Majka, C. G., Michaud J. P. & Moreau G. (2009b). Adventive species of *Quedius* (Coleoptera: Staphylinidae) in North America: A survey and new Canadian record. *ZooKeys*, *22*, 341-345.

Majka, C. G., Michaud J. P., Moreau G. & Smetana A. (2009c). *Philonthus hepaticus* (Coleoptera: Staphylinidae) in eastern Canada: Are distribution gaps distinctive features or collecting artifacts? *ZooKeys*, *22*, 347-354.

Marchiori, C. H., Caldas, E. R. & Almeida, K. G. S. (2003). Succession of Scarabaeidae on bovine dung in Itumbiara, Goiás, Brazil. *Neotropical Entomology*, *32*, 173-176.

Márquez, J. (2003). Ecological patterns in necrophilous Staphylinidae (Insecta: Coleoptera) from Tlayacapan, Morelos, México. *Acta Zoológica Mexicana*, *89*, 69-83.

Martikainen, P., Siitonen, J., Punttila, P., Kaila, L. & Rauh, J. (2000). Species richness of Coleoptera in mature managed and old-growth boreal forests in southern Finland. *Biological Conservation*, *94*, 199-209.

Martínez, M. I., Cruz, R. M. & Lumaret, J. P. (2000). Efecto del diferente manejo de los pastizales y del ganado sobre los escarabajos coprófagos *Ataenius apicalis* Hinton y *Ataenius sculptor* Harold (Scarabaeidae: Aphodiinae: Eupariini). *Acta Zoológica Mexicana*, *80*, 185-196.

Martinson, H. M. & Raupp, M. J. (2013). A meta-analysis of the effects of urbanization on ground beetle communities. *Ecosphere*, *4*, 60.

Maser, C., Anderson, R., Cromack, Jr., K., Williams, J. T. & Martin, R. E. (1979). Dead and down woody material. In: Thomas, J.W. (Ed.). *Wildlife Habitats in Managed Forests: the Blue Mountains of Oregon and Washington.*, 78-95. *USDA Agricultural Handbook*, *553*, Washington, DC.

Matuszewski, S. (2011). Estimating the pre-appearance interval from temperature in *Necrodes littoralis* L. (Coleoptera: Silphidae). *Forensic Science International*, *212*, 180-88.

Matuszewski, S. (2012). Estimating the preappearance interval from temperature in *Creophilus maxillosus* L. (Coleoptera: Staphylinidae). *Journal of Forensic Sciences*, *57*, 136-145.

Matuszewski, S. & Szafałowicz, M. (2013). Temperature-dependent appearance of forensically useful beetles on carcasses. *Forensic Science International*, *229*, 92-99.

Mayor, X. & Rodà, F. (1993). Growth response of holm oak (*Quercus ilex* L) to commercial thinning in the Montseny mountains (NE Spain). *Annales des Sciences Forestières*, *50*, 247-256.

McDonnell, M. J., Pickett, S. T., Groffman, P., Bohlen, P., Pouyat, R. V., Zipperer, W. C., Parmelee, R. W., Carreiro, M. M. & Medley, K. (1997). Ecosystem processes along an urban-to-rural gradient. *Urban Ecosystems*, *1*, 21-36.

McKinney, M. L. (2008). Effects of urbanization on species richness: A review of plants and animals. *Urban Ecosystems*, *11*, 161-176.

Mégnin, P. (1883). L'application de l'entomologie à la médecine légale. *Gazette Hebdomadaire de Médecine et de Chirurgie*, *29*, 480-482.

Mégnin, P. (1887). La faune des tombeaux. *Compte Rendu Hebdomadaire des Séances de l'Académie des Sciences*, *105*, 948-951.

Mégnin, P. (1894). *La faune des cadavres. Application de l'Entomologie à la Médecine Légale*. G. Masson, Gauthier-Villars et Fils, Paris, France.

Menéndez, R. & Gutiérrez, D. (1999). Heterotrophic succession within dung-inhabiting beetle communities in northern Spain. *Acta Oecologica*, *20*, 527-535.

Michaud, J. P., Majka C. G., Privé J. P. & Moreau G. (2010). Natural and anthropogenic changes in the insect fauna associated with carcasses in the North American Maritime lowlands. *Forensic Science International*, *202*, 64-70.

Michaud, J. P. & Moreau G. (2009). Predicting the visitation of carcasses by carrion-related insects under different rates of degree-day accumulation. *Forensic Science International*, *185*, 78-83.

Michaud, J. P. & Moreau, G. (2011). A statistical approach based on accumulated degree-days to predict decomposition-related processes in forensic studies. *Journal of Forensic Sciences*, *56*, 229-232.

Michaud, J. P. & Moreau, G. (2013). Effect of variable rates of daily sampling of fly larvae on decomposition and carrion-insect community assembly: Implications for forensic entomology field study protocols. *Journal of Medical Entomology*, *50*, 890-897.

Michaud, J. P., Schoenly, K. G. & Moreau, G. (2012). Sampling flies or sampling flaws? Experimental design and inference strength in forensic entomology. *Journal of Medical Entomology*, *49*, 1-10.

Micozzi, M. S. (1986). Experimental study of postmortem change under field conditions: Effects of freezing, thawing, and mechanical injury. *Journal of Forensic Science*, *31*, 953-961.

Milad, M., Schaich, H., Bürgi, M. & Konold, W. (2011). Climate change and nature conservation in Central European forests: A review of consequences, concepts and challenges. *Forest Ecology and Management*, *261*, 829-843.

Miller, H. G. (1984). Dynamics of nutrient cycling in platation ecosystems. In: Boven, G. D., Nambiar, E. K. S. (Eds.) *Nutrition of Plantation Forest.*, 53-78. Acadamic Press, London, UK.

Miller, M. L., Lord, W. D., Goff, L. M. & Donnelly, B. (1994). Isolation of amitriptyline and nortriptyline from fly puparia (Phoridae) and beetle exuviae (Dermestidae) associated with mummified human remains. *Journal of Forensic Sciences*, *39*, 1305-1305.

Milne, L. J. & Milne, M. (1976). The social behavior of the burying beetles. *Scientific American*, *235*, 84-89.

Mohr, C. O. (1943). Cattle droppings as ecological units. *Ecological Monographs*, *13*, 275-298.

Montes de Oca, E. & Halffter, G. (1995). Daily and seasonal activities of a guild of the coprophagous, burrowing beetle (Coleoptera Scarabaeidae Scarabaeinae) in tropical grassland. *Tropical Zoology*, *8*, 159-180.

Monthei, D. R. (2009). Entomotoxicological and thermal factors affecting the development of forensically important flies. PhD dissertation, Virginia Polytechnic Institute and State University, Virginia.

Moreau, G., Eveleigh, E. S., Lucarotti, C. J. & Quiring, D. T. (2006). Ecosystem alteration modifies the relative strengths of bottom-up and top-down forces in a herbivore population. *Journal of Animal Ecology*, *75*, 853-861.

Moreau, G., Michaud, J.-P., & Schoenly, K. G. (2015). Experimental Design, Statistics and Computer Modeling. In: Tomberlin, J. K., Benbow, E. (Eds.) *International Dimensions and Frontiers of Forensic Entomology.*, 205-229. Taylor & Francis Books Inc, NY.

Motter, M. G. (1898). A contribution to the study of the fauna of the grave. A study of one hundred and fifty disinterments, with some additional experimental observations. *Journal of the New York Entomological Society*, *6*, 201-231.

Moura, M. O., Monteiro-Filho E. L. A. & Carvalho, C. J. B. (2005). Heterotrophic succession in carrion arthropod assemblages. *Brazilian Archives of Biology and Technology*, *48*, 477-86.

Müller, J. & Bütler, R. (2010). A review of habitat thresholds for dead wood: A baseline for management recommendations in European forests. *European Journal of Forest Research*, *129*, 981-992.

Müller, J., Hothorn, T. & Pretzsch, H. (2007). Long-term effects of logging intensity on structures, birds, saproxylic beetles and wood-inhabiting fungi in stands of European beech *Fagus sylvatica* L. *Forest Ecology and Management*, *242*, 297-305.

Murcia, C. (1995). Edge effects in fragmented forests: Implications for conservation. *Trends in Ecology & Evolution*, *10*, 58-62.

Murphy, J. F., Giller, P. S. & Horan, M. A. (1998). Spatial scale and the aggregation of stream macroinvertebrates associated with leaf packs. *Freshwater Biology*, *39*, 325-337.

Murray, S. N. & Littler, M. M. (1978). Patterns of algal succession in a perturbated marine intertidal community. *Journal of Phycology*, *14*, 506-512.

Myers, L. (2004). *Sap Beetles of Florida, Nitidulidae (Insecta: Coleoptera: Nitidulidae).* University of Florida Cooperative Extension Service, Institute of Food and Agricultural Sciences, EDIS.

Nadeau, P., Majka, C. G., Moreau, G. (*in press*). Short-term response of coleopteran assemblages to thinning-induced differences in dead wood volumes. *Forest Ecology and Management.*

Newberry, J. E., Lewis, K. J. & Walters, M. B. (2004). Estimating time since death of *Picea glauca* × *P. engelmannii* and *Abies lasiocarpa* in wet cool sub-boreal spruce forest in east-central British Columbia. *Canadian Journal of Forest Research*, *34*, 931-938.

Nichols, E., Gardner, T. A., Peres, C. A., Spector, S. & The Scarabaeinae Research Network (2009). Co- declining mammals and dung beetles: An impending ecological cascade. *Oikos*, *118*, 481-487.

Nichols, E., Larsen, T., Spector, S., Davis, A. L., Escobar, F., Favila, M., Vulinec, K. & The Scarabaeinae Research Network (2007). Global dung beetle response to tropical forest modification and fragmentation: A quantitative literature review and meta-analysis. *Biological Conservation*, *137*, 1-19.

Nichols, E., Spector, S., Louzada, J., Larsen, T., Amezquita, S., Favila, M. E. The Scarabaeinae Research Network (2008). Ecological functions and ecosystem services provides by Scarabaeinae dung beetles. *Biological Conservation*, *141*, 1461-1474.

Niemelä, J., Langor, D. & Spence, J. R. (1993). Effects of Clear-Cut Harvesting on Boreal Ground-Beetle Assemblages (Coleoptera: Carabidae) in Western Canada. *Conservation Biology*, *7*, 551-561.

Nieto, A. & Alexander, K. N. A. (2010). *European Red List of Saproxylic Beetles*. Publications Office of the European Union, Luxembourg.

Odeyemi, O. O. (1997). Interspecific competition between the beetles *Dermestes maculatus* DeGeer and *Necrobia rufipes* DeGeer in dried fish. *International Journal of Tropical Insect Science*, *17*, 213-220.

O'Hea, N. M., Kirwan, L., Giller, P. S. & Finn, J. A. (2010). Lethal and sub-lethal effects of ivermectin on north temperate dung beetles, *Aphodius ater* and *Aphodius rufipes* (Coleoptera: Scarabaeidae). *Insect Conservation and Diversity*, *3*, 24-33.

Ohno, M., Arai, S. & Watanabe, M. (1990). Seaweed succession on artificial reefs on different bottom substrata. *Journal of applied phycology*, *2*, 327-332.

Økland, B., Skarpaas O. & Kausrud, K. (2009). Threshold facilitations of interacting species. *Population Ecology*, *51*, 513-523.

Olff, H., Huisman, J. & Van Tooren, B. F. (1993). Species dynamics and nutrient accumulation during early primary succession in coastal sand dunes. *Journal of Ecology*, *81*, 693-706.

Paarmann, W., Adis, J., Stork, N., Gutzmann, B., Stumpe, P., Staritz, B., Bolte, H., Küppers, S., Holzkamp, K., Niers, C. & da Fonseca, C. R. (2001). The structure of ground beetle assemblages (Coleoptera: Carabidae) at fig fruit falls (Moraceae) in a terra firme rain forest near Manaus (Brazil). *Journal of Tropical Ecology*, *17*, 549-561.

Padayachi, Y., Procheş, Ş. & Ramsay, L. F. (2014). Beetle assemblages of indigenous and alien decomposing fruit in subtropical Durban, South Africa. *Arthropod-Plant Interactions*, *8*, 135-142.

Pawson, S. M., Brockerhoff, E. G., Meenken, E. D. & Didham, R. K. (2008). Non-native plantation forests as alternative habitat for native forest beetles in a heavily modified landscape. *Biodiversity and Conservation*, *17*, 1127-1148.

Payne, J. A. & Mason, W. R. M. (1971). Hymenoptera associated with pig carrion. *Proceedings of the Entomological Society of Washington*, *73*, 132-141.

Payne, J. A. & King, E. W. (1968). Arthropod succession and decomposition of buried pigs. *Nature*, *219*, 1180-1181.

Payne, J. A. (1965). A summer carrion study of the baby pig *Sus scrofa* Linnaeus. *Ecology*, *46*, 592-602.

Pedley, S. M., Martin, R. D., Oxbrough, A., Irwin, S., Kelly, T. C. & O'Halloran, J. (2014). Commercial spruce plantations support a limited canopy fauna: Evidence from a multi taxa comparison of native and plantation forests. *Forest Ecology and Management*, *314*, 172-182.

Perez, A. E., Haskell, N. H. & Wells, J. D. (2014). Evaluating the utility of hexapod species for calculating a confidence interval about a succession based postmortem interval estimate. *Forensic Science International*, *241*, 91-95.

Perez, S. P. , Duque, P. & Wolff, M. (2005). Successional behavior and occurrence matrix of carrion-associated arthropods in the urban area of Medellín, Colombia. *Journal of Forensic Sciences*, *50*, 448-454.

Poole, R. W. (1974). *An Introduction to Quantitative Ecology*. McGraw-Hill, New York, New York.

Posamentier, H. G., Clark, S. S., Hain, D. L. & Recher, H. F. (1981). Succession following wildfire in coastal heathland (Nadgee Nature Reserve NSW). *Australian Journal of Ecology*, *6*, 165-175.

Potter, D. A. & Held, D. W. (2002). Biology and management of the Japanese beetle. *Annual Review of Entomology*, *47*, 175-205.

Pukowski, E. (1933). Ökologische untersuchungen an *necrophorus* F. *Zeitschrift für Morphologie und Ökologie der Tiere*, *27*, 518-586.

Putman R. J. (1978). The role of carrion-frequenting arthropods in the decay process. *Ecological Entomology*, *3*, 133-139.

Putman, R. J. (1983). *Carrion and Dung: The Decomposition of Animal Wastes*. The Institute of Biology's Studies in Biology No. 156. Edward Arnold, London, UK.

Reed, H. B. (1958). A study of dog carcass communities in Tennessee, with special reference to the insects. *American Midland Naturalist*, *59*, 213-245.

Renvall, P. (1995). Community structure and dynamics of wood-rotting Basidiomycetes on decomposing conifer trunks in northern Finland. *Karstenia*, *35*, 1-51.

Richards, O. W. & Davies, R. G. (1977). *Imm's General Textbook of Entomology: Classification and Biology, Volume 2*. Tenth edition. Chapman & Hall, London, UK.

Richardson, M. S. & Goff, M. L. (2001). Effects of temperature and intraspecific interaction on the development of *Dermestes maculatus* (Coleoptera: Dermestidae). *Journal of medical entomology*, *38*, 347-351.

Ridgeway, J. A., Midgley, J. M., Collett, I. J. & Villet, M. H. (2014). Advantages of using development models of the carrion beetles *Thanatophilus micans* (Fabricius) and *T. mutilatus* (Castelneau)(Coleoptera: Silphidae) for estimating minimum post mortem intervals, verified with case data. *International journal of legal medicine*, *128*, 207-220.

Riffell, S., Verschuyl, J., Miller, D. & Wigley, T. B. (2011). Biofuel harvests, coarse woody debris, and biodiversity–a meta-analysis. *Forest Ecology and Management*, *261*, 878-887.

Rodriguez, W. C. & Bass, W. M. (1983). Insect activity and its relationship to decay rates of human cadavers in East Tennessee. *Journal of Forensic Science*, *28*, 423-432.

Rondon, S. I., Price, J. F. & Cantliffe, D. J. (2011). *Sap Beetle (Coleoptera: Nitidulidae) Management in Strawberries*. University of Florida Cooperative Extension Service, Institute of Food and Agricultural Sciences, EDIS.

Rozen, D. E., Engelmoer, D. J. P. & Smiseth, P. T. (2008). Antimicrobial strategies in burying beetles breeding on carrion. *Proceedings of the National Academy of Sciences*, *105*, 17890-17895.

Sabu, T. K., Vinod, K. V. & Vineesh, P. J. (2006). Guild structure, diversity and succession of dung beetles associated with Indian elephant dung in South Western Ghats forests. *Journal of Insect Science*, *6*, 1-12.

Saint-Germain, M., Drapeau, P. & Buddle, C. (2007). Host-use patterns of saproxylic phloeophagous and xylophagous Coleoptera adults and larvae along the decay gradient in standing dead black spruce and aspen. *Ecography*, *30*, 737-748.

Saint-Germain, M., Larrivée, M., Drapeau, P., Fahrig, L. & Buddle, C. M. (2005). Short-term response of ground beetles (Coleoptera: Carabidae) to fire and logging in a spruce-dominated boreal landscape. *Forest Ecology and Management*, *212*, 118-126.

Sallabanks, R. & Courtney, S. P. (1993). On fruit-frugivore relationships: Variety is the spice of life. *Oikos*, *68*, 567-570.

Savely, Jr., H. E.. (1939). Ecological relations of certain animals in dead pine and oak logs. *Ecological Monographs*, *9*, 322-385.

Schaefer, M. (1996). Die Artenzahl von Waldinsekten: muster und mögliche Ursachen der Diversität. *Mitteilungen der Deutschen Gesellschaft für Allgemeine und Angewandte Entomology*, *10*, 387-395.

Schelhaas, M. J., Nabuurs, G. J. & Schuck, A. (2003). Natural disturbances in the European forests in the 19th and 20th centuries. *Global Change Biology*, *9*, 1620-1633.

Schoenly, K. G., Shahid, S. A., Haskell, N. H. & Hall, R. D. (2005). Does carcass enrichment alter community structure of predaceous and parasitic arthropods? A second test of the arthropod saturation hypothesis at the Anthropology Research Facility in Knoxville, Tennessee. *Journal of Forensic Science*, *50*, 134-142.

Schoenly, K. Haskell, N. H., Hall, R. D. & Gbur, R. J. (2007). Comparative performance and complementarity of four sampling methods and arthropod preference tests from human and porcine remains at the Forensic Anthropology Center in Knoxville, Tennessee. *Journal of Medical Entomology*, *44*, 881-894.

Schoenly, K. & Reid, W. (1987). Dynamics of heterotrophic succession in carrion arthropod assemblages: Discrete seres or a continuum of change? *Oecologia*, *73*, 192-202.

Schoenly, K. & Reid, W. (1989). Dynamics of heterotrophic succession in carrion revisited: A reply to Boulton and Lake, (1988). *Oecologia*, *79*, 140-142.

Schroeder, H., Klotzbach, H., Oesterhelweg, L. & Püschel, K. (2002). Larder beetles (Coleoptera, Dermestidae) as an accelerating factor for decomposition of a human corpse. *Forensic Science International*, *127*, 231-236.

Schroeder, L. M. (1997). Oviposition behavior and reproductive success of the cerambycid *Acanthocinus aedilis* in the presence and absence of the bark beetle *Tomicus Piniperda*. *Entomologia Experimentalis et Applicata*, *82*, 9-17.

Schulze, E. D. (2000). *Carbon and Nitrogen Cycling in European Forest Ecosystems, Ecological Studies, Volume, 142*. Springer, New York, New York.

Schwarz, H. H. & Koulianos, S. (1998). When to leave the brood chamber? Routes of dispersal in mites associated with burying beetles. *Experimental and Applied Acarology*, *22*, 621-631.

Scott, M. P. (1998). The ecology and behavior of burying beetles. *Annual Review of Entomology*, *43*, 595-618.

Scott, M. P., Lee, W. J. & van Der Reijden, E. D. (2007). The frequency and fitness consequences of communal breeding in a natural population of burying beetles: A test of reproductive skew. *Ecological Entomology*, *32*, 651-661.

Scott, M. P. & Traniello, J. F. A. (1990). Behavioral and ecological correlates of male and female parental care and reproductive success in the burying beetles (*Nicrophorus* spp.). *Animal Behavior*, *39*, 247-283.

Segura, N. A., Usaquén, W., Sánchez, M. C., Chuaire, L. & Bello, F. (2009). Succession pattern of cadaverous entomofauna in a semi-rural area of Bogotá, Colombia. *Forensic Science International*, *187*, 66-72.

Seidl, R., Rammer, W., Jaeger, D. & Lexer, M. J. (2008). Impact of bark beetle (*Ips typographus* L.) disturbance on timber production and carbon sequestration in different management strategies under climate change. *Forest Ecology and Management*, *256*, 209-220.

Sevenster, J. G. & van Alphen, J. J. M. (1993). A life history trade-off in *Drosophila* species and community structure in variable environments. *Journal of Animal Ecology*, *62*, 720-736.

Sevenster, J. G. & van Alphen, J. J. M. (1996). Aggregation and coexistence. II A neotropical *Drosophila* community. *Journal of Animal Ecology*, *65*, 308-324.

Shahid, S. A., Schoenly, K., Haskell, N. H., Hall, R. D. & Zhang, W. (2003). Carcass enrichment does not alter decay rates or arthropod community structure: A test of the arthropod saturation hypothesis at the Anthropology Research Facility in Knoxville, Tennessee. *Journal of Medical Entomology*, *40*, 559-569.

Shelford, V. E. (1912). Ecological succession IV. Vegetation and the control of land animal communities. *The Biological Bulletin*, *23*, 59-99.

Shorrocks, B. & Rosewell, J. (1987). Spatial patchiness and community structure: Coexistence and guild size in Drosophilids in ephemeral resources. In: Gee, J. H. R., Giller, P. S. (Eds.) *Organisation of Communities: Past and Present.*, 29-51. Blackwell Scientific, Oxford, UK.

Siitonen, J. (2001). Forest management, coarse woody debris and saproxylic organisms: Fennoscandian boreal forests as an example. *Ecological Bulletins*, *49*, 11-41.

Siitonen, J. & Saaristo, L. (2000). Habitat requirements and conservation of *Pytho kolwensis*, a beetle species of old-growth boreal forest. *Biological Conservation*, *94*, 211-220.

Similä, M., Kouki, J. & Martikainen, P. (2003). Saproxylic beetles in managed and seminatural Scots pine forests: Quality of dead wood matters. *Forest Ecology and Management*, *174*, 365-381.

Simmons, L. W. & Ridsdill-Smith, J. (2011). Reproductive competition and its impact on the evolution and ecology of dung beetles. In: Simmons, L. W., Ridsdill-Smith, T. J. (Eds.) *Ecology and Evolution of Dung Beetles.*, 1-20. John Wiley & Sons Ltd., Chichester, UK.

Slade, E. M., Mann, D. J., Villanueva, J. F. & Lewis, O. T. (2007). Experimental evidence for the effects of dung beetle functional group richness and composition on ecosystem function in a tropical forest. *Journal of Animal Ecology*, *76*, 1094-1104.

Smith, C. R. & Baco, A. R. (2003). Ecology of whale falls at the deep-sea floor. *Oceanography and Marine Biology: An Annual Review*, *41*, 311-354.

Sousa, W. P. (1979). Experimental investigations of disturbance and ecological succession in a rocky intertidal algal community. *Ecological Monographs*, *49*, 227-254.

Speight, M. C. D. (1989). *Saproxylic Invertebrates and their Conservation.* Nature and Environment Series, 42. Council of Europe, Strasbourg, France.

Springett, B. P. (1968). Aspects of the relationship between burying beetles, *Necrophorus* spp., and the mite, *Poecilochirus necrophori* Vitz. *Journal of Animal Ecology*, *37*, 417-424.

Stenbacka, F., Hjältén, J., Hilszczanski, J. & Dynesius, M. (2010). Saproxylic and non-saproxylic beetle assemblages in boreal spruce forests of different age and forestry intensity. *Ecological Applications*, *20*, 2310-2321.

Stephens, S. & Wagner, M. R. (2007). Forest plantations and biodiversity: A fresh perspective. *Journal of Forestry*, *105*, 307-313.

Stokland, J. N., Siitonen, J. & Jonsson, B. G. (2012). *Biodiversity in Dead Wood*. Cambridge University Press, New York, New York.

Stone, D., Jepson, P., Kramarz, P. & Laskowski, R. (2001). Time to death response in carabid beetles exposed to multiple stressors along a gradient of heavy metal pollution. *Environmental Pollution*, *113*, 239-244.

Strümpher, W. P., Farrell, J. & Scholtz, C. H. (2014). (In Press) Trogidae (Coleoptera: Scarabaeoidea) in forensic entomology: occurrence of known and new species in Queensland, Australia. *Austral Entomology*.

Stubbs, A. E. (1982). Hoverflies as primary woodland indicators with reference to Warncliffe Wood. *Sorby Record*, *20*, 62-67.

Su, N. Y. & Scheffrahn, R. H. (1990). Efficacy of sulfuryl fluoride against four beetle pests of museums (Coleoptera: Dermestidae, Anobiidae). *Journal of Economic Entomology*, *83*, 879-882.

Sverdrup-Thygeson, A., Gustafsson, L. & Kouki, J. (2014). Spatial and temporal scales relevant for conservation of dead-wood associated species: Current status and perspectives. *Biodiversity and Conservation*, *23*, 513-535.

Swift, M. J. (1977). The roles of fungi and animals in the immobilization and release of nutrient elements from decomposing branchwood. *Ecological Bulletins*, *25*, 193-202.

Swift, M. J., Heal, O. W. & Anderson, M. J. (1979). *Decomposition in Terrestrial Ecosystems*. Blackwell Scientific, Oxford, UK.

Swinburne, T. R. (1973). The resistance of immature bramley's seedling apples to rotting by *Nectria galligena* Bres. In: Byrde, R. J. W., Cutting, C.V. (Eds.) *Fungal Pathogenicity and the Plant's Response*., 365-379. Academic Press Inc., London, UK.

Tabor, K., Fell, R. & Brewster, C. (2005a). Insect fauna visiting carrion in Southwest Virginia. *Forensic Science International*, *150*, 73-80.

Tabor, K. L., Fell, R. D., Brewster, C. C., Pelzer, K. & Behonick, G. S. (2005b). Effects of antemortem ingestion of ethanol on insect successional patterns and development of *Phormia regina* (Diptera: Calliphoridae). *Journal of Medical Entomology*, *42*, 481-489.

Tang, A. M., Corlett, R. T. & Hyde, K. D. (2005). The persistence of ripe fleshy fruits in the presence and absence of frugivores. *Oecologia*, *142*, 232-237.

Tibbett, M. & Carter, D. O. (2009). Research in forensic taphonomy: A soil-based perspective. In: Ritz, K., Dawson, L., Miller, D. (Eds.) *Criminal and Environmental Soil Forensics*., 317-331. Springer Science, New York, New York.

Tinker, D. B. & Knight, D. H. (2000). Coarse woody debris following fire and logging in Wyoming lodgepole pine forests. *Ecosystems*, *3*, 472-483.

Topp, W., Kappes, H., Kulfan, J. & Zach, P. (2006). Litter-dwelling beetles in primeval forests of Central Europe: Does deadwood matter? *Journal of Insect Conservation*, *10*, 229-239.

Trumbo, S. T. & Fiore, A. J. (1994). Interspecific competition and the evolution of communal breeding in burying beetles. *American Midland Naturalist*, *131*, 169-174.

Ulyshen, M. D. & Hanula, J. L. (2009). Responses of arthropods to large-scale manipulations of dead wood in loblolly pine stands of the southeastern United States. *Environmental Entomology*, *38*, 1005-1012.

Ulyshen, M. D. & Hanula, J. L. (2010). Patterns of saproxylic beetle succession in loblolly pine. *Agricultural and Forest Entomology*, *12*, 187-194.

Vanderwel, M. C., Malcolm, J. R., Smith, S. M. & Islam, N. (2006). Insect community composition and trophic guild structure in decaying logs from eastern Canadian pine-dominated forests. *Forest Ecology and Management*, *225*, 190-199.

Vass, A. A. (2012). Odor mortis. *Forensic Science International*, *222*, 234-241.

Velásquez, Y. & Viloria, A. L. (2009). Effects of temperature on the development of the Neotropical carrion beetle *Oxelytrum discicolle* (Brullé, 1840)(Coleoptera: Silphidae). *Forensic Science International*, *185*, 107-109.

Victorsson, J. (2012). Semi-field experiments investigating facilitation: Arrival order decides the interrelationship between two saproxylic beetle species. *Ecological Entomology*, *37*, 395-401.

VijayaVenkataRaman, S., Iniyan, S. & Goic, R. (2012). A review of climate change, mitigation and adaptation. *Renewable and Sustainable Energy Reviews*, *16*, 878-897.

Voss, S. C., Cook, D. F. & Dadour, I. R. (2011). Decomposition and insect succession of clothed and unclothed carcasses in Western Australia. *Forensic Science International*, *211*, 67-75.

Wang, B. C. & Smith, T. B. (2002). Closing the seed dispersal loop. *Trends in Ecology and Evolution*, *17*, 379-386.

Wasti, W. S. (1972). A study of the carrion of the common fowl, *Gallus domesticus*, in relation to arthropod succession. *Journal of the Georgia Entomological Society*, *7*, 221-229.

Waterhouse, D. F. (1974). The biological control of dung. *Scientific American*, *230*, 100-109.

Wermelinger, B., Flückiger, P. F., Obrist, M. K. & Duelli, P. (2007). Horizontal and vertical distribution of saproxylic beetles (Col., Buprestidae, Cerambycidae, Scolytinae) across sections of forest edges. *Journal of Applied Entomology*, *131*, 104-114.

Wertheim, B., Sevenster, J. G., Eijs, I. E. & van Alphen, J. J. M. (2000). Spatial diversity in a mycophagous insect community: A case study of spatial aggregation vs. resource partitioning. *Journal of Animal Ecology*, *69*, 335-351.

Weslien, J., Djupström, L. B., Schroeder, M. & Widenfalk, O. (2011). Long-term priority effects among insects and fungi colonizing decaying wood. *Journal of Animal Ecology*, *80*, 1155-1162.

Westoby, M. (1984). The self-thinning rule. *Advances in Ecological Research*, *14*, 167-225.

Whipple, S. D. & Hoback, W. W. (2012). A comparison of dung beetle (Coleoptera: Scarabaeidae) attraction to native and exotic mammal dung. *Environmental Entomology*, *41*, 238-244.

Wilson, E. O. (1971). *The Insect Societies*. Harvard Belknap, Cambridge, UK.

Wilson, E. O. (1999). *The Diversity of Life*. WW Norton and Company, New York, New York.

Wilson, D. S. & Knollenberg, W. G. (1987). Adaptive indirect effects: The fitness of burying beetles with and without their phoretic mites. *Evolutionary Ecology*, *1*, 139-159.

Wolff, M., Builes, A., Zapata, G., Morales, G. & Benecke, M. (2004). Detection of Parathion (O, O-diethyl O-(4-nitrophenyl) phosphorothioate) by HPLC in insects of forensic importance in Medellín, Colombia. *Annual Aggrawal's Internet Journal of Forensic Medicine and Toxicology*, *5*, 6-11.

Wright, E. J., Sinclair, E. A. & Annis, P. C. (2002). Laboratory determination of the requirements for control of *Trogoderma variabile* (Coleoptera: Dermestidae) by heat. *Journal of Stored Products Research*, *38*, 147-155.

von Hoermann, C., Ruther, J., Reibe, S., Madea, B. & Ayasse, M. (2011). The importance of carcass volatiles as attractants for the hide beetle *Dermestes maculatus* (De Geer). *Forensic Science International*, *212*, 173-179.

von Hoermann, C., Ruther, J. & Ayasse, M. (2012). The attraction of virgin female hide beetles (*Dermestes maculatus*) to cadavers by a combination of decomposition odour and male sex pheromones. *Frontiers in Zoology*, *9*, 18.

Wood, C. M. (2012). Saproxylic beetles (Coleoptera) associated with aspen deadwood in broad-leaved boreal mixedwood stands. Doctoral dissertation, University of Alberta, Canada.

Woodcock, B. A., Watt, A. D. & Leather, S. R. (2002). Aggregation, habitat quality and coexistence: A case study of carrion fly communities in slug cadavers. *Journal of Animal Ecology*, *71*, 131-140.

Work, T. T., Jacobs, J. M., Spence, J. R. & Volney, W. J. (2010). High levels of green-tree retention are required to preserve ground beetle biodiversity in boreal mixedwood forests. *Ecological Applications*, *20*, 741-751.

Zeide, B. (2001). Thinning and growth: A full turnaround. *Journal of Forestry*, *99*, 20-25.

Zhong, H. & Schowalter, T. D. (1989). Conifer bole utilization by wood-boring beetles in western Oregon. *Canadian Journal of Forest Research*, *19*, 943-947.

Zimmer, M., Kautz, G. & Topp, W. (2005). Do woodlice and earthworms interact synergistically in leaf litter decomposition? *Functional Ecology*, *19*, 7-16.

In: Beetles: Biodiversity, Ecology and Role in the Environment ISBN: 978-1-63463-380-2
Editor: Camilla Stack

Chapter 7

How a Locality Can Have so Many Species? A Case Study with Dung Beetles (Coleoptera: Scarabaeinae) In a Tropical Rain Forest in Colombia

Jorge Ari Noriega A. *

Laboratorio de Zoología y Ecología Acuática - LAZOEA,
Universidad de Los Andes, Bogotá, Colombia

Abstract

Dung beetles of the subfamily Scarabaeinae (Coleoptera: Scarabaeidae) are one of the most important insect groups in tropical regions because their relevant ecosystem role as organic recycling agents, biological pest controllers and secondary seed dispersers. However, our knowledge of the different mechanisms that maintain assemblage structure and allow species coexistence is poor. We lack such information because dung beetle assemblages are often sampled over the short term, there are few studies that cover extended periods of time and broad spatial areas, and it is common to use just one type of trap and bait. To better explore the mechanisms underlying dung beetle coexistence, it was studied a beetle assemblage found in a tropical rain forest located in Tinigua National Park in Colombia. Sampling was conducted in August 1992, May-July and December 1994, January and December 1995, January and July 1996, January-July 1997, and January and July 1998. Dung beetles were sampled in the three main habitats found in the study area: mature mainland forest, flooded lowland forest, and river beaches. Specimens were collected using baited pitfall traps, flight interception traps, and baited elevated traps placed at different heights using different type of baits. In addition, feces from different vertebrates, vegetation, logs, fungi, litter, and fallen fruit were checked for the presence of dung beetles. A total of 70 species belonging to 19 genera and 7 tribes were collected. Compared to other areas, previously sampled in Colombia, this area demonstrates a remarkable diversity and has a very high level of species richness (one of the highest in the country). Nevertheless, it cannot be considered a hotspot for endemism because there is a strong influence and contribution of species of nearby biogeographic

* Email: jnorieg@hotmail.com

provinces to its own fauna. Some of the mechanisms that may affect the richness and structure of the assemblage, favoring the coexistence of species are: spatial partitioning (habitats and microhabitats), temporal partitioning (seasonality and dial activity), resource partitioning, and guild-specific resource relocation behavior. The values of the Jaccard similarity index reveal low levels of species overlap. Therefore, these coexistence mechanisms appear to act in tandem to partition the ecosystem in a way that decreases overlap among similar species, thus allowing a large number of dung beetle species to coexist in the same area. These results demonstrate that it is necessary to carry out broad-ranging and exhaustive sampling to fully characterize the structure of tropical dung beetle assemblages. Finally, to completely describe the assemblage composition at this locality, the following additional improvements should be made to the sampling protocol: carrying out exhaustive sampling over one full year; using more specialized baits; studying the forest canopy; searching in specialized microhabitats, such as bromeliads and ant and termite nests; and characterizing potential phoretic interactions with mammals.

Keywords: Assemblage structure, coexistence mechanisms, Macarena, richness, Scarabaeidae, Tinigua National Park

INTRODUCTION

The study and inventory of diverse areas is a global priority that should be supported with as much research infrastructure as possible, especially when it comes to extremely diverse groups that could be useful bioindicators in monitoring and conservation studies (Maguran 1988, Halffter & Favila 1993, Gaston & Hudson 1994, Hammond 1994, Myers et al. 2000). Dung beetles of the subfamily Scarabaeinae (Coleoptera: Scarabaeidae) are one of the most commonly used groups of insect bioindicators because of the important ecological roles they play (Nichols et al. 2008). Adults and larvae feed on dung produced by mammals, thus helping to recycle organic matter (Halffter & Matthews 1966, Howden & Young 1981, Halffter & Halffter 1989, Gill 1991). Besides this function, dung beetles act as parasitic controllers of flies and nematodes (Waterhouse 1974) and also as secondary seed dispersers (Estrada & Coates 1991, Andresen 2001).

Among the countries located in the Neotropics, Colombia has one of the highest levels of dung beetles diversity. Knowledge about this group has been growing over the last two decades thanks to the publication of several important articles (Escobar & Medina 1996, Amat et al. 1997, Amezquita et al. 1999, Escobar 2000, Medina et al. 2001, 2002, Escobar 2004, Escobar et al. 2005, Fuentes & Camero 2006, Noriega et al. 2007, Martinez et al. 2009, Giraldo et al. 2011, Solis et al. 2011, Cultid et al. 2012, Delgado-Gómez et al. 2012, Noriega et al. 2012, Otavo et al. 2013). However, some of these studies cover only short time periods of sampling and use only a few different types of traps, frequently with just one kind of bait, which means that they are little more than rapid surveys of biodiversity; consequently, they provide an incomplete characterization of dung beetle assemblage richness.

In ecological classic literature is commonly assumed that the dynamic of the structure of an assemblage is influenced mainly by interspecific competition of a limiting resource in space and time, affecting the coexistence of more then two species (Hairston et al. 1960, Levins & Culver 1971, Hastings 1980, Tilman 1980, Atkinson & Shorrocks 1981, Connor & Simberloff 1983, Schoener 1983). Following the principal of competitive exclusion, two species cannot occupy the same ecological niche and if the resource they share is a limiting factor one of them will exclude the other one (Hardin 1960). In this sense, several studies with dung beetles have found that competition is an important factor that affects the composition, richness and abundance in assemblages (Hanski 1981, Holter 1982, Peck & Forsyth 1982, Giller & Doube 1989, Hanski & Cambefort 1991, Giller & Doube 1994, Montes de Oca & Halffter 1995, Hirschberger 1998, Krell-Westerwalbesloh et al. 2004, Feer & Pincebourde 2005, Horgan 2005, Horgan & Fuentes 2005, Vernes et al. 2005, Horgan 2006, Hernández et al. 2011). However, there is so few information that characterizes the mechanisms that support the coexistence in rich-species assemblages localities.

This study aimed to provide greater insight into this group by conducting a complete species inventory at one of the most biologically interesting areas in the country (Noriega et al. 2015). The study's main question was the following: how can a given area support such a large number of species? The goal was also to understand the coexistence mechanisms operating in the dung beetles assemblage that minimize interspecific competition, including spatial, temporal, and resource partitioning as well as resource relocation behavior.

Materials and Methods

Study Area

The study was carried out at the Center for Ecological Research of La Macarena (Spanish abbreviation: CIEM); the study area (2°40' N - 74°10' W, 350-400 m.a.s.l.; Figure 1) was situated in a tropical rainforest found to the west of the Duda river, 13 km upstream from the Duda's junction with the Guayabero, and located near the eastern border of Tinigua National Park (201.875 ha; Meta Department, Colombia). The park was created in 1989 to establish a biological corridor between the parks of La Macarena and Picachos. The Serrania of La Macarena is part of the biogeographic province of the Guayana presenting a complex mosaic of faunal and floral elements (Hernandez et al. 1992).

The region is characterized by a high degree of seasonality (Kimura et al. 1994). The dry season lasts from December to March, and rainfall varies the rest of the year (>100 mm per month). Average annual rainfall is 2600 mm; monthly rainfall is lowest in January (as low as 0 mm) and highest in May (up to 530 mm). Average annual temperature is 25°C (Kimura et al. 1994). The area has soils with different and complex geological origins. They were largely created by Plio-Pleistocene alluvial plates, experiencing recent tectonic fracturing and water erosion; consequently, they have low fertility and a poor cationic exchange capacity, are oligotrophic and deficient in mineral elements (Hirabuki 1990).

Sampling was carried out in the study area's three main habitats (described by Hirabuki 1990; Figure 2):

- Mature mainland tropical rain forest (MF): primary tropical rainforest with a continuous canopy containing trees of 25–30 m and emergent trees that reach 35 m. This type of forest is associated with the highest diversity of tree species (Barbosa & Hirabuki 1992) and has the highest vegetative cover in the area (53%) (Hirabuki 1990, Stevenson et al. 2004).

- Flooded lowland tropical rain forest (FF): lowland forest located on flatlands that are partially flooded during the rainy season by the river. It has a discontinuous canopy that is dominated by *Ficus* spp., *Inga* spp., and *Cecropia* sp. and an underbrush that contains mainly *Heliconia* spp. In this area, vegetative cover is 11% in this type of forest (Hirabuki 1990).
- River beaches (RB): beaches located on the banks of the Duda river. Their number and size vary throughout the year depending on rainfall seasonality. During the rainy months, the river increases in volume and covers the beaches completely within a few days. In the dry season, the beaches reappear, although the river never dries up completely. During the dry months, sand surface temperatures reach 50–55°C. These beaches are characterized by a type of early successional riparian forest containing *Tessaria integrifolia* and young trees of *Cecropia* sp.

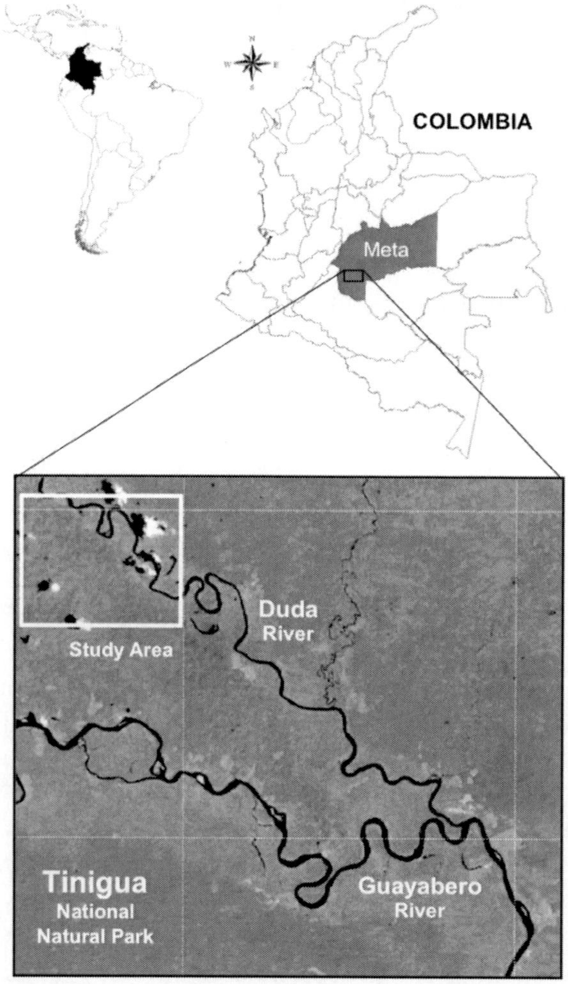

Figure 1. Geographical location of the study area: CIEM - Tinigua National Park, Meta Department, Colombia.

Figure 2. Habitats sampled at CIEM - Tinigua National Park, Meta Department, Colombia. A) Mature mainland tropical rain forest - MF, B) Flooded lowland tropical rain forest – FF and C) River beaches – RB.

In terms of vertebrates this area is a diverse spot with high species richness in different orders as Artiodactyla (3 spp.), Carnivora (9 spp.), Lagomorpha (1 sp.), Perissodactyla (1 sp.), Rodentia (6 spp.) and Xenarthra (5 spp.) (Stevenson 2002). In addition, it is home to seven coexisting species of primates: *Lagothrix lagothricha* (Humboldt, 1812 – common woolly monkey), *Sapajus apella* (Linnaeus, 1758 - brown capuchin), *Ateles belzebuth* Geoffroy, 1806 - spider monkey, *Alouatta seniculus* (Linnaeus, 1766 - red howler monkey), *Saimiri sciureus* (Linnaeus, 1758 - squirrel monkey), *Callicebus cupreus* (Spix, 1823 - red titi) and *Aotus brumbacki* Hershkovitz, 1983 - brumback´s night monkey (Stevenson 2002). Woolly monkeys and howler monkeys have the highest densities and levels of biomass (Stevenson & Quiñones 1993).

Dung Beetles Sampling, Data Review and Food Relocation Guilds

The author collected specimens in the habitats described above between June to July 1994, December 1994 to January 1995, December 1995 to January 1996, as well as in July 1996 and January 1998. Besides, other researchers studying dung beetles in the same area also collected some samples (from January to July 1997 and in July 1998) that were included in the analyses. In addition, it was included data from specimens that were found while going through the review of the largest entomological collections in Colombia: Entomological Collection Instituto Alexander von Humboldt (IAvH), Entomological Collection Instituto de Ciencias Universidad Nacional de Colombia (ICN-MHN), Entomological Collection Universidad de Antioquia (CEUA), Entomological Collection Universidad del Cauca (MHN-UC), Entomological Collection Pontificia Universidad Javeriana (MPUJ), Entomological Collection Universidad de Nariño (PSO), Entomological Collection Universidad de Tunja (CCM-FM) and Entomological Collection Natural History Museum of Universidad de Los Andes (EANDES). These collections contain some specimens that were collected during some of the same time periods as those mentioned above as well as some specimens collected in August 1992 and May 1994. Furthermore, data were added from any published reports or studies for which dung beetles in this area were sampled (Castellanos et al. 1999, Laverde et al. 2002, Noriega 2002b, 2004, Noriega & Calle 2008, Noriega 2011, Noriega & Acosta 2011, Noriega 2012).

Specimens were collected using pitfall trap arrays (Noriega & Fagua 2009); traps were baited with human excrement, carrion (rotten fish, squid, calamari, and chicken), or fruit (banana) and were left at least 48 hours (20160 hours/traps; Noriega et al. 2005). Flight interception traps were also used. Arrays of elevated traps placed at different heights were baited with human excrement, rotten fish, mushrooms, and decaying fruit (Noriega, unpubl. data). Additionally, dung from different vertebrates—such as *L. lagothricha, A. seniculus, Hydrochaeris hydrochaeris* (Linnaeus, 1766) (capybara), *Crocodylus intermedius* (Graves, 1819) (Orinoco crocodile), and *Tapirus terretris* (Linnaeus, 1758) (brazilian tapir)—was visually inspected for the presence of beetles. In addition, individuals were manually collected from vegetation, logs, flowers, fungi, leaf litter, and fallen fruit along the CIEM's trail system.

All specimens were preserved in 70% ethanol upon collection. They were later dried and identified to species using different taxonomic keys (Halffter & Martinez 1977, Howden & Young 1981, Kohlmann 1984, Jessop 1985, Edmonds 1994, Genier 1996, Kohlmann & Solis 1997, Cook 1998, Edmonds 2000, Medina & Lopera 2000, Kohlmann & Solis 2001, Solis & Kohlmann 2002, Edmonds & Zidek 2004, Solis & Kohlmann 2004, Kohlmann & Solis 2006, Vaz-de-Mello 2008, González et al. 2009, Camero 2010, Edmonds & Zidek 2010, Molano & Medina 2010, Vaz-de-Mello et al. 2011, Cupello & Vaz-de-Mello 2013, Solis & Kohlmann 2013), reference specimens in entomological collections, and the assistance of taxonomic experts. A reference collection was deposited at the EANDES in Bogotá, Colombia, and specimens from almost all the species collected were distributed to various entomological collections in Colombia: IAvH, ICN-MHN, and Universidad Pedagógica Nacional (MHNUPN). Specimens were also placed in the personal reference collections of Alejandro Lopera (CAL) and the author (CJAN).

Dung beetles food-relocation groups incorporated in the analysis where defined using the guilds described by Bornemissza (1976) and Halffter & Edmonds (1982): paracoprids (T, tunnelers)—species that dig a vertical tunnel below the food resource and transport a fraction of it into the bottom of the burrow to build their nests; telecoprids (R, rollers)—species that make a ball from the resource, roll it horizontally and later bury it into the ground building a nests; and endocoprids (D, dwellers)—species that occur within the food resource and make their nests there or in the interphase soil – resource. Since, for some species, it was impossible to directly observe specific relocation behaviors in the field, the distinctive behavior of the genus was assumed to occur.

In addition, the beetles were placed in three researcher-defined size categories: small (<10 mm), medium (10–18 mm) and large individuals (>18 mm). This was an adaptation of the system proposed by Doube (1990).

Data Analysis

An Excel matrix was constructed using the information on the niche characteristics of all of the 70 species sampled (habitat, season, dial activity, resource preferences, and guild-resource relocation behavior). These niche characteristics (coded as 14 grouped binomial variables placed in 6 ecological categories) were transformed into orthogonal components by performing a principal coordinates analysis on a matrix of Jaccard similarities. The five first components that accounted for more than 5% of the variability of the original matrix (70 species x 14 variables) and that, in addition, fulfilled the "broken stick" criterion were retained (Peres-Neto et al. 2006). Ecological patterns were interpreted by examining the five components obtained and comparing the positions of each species in each component based on the variable states (0-1); for each component, variables were selected using ANOVAs ($p<0.001$).

Using all of the species positions (derived from the principal coordinates analysis), a matrix of the Euclidean distances between the different pairs of species was obtained. Using this new matrix, it was possible to calculate the overall average distance between species as well as the average of the top 1% of distances between species ($C_{70,2}=2415$). These two measurements reveal the average and maximum degree of ecological divergence between species, respectively.

The matrix of 70 species x 5 components, which defined the ecological niches of the species in the study area, was used to generate 9999 neutral matrices of the "ecological positions" of the species. The neutral positions of the 70 species in each of the five orthogonal components were determined by randomly redistributing the values of the species within each component; consequently, because species values were not correlated across components, it was ensured that the different components would not be related (i.e., $r\approx0$).

These neutral matrices had the same dimensions (70 x 5) and contained the same amount of information as the original matrix of observed values. However, because they described randomly generated species, whose niches were not determined by real-life interspecific ecological interactions, they served as controls that could be used to determine if the observed degree of ecological packing was greater or less than what would be expected according to a neutral model based on random distributions. The overall average distance between the 70 species as well as the average of the top 1% of distances between species were calculated for each of the 9999 neutral matrices.

These values were used to build two distributions, which revealed whether the observed average and maximum degrees of species divergence were greater or smaller than what would be expected by chance. In addition, a species similarity analysis based on the Jaccard index was performed using all the ecological characteristics examined. All the analyses were done using the free statistical software packages PAST (v. 3.02, 2014; Hammer et al. 2001) and PopTools (v. 3.2; Hood 2010).

RESULTS AND DISCUSSION

Assemblage Structure

A total of 70 species—belonging to 19 genera and 7 tribes—were collected (Table 1, Figure 3). In terms of species richness, the dominant genus in the area was *Dichotomius* (11 species), followed by *Canthon* (9 species) and *Canthidium* (8 species). The most abundant species in the region were *O. haematopus*, *C. cupreum*, *C. aequinoctialis*, *D. parile*, and *O. conspicillatum*; they were the most commonly sampled species in both this study and some other studies conducted in the area (Castellanos et al. 1999). It is particularly interesting to note that *S. leander* and *G. lemoinei* were present; they are the only two species found on river beaches (Figure 3; Noriega 2002).

Table 1. List of the species of dung beetles (Scarabaeinae) collected at the CIEM - Tinigua National Park, Meta Department, Colombia. Code Spp: species code

Tribe	Species	Code Spp.
Ateuchini	*Ateuchus murrayi* (Harold, 1868)	*Ate mur*
	Ateuchus pygidialis (Harold, 1939)	*Ate pyg*
	Ateuchus cf. *scatimoides* (Balthasar, 1939)	*Ate sca*
	Eutrichillum cf. *hirsutum* (Boucomont, 1928)	*Eut hir*
	Uroxys bidentis Howden & Young, 1981	*Uro bid*
	Uroxys micros Bates, 1887	*Uro mic*
	Uroxys sp. 1	*Uro sp1*
	Uroxys sp. 2	*Uro sp2*
Coprini	*Canthidium euchalceum* Balthasar, 1939	*Can euc*
	Canthidium funebre Balthasar, 1939	*Can fun*
	Canthidium gerstaeckeri Harold, 1867	*Can ger*
	Canthidium onitoides (Perty, 1830)	*Can oni*
	Canthidium ruficolle (Germar, 1824)	*Can ruf*
	Canthidium splendidum Preudhomme de Borre, 1886	*Can spl*
	Canthidium cf. *centrale* Boucomont, 1928	*Can cen*
	Canthidium cf. *cupreum* (Blanchard, 1845)	*Can cup*
	Dichotomius belus (Harold, 1880)	*Dic bel*
	Dichotomius boreus (Olivier, 1789)	*Dic bor*
	Dichotomius compressicollis (Luederwaldt, 1929)	*Dic com*
	Dichotomius deyrollei (Harold, 1869)	*Dic dey*
	Dichotomius mamillatus (Felsche, 1901)	*Dic mam*
	Dichotomius ohausi (Luederwaldt, 1923)	*Dic oha*

Tribe	Species	Code Spp.
	Dichotomius podalirius (Felsche, 1901)	*Dic pod*
	Dichotomius worontzowi (Pereira, 1942)	*Dic wor*
	Dichotomius cf. *problematicus* (Luederwaldt, 1922)	*Dic prb*
	Dichotomius cf. *protectus* (Harold, 1867)	*Dic prt*
	Dichotomius aff. *inachus* (Erichson, 1847)	*Dic ina*
	Ontherus azteca Harold, 1869	*Ont azt*
	Ontherus pubens Génier, 1996	*Ont pub*
Deltochilini	*Canthon aequinoctialis* Harold, 1868	*Can aeq*
	Canthon angustatus Harold, 1867	*Can ang*
	Canthon cyanellus Harold, 1863	*Can cya*
	Canthon femoralis (Chevrolat, 1834)	*Can fem*
	Canthon fulgidus Redtenbacher, 1867	*Can ful*
	Canthon lituratus (Germar, 1813)	*Can lit*
	Canthon luteicollis Erichson, 1847	*Can lut*
	Canthon mutabilis Lucas, 1859	*Can mut*
	Canthon cf. *gutierrezi* Martínez, 1950	*Can gut*
	Deltochilum amazonicum Bates, 1887	*Del ama*
	Deltochilum orbiculare van Lansberge, 1874	*Del orc*
	Deltochilum aff. *orbignyi* (Blanchard, 1845)	*Del org*
	Deltochilum aff. *parile* Bates, 1887	*Del par*
	Scatonomus cf. *insignis* Harold, 1867	*Sca ins*
	Scybalocanthon pygidialis (Schmidt, 1922)	*Scy pyg*
	Scybalocanthon cf. *imitans* (Harold, 1868)	*Scy imi*
Demarziellini	*Bdelyrus* cf. *metaensis* Cook, 1998	*Bde met*
Oniticellini	*Eurysternus caribaeus* (Herbst, 1789)	*Eur car*
	Eurysternus contractus Génier, 2009	*Eur con*
	Eurysternus foedus Guérin-Méneville, 1844	*Eur foe*
	Eurysternus hamaticollis Balthasar, 1939	*Eur ham*
	Eurysternus hypocrita Balthasar, 1939	*Eur hyp*
	Eurysternus mexicanus Harold, 1869	*Eur mex*
	Eurysternus plebejus Harold, 1880	*Eur ple*
	Eurysternus wittmerorum Martínez, 1988	*Eur wit*
Onthophagini	*Onthophagus buculus* Mannerheim, 1829	*Ont buc*
	Onthophagus haematopus Harold, 1875	*Ont hae*
	Onthophagus rubrescens Blanchard, 1845	*Ont rub*
	Onthophagus cf. *curvicornis* (Latreille, 1811)	*Ont cur*
	Onthophagus aff. *clypeatus* Blanchard, 1846	*Ont cly*
Phanaeini	*Coprophanaeus telamon* (Erichson, 1847)	*Cop tel*
	Coprophanaeus ohausi (Felsche, 1911)	*Cop oha*
	Dendropaemon cf. *waterhousei* d´Olsoufieff, 1924	*Den wat*
	Gromphas lemoinei Waterhouse, 1891	*Gro lem*
	Oxysternon conspicillatum (Weber, 1801)	*Oxy con*
	Oxysternon silenus Castelnau, 1840	*Oxy sil*
	Phanaeus bispinus Bates, 1868	*Pha bis*
	Phanaeus cambeforti Arnaud, 1982	*Pha cam*
	Phanaeus chalcomelas (Perty, 1830)	*Pha cha*
	Sulcophanaeus faunus (Fabricius, 1775)	*Sul fau*
	Sulcophanaeus leander (Waterhouse, 1891)	*Sul lea*

One of the least common species was *S. faunus*: it never came to baited pitfall traps, even those baited with carrion (rotten fish, squid, calamari, and chicken). The only two individuals that were collected (one male and one female) were captured in a mist net intended for birds. It is also interesting to note that individuals of the genus *Ontherus* (*O. azteca* and *O. pubens*) occurred in low numbers. Genier (1996) says that some species of this genus have very specific habitat restrictions and resource preferences that may explain these low numbers. For instance, in the study area, there are ant and termite nests that could serve as microhabitats for these two species (Kistner 1982, Krikken 2008, Krell & Phillips 2010, Vårdal & Forshage 2010, Maruyama 2012), which may explain why *O. azteca* was frequently observed in interception traps. Other species that also occurred in low numbers were *Scatonomus* cf. *insignis*, *Bdelyrus* cf. *metaensis*, and *Eutrichilum* cf. *hirsutum*; little information is available on the ecology, behavior, and habitat preferences of these three genera (Pereira 1954, Martínez 1967, Vulcano & Pereira 1973, Ratcliffe 1980, Cook 1998, Medina et al. 2001, Solis et al. 2011).

Table 2. Different numbers of Scarabaeinae found in five biogeographical regions in Colombia (Ama: Amazon, And: Andean, Car: Carribean, Cho: Choco – Pacific, Ori: Orinoquia – Plain grasslands and Guy: Guayana) and at the CIEM - Tinigua National Park, Meta Department, Colombia

Region	Department	Locality	Altitude (m.a.s.l.)	Genera/ Species	Reference
Ama	Amazonas	Leticia	80	15/60	Howden & Nealis (1975)
	Caquetá	PNN Serranía de Chiribiquete	300	14/61	Pulido et al. (2003)
And	Caldas	Cuenca Río La Miel	160, 2600 - 2750	12/55	Arango & Montes (2009)
	Tolima	Mariquita	690	13/30	Fuentes & Camero (2006)
Car	Atlántico	Barranquilla	0 - 500	17/35	Solís et al. (2011)
	Magdalena	PNN Sierra Nevada de Santa Marta	50 - 940	15/29	Martínez et al. (2009)
Cho	Choco	Salero	115	13/23	Neita et al. (2003)
	Choco	Lloro	90	13/19	Neita & Escobar (2011)
Ori	Meta	Puerto Colombia	200	14/32	Amézquita et al. (1999)
	Guaviare	San José del Guaviare, RN Nukak	200 - 360	14/48	Escobar (2000)
Guy	Meta	CIEM	350	19/70	(this work)

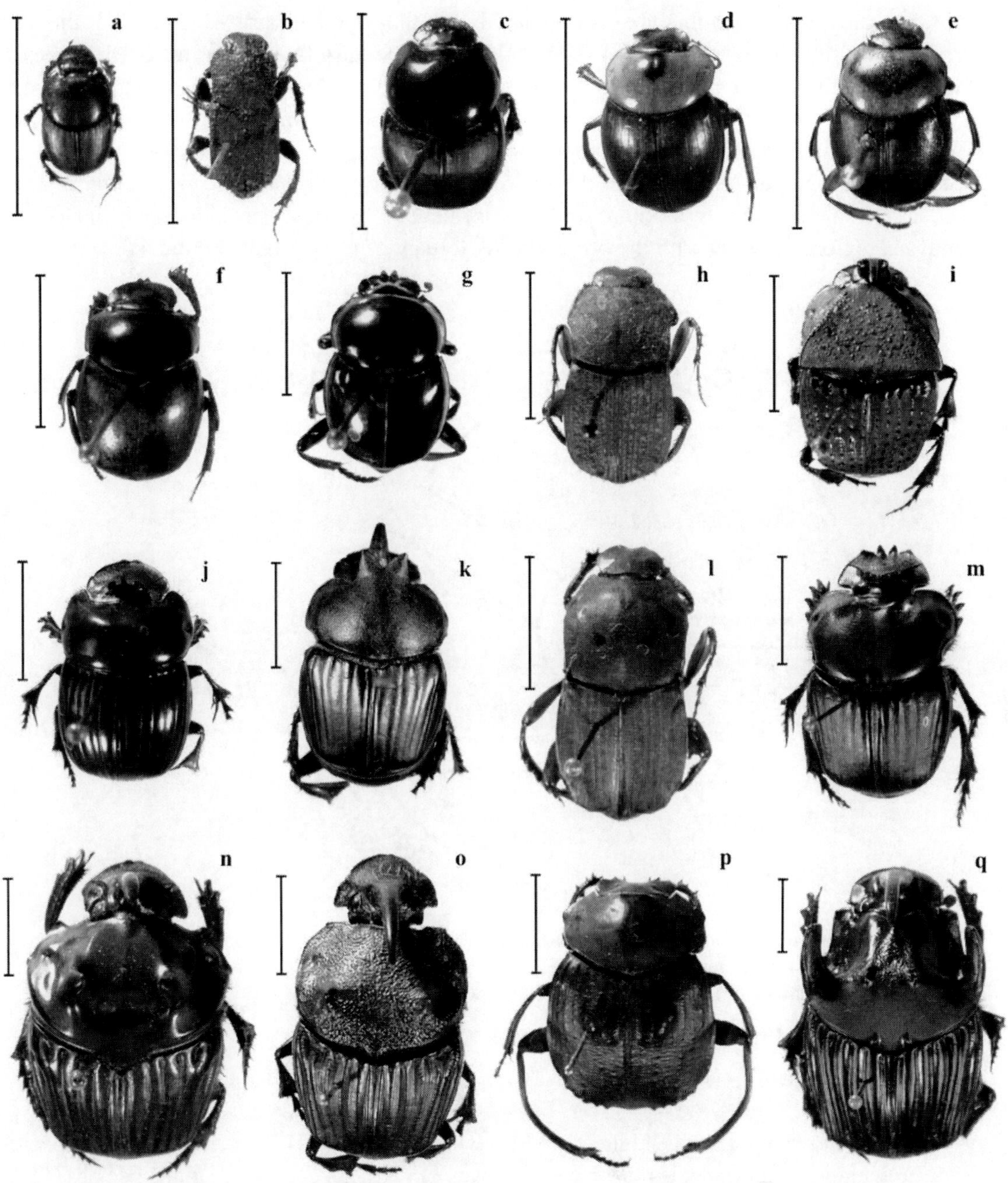

Figure 3. Dorsal view of a sample of dung beetle species in the CIEM - Tinigua National Park, Meta Department, Colombia. a) *Onthophagus haematopus* (♀), b) *Eurysternus plebejus* (♂), c) *Canthidium funebre* (♀), d) *Scybalocanthon pygidialis* (♀), e) *Canthon luteicollis* (♀), f) *Canthon aequinoctialis* (♂), g) *Canthon fulgidus* (♂), h) *Eurysternus caribaeus* (♀), i) *Phanaeus chalcomelas* (♂), j) *Dichotomius belus* (♂), k) *Gromphas lemoinei* (♂), l) *Eurysternus hamaticollis* (♂), m) *Coprophanaeus telamon* (♂), n) *Oxysternon conspicillatum* (♂), o) *Sulcophanaeus leander* (♂), p) *Deltochilum orbignyi* (♀), q) *Sulcophanaeus faunus* (♂). Bar scale = 1 cm.

If species richness in this area is compared with that in other parts of Colombia, the area around the CIEM is one of the most species rich in the country (Noriega et al. 2015; Table 2). Given the high number of species it shares with neighboring biogeographical regions (n=56; 79.8%) and its low number of endemic species (n=4; 5.7%) compared to other areas of the country (Escobar 2000), faunal composition in this area is likely greatly determined by species contributions distributed in the entire country (n=8; 11.4%), in surrounding regions (Andes, Amazonia, and Orinoquia) and less influenced by local speciation (i.e., inside the Guayana region, agreeing with the proposed by Hernandez et al. 1992) (Figure 4).

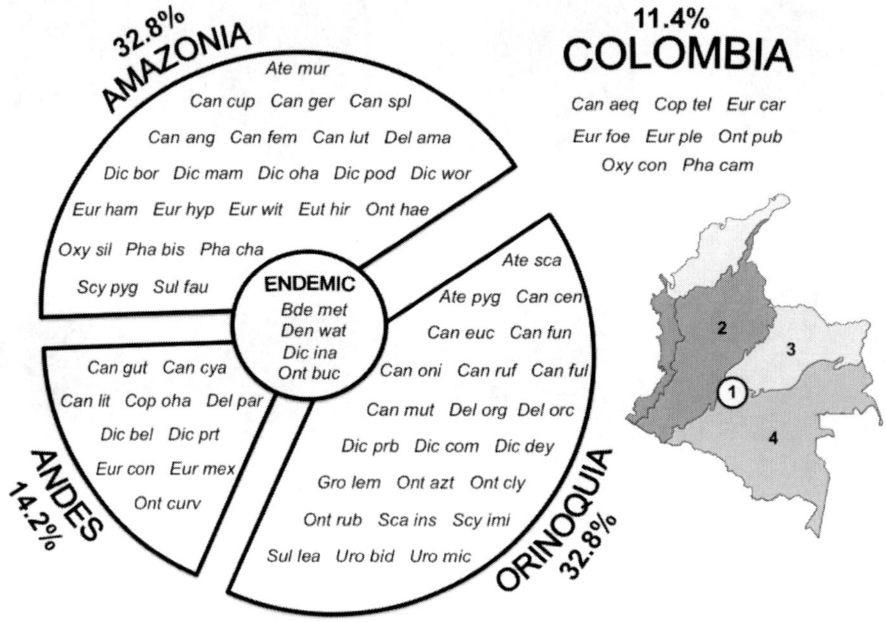

Figure 4. Composition of dung beetle species assemblage at the CIEM - Tinigua National Park, Meta Department, Colombia, in terms of the species biogeographical origin: Endemic (1), Andes (2), Orinoquia (3), Amazonia (4) or present throughout the entire country.

Spatial Coexistence Mechanisms

Taxonomic richness was highest in MF, which harbored 17 genera, 53 total species (75.7%), and 41 unique species (58.6%). Next was FF with 12 genera, 28 total species (40%), and 15 unique species (21.4%). Last were the RB with just 2 genera, 2 total species (2.9%), and 1 unique species (*S. leander*). There are no species that were present in all the CIEM's main habitats. However, MF and FF shared 13 species (18.6%), and FF and the RBs shared just one species (*G. lemoinei*) (Figure 5).

Vertical partitioning of forest habitat appears to be another important spatial coexistence mechanism. Given the results obtained from the linear transect of elevated traps (Noriega 2011) placed at heights of 3, 6, 9, and 15 m in MF, it is clear that some species are using different strata within the forest (floor - 0 m vs. arboreal level – 6 m) and are especially abundant between 6 and 9 m (Noriega unpub. data). Of the 70 species found in this area, 10

(14.7%; *B.* cf. *metaensis*, *C. funebre*, *C. gerstaeckeri*, *C.* cf. *cupreum*, *C. aequinoctialis*, *C. fulgidus*, *C. luteicollis*, *O. buculus*, *O. haematopus* and *S. pygidialis*) came to the elevated traps.

In particular, *B.* cf. *metaensis* was collected exclusively in one such trap, and *C. fulgidus* was by far more common in traps located 6 m above the ground than in ground-level pitfall traps. Other species like *C. splendidum*, *C. angustatus*, and *O. conspicillatum* were found perching on leaves in the forest but never occurred in the elevated traps (Noriega unpub. data). It is recommended to install elevated traps in the canopy and studying this as yet unexamined stratum of the forest; it is possible that some species use the canopy microhabitat.

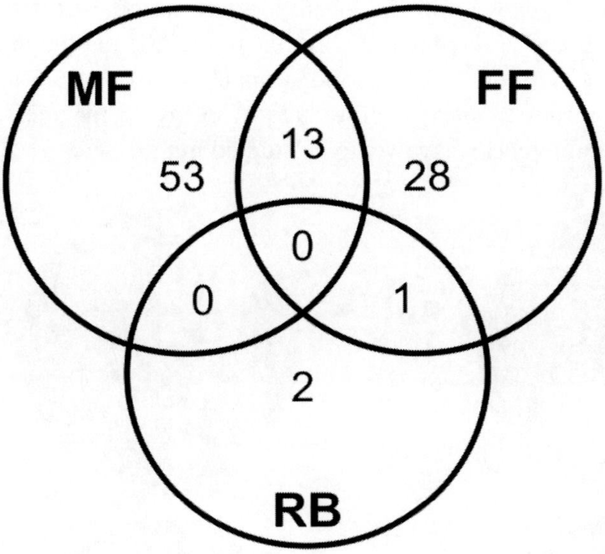

Figure 5. Shared and unique species in the three main habitats (MF=Mature mainland forest, FF=Flooded lowland forest and RB=River beaches) of the CIEM - Tinigua National Park, Meta Department, Colombia.

Temporal Coexistence Mechanisms

The wet season was more taxonomically rich (17 genera; 57 total species, 81.4%, 60.2% in contrast; 32 unique species, 45.7%) than the dry season (13 genera; 38 total species, 54.3%, 39.8% in contrast; 13 unique species, 18.6%; Figure 6). Twenty-five species (35.7%) occurred during both seasons. However, sampling during the transition season was incomplete, and it is certain that there are no species that were present throughout the entire year at the CIEM. It is noteworthy that five of the eight species of the genus *Eurysternus* were present during both seasons. In general, the arrival of the rainy season produces an increase in resources in the forest that favors the coexistence of a greater number of species; the opposite occurs during the dry season, as it was registered in other studies (Montes de Oca & Halffter 1995, Morelli et al. 2002, Vernes et al. 2005).

As for dial activity, slightly more species were out during the day (14 genera; 40 total species, 57.1%, 52.4% in contrast; 31 unique species, 44.3%) than at night (13 genera, 39 total species, 55.7%, 47.6% in contrast; 30 unique species, 42.9%; Figure 6). Nine species (12.9%) occurred both during the day and at night hours. However, no species was active all day (24 hours), and most had very short windows of activity during the crepuscular hours—for instance, *S. leander* is active for less than an hour at dusk and dawn (Noriega 2002b). These differences in periods of activity decrease the competition for limited resources in dung beetles (Hanski & Cambefort 1991, Estrada et al. 1993, Giller & Doube 1994, Montes de Oca & Halffter 1995, Hernández 2002, Krell-Westerwalbesloh et al. 2004).

This crepuscular peak in activity is probably related to the fact that mammals, and especially primates, produce much more dung at dawn and during the early morning, as it has been registered for *L. lagotricha* in this locality, after one hour of rest (P. Stevenson pers. comm.). During the middle of the day (12–2 pm) and after the middle of the night (12–2 am), species number and activity are lower (Noriega unpub. data). In addition, it is possible that some species prefer resources in a specific state of decay or freshness (e.g., fresh vs. old; Hanski 1980b); such preferences have yet to be studied in this area.

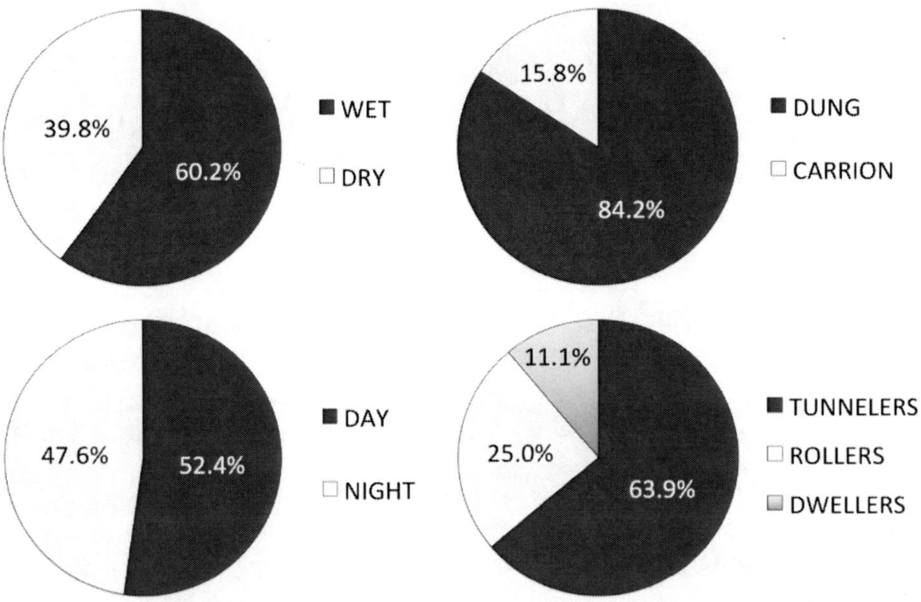

Figure 6. Structure of the dung beetle assemblage found at the CIEM - Tinigua National Park, Meta Department, Colombia. Four different categories of niche partitioning were examined: season (wet vs dry), resource preference (dung vs carrion), dial activity (day vs night) and resource-relocation behavior (tunnelers vs rollers vs dwellers).

Resource Coexistence Mechanisms

More species were found on dung (19 genera; 63 total species, 90%, 84.2% in contrast; 58 unique species, 82.9%) than on carrion (7 genera; 12 total species, 17.1%, 15.8% in contrast; 7 unique species, 10%; Figure 6). Five species (7.1%) occurred on both resources (*C. aequinoctialis, C. telamon, D. amazonicum, E. caribaeus* and *E. plebejus*), which show that they are generalists in terms of their food preferences. The differences in the spatial distributions and abundances of these two resources have a clear effect on the number of species associated with each. The high richness and abundance of dung beetles in this area is supported by the high diversity and density of mammals. Castellanos et al. (1999) estimated that woolly monkey populations in this area might produce 400–600 fecal deposits/km^2 per day. However, this study found that a large number of species use carrion, a result that contrasts with those of other studies in Colombia (Howden & Nealis 1975, Bustos-Gómez & Lopera 2003, Pardo et al. 2004, Esparza-Leon & Amat-García 2007, Murillo et al. 2010, Delgado-Gomez et al. 2012) and in other regions (Hanski 1983, Klein 1989, Hill 1996, Boonrotpong et al. 2004, Larsen et al. 2006, Damborsky et al. 2008, Almeida & Louzada 2009, Davis et al. 2010, Barragan et al. 2011). The exploitation of carrion could be related to the Neotropical absence of large mammals (Halffter & Matthews 1971, Louzada & Lopes 1997), which can produce large quantities of excrement; consequently, some species of dung beetles may have turned to other available resources.

In addition, some studies in Colombia (Santos-Heredia et al. 2010) and in other regions (Howden & Young 1982, Estrada & Coates 1991, Gill 1991, Estrada et al. 1993) suggest that dung beetles might be highly specialized in the type of mammal excrement (carnivorous, herbivorous or omnivorous) they use. Another study conducted in this area (Noriega 2012) compared species preferences for excrement produced by two different primates (*A. seniculus* and *L. lagotricha*); it found that some beetle species were more attracted to the dung produced by one primate and not to that produced by the other. It is also important to note that most of the baits used in this study were small in size (20–30 g), and it is possible that some species could be more attracted to smaller or larger baits (Gill 1991). For instance, based on data obtained in other areas and from the literature (Edmonds & Zidek 2010), it is proposed that *S. faunus* is not attracted to small baits but rather prefers large carcasses in highly conserved forests.

Furthermore, numerous records show that some dung beetles use other, very specific resources that are completely different from vertebrate dung and carrion, such as invertebrate feces, flowers, fungi, fruits, eggs, and diplopods, among others (Young 1980, Monteith & Storey 1981, Howden & Young 1982, Gill 1991, Navarrete & Galindo 1997, Cano 1998, Pfrommer & Krell 2004, Noriega & Calle 2008, Halffter & Halffter 2009). The fact that these species exploit such a wide range of non-fecal resources raises the question as to whether this group of insects should even be called "dung beetles" at all.

Guild-Resource Relocation Coexistence Mechanisms

In terms of the dung beetle guilds that show resource relocation behavior, the paracoprids (T) were the most common (14 genera; 46 species, 63.9%), followed by the telecoprids (R) (4 genera; 16 species, 25.0%) and the endocoprids (D) (1 genus, *Eurysternus*; 8 species, 11.6%; Figure 6). Some additional resource relocation behaviors were observed at the CIEM, such as pellet- or fragment-pushing, but they were exhibited by very few species and not commonly.

If beetle size (small, medium and large) is examined in tandem, it is possible to gain a better idea of how effectively relocation behavior limits direct competition and partitions species in this area. Indeed, the results indicate that, in some categories like medium and small rollers and large, medium, and small dwellers, there were no more than three species sharing the resource (Table 3). No kleptoparasitism was directly observed in the field in this area, but since this behavior is very common at other localities (Cambefort 1991, Gill 1991, Rougon & Rougon 1991), it seems highly probable that it occurred between small and large species.

Table 3. Matrix of beetle species belonging to different resource-relocation guilds (tunnelers, rollers and dwellers) and of different sizes (large, medium and small) found at CIEM - Tinigua National Park, Meta Department, Colombia

	TUNNELERS		ROLLERS		DWELLERS
LARGE (> 18 mm)	*Cop tel* *Cop oha* *Dic bor* *Dic mam* *Dic pod*	*Dic wor* *Oxy con* *Sul fau* *Sul lea*	*Del ama* *Del orc* *Del org*		*Eur con* *Eur ham* *Eur hyp*
MEDIUM (10 – 18 mm)	*Can cen* *Can ful* *Den wat* *Dic bel* *Dic com* *Dic dey* *Dic ina* *Dic oha* *Dic prt*	*Dic prb* *Gro lem* *Ont pub* *Ont azt* *Oxy sil* *Pha bis* *Pha cam* *Pha cha*	*Can aeq*		*Eur car* *Eur foe*
SMALL (< 10 mm)	*Ate mur* *Ate pyg* *Ate sca* *Can cup* *Can euc* *Can fun* *Can ger* *Can lut* *Can oni* *Can ruf* *Can spl*	*Eut hir* *Ont buc* *Ont cly* *Ont cur* *Ont hae* *Ont rub* *Uro bid* *Uro mic* *Uro sp1* *Uro sp2* *Bde met*	*Can ang* *Can cya* *Can fem* *Can gut* *Can lit*	*Can mut* *Del par* *Sca ins* *Scy pyg* *Scy imi*	*Eur mex* *Eur ple* *Eur wit*

Ecological Segregation of Species

The five components retained in the principal coordinates analysis accounted for 71% of the interspecific variability in species niche characteristics. The first component (Table 4) accounted for 27.76% of the niche information (species that were found in MF, active during wet season, with nocturnal activity and they use carrion resources).

The second component included the 20.57% of the niche (species that were found in FF, active during dry season, with nocturnal activity and with paracoprid behavior). These two components alone accounted for nearly the 50% of the ecological divergence between species. The third, fourth, and fifth components, which explained the 22.9% of the variability, are described in Table 4. Based on these five components, the average degree of divergence between species was 0.363 (the maximum theoretical value is 1, and the maximum observed Euclidean distance was 0.723).

Table 4. Analysis of the principal coordinates matrix for all the species sampled at CIEM - Tinigua National Park, Meta Department, Colombia. The signs + and – indicates respectively, the positive and negative relationships that exist between the binomial variables (0-1), which describe the species niche characteristics and the five principal coordinates components

	Comp 1	Comp 2	Comp 3	Comp 4	Comp 5
MF	+	-			-
FF	-	+	-	-	
RB					
Dry	-	+	-	-	-
Wet	+	-		+	-
Day	-	-	+	-	
Night	+	+	-	+	
Dung	-			+	+
Carrion	+		-	-	
Tunneler		+	+	-	+
Roller			-		+
Dweller					-
Floor	-			-	+
Arboreal				+	+
Autovalue	3.90	2.89	1.49	0.94	0.79
% variance	27.76	20.57	10.60	6.71	5.59

This observed average value was significantly greater than the average of the neutral distribution (P=0.050; Figure 7). This result means that the 70 species sampled were slightly more segregated than expected by chance based on a neutral model of ecological segregation. In contrast, average maximum divergence was significantly smaller (p=0.036; Figure 7) than expected by chance (neutral model: average maximum value of 0.970). This result indicates that there are no extreme ecological positions within the niche space of the group of species and therefore not observed degrees of ecological segregation that did not have representation in the 70 species registered.

On average, species were more ecologically segregated than expected by chance, but species with extreme ecological niches were not detected, which generated maximum values of interspecific divergence within the niche space available for the 70 species.

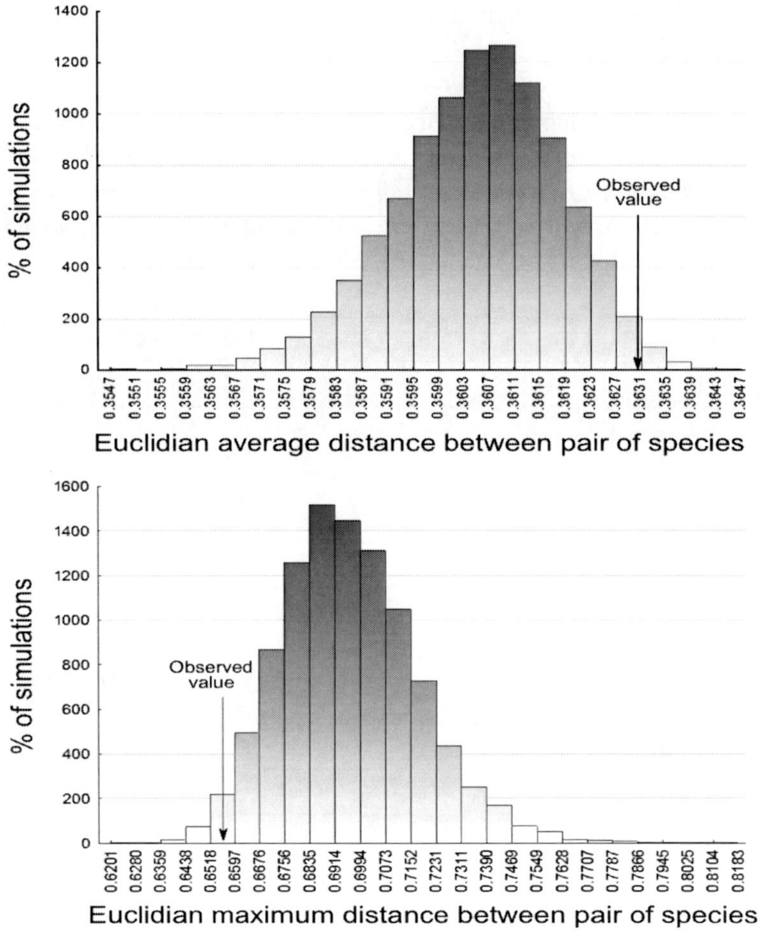

Figure 7. Euclidian average and maximum distance between pair of species sampled at the CIEM - Tinigua National Park, Meta Department, Colombia.

The analysis of similarity using Jaccard index shows that there are 20 different groups of species or single species with a 75% of similarity (25% of dissimilarity) that do not use and share the niche in the same way (Figure 8).

Relative to the high level of species richness in this area, the degree of interspecific niche overlap is low, meaning that few species interact directly in the same space, at the same time, and on the same resource (Figure 9). Indeed, this assertion is supported by the presence of all of the coexistence mechanisms described above.

If the information of food relocation guilds and categories of size (Table 3) is added with the coexistence scheme of niche segregation structure (Figure 9) there were very few species sharing the same space, time and resource decreasing radically the interspecific competition. In addition, it is important to notice that at least there is one species using the available resource in each "niche-space" in the coexistence scheme (Figure 9).

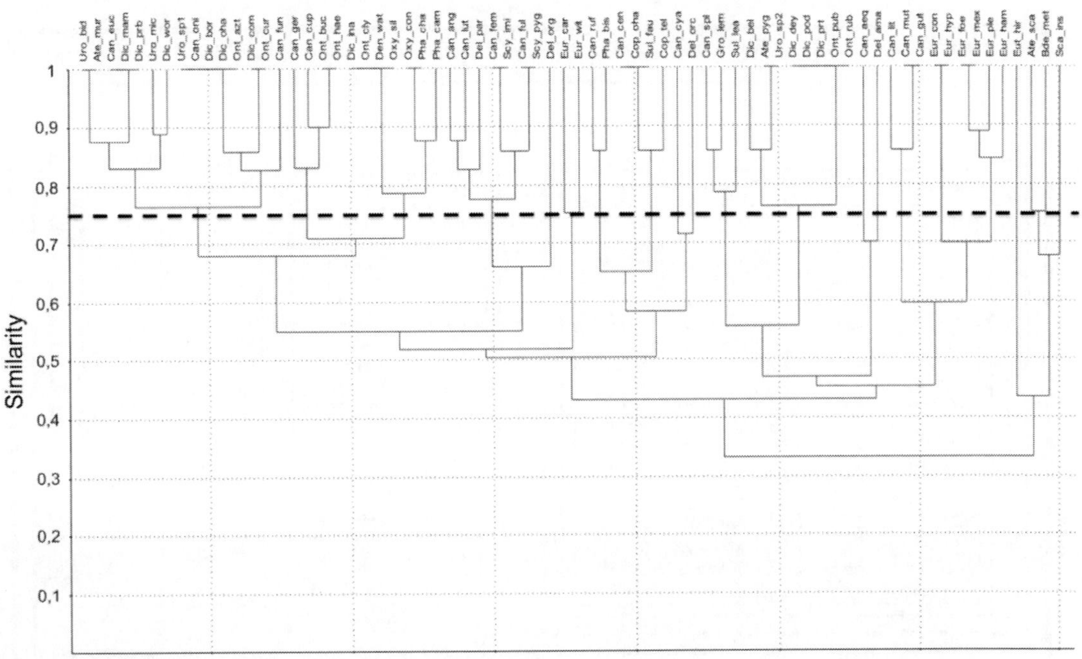

Figure 8. Jaccard index of similarity (dashed line=75%) for the species sampled at the CIEM - Tinigua National Park, Meta Department, Colombia.

All these mechanisms are the result of interactions between historical and ecological factors (Davis & Scholtz 2001) and promote the coexistence of different species by limiting the effect of the factors that are primarily responsible for high levels of interspecific competition (Hanski 1980a, b, Giller & Doube 1989, Lumaret et al. 1992, Finn & Gittings 2003, Hutton & Giller 2004).

"Rare Species" and Sampling Recommendations

Over the course of the sampling process, some species appeared to be rare because of their low abundances. However, when a new habitat or microhabitat was sampled, a new bait or trap type was used, or sampling was conducted at a new time of the year, it was discovered that that these "rare" species were not rare at all, instead, their presence was underestimated as a result of sampling inaccuracy. For example, this was the case for *C. fulgidus*, which rarely occurred in ground-level pitfall traps but was common in 6 m elevation traps, for *S. faunus*, which was collected when mist nets were used, and for *S. leander*, which was only found on river beaches.

These examples of "rare" species reveal how imprecise sampling can be and underscore the necessity of using a sampling approach that includes a wide variety of trap types, a wide range of bait types and sizes, and all possible habitats and microhabitats; that is carried out over the entire year; and that incorporates more natural history studies.

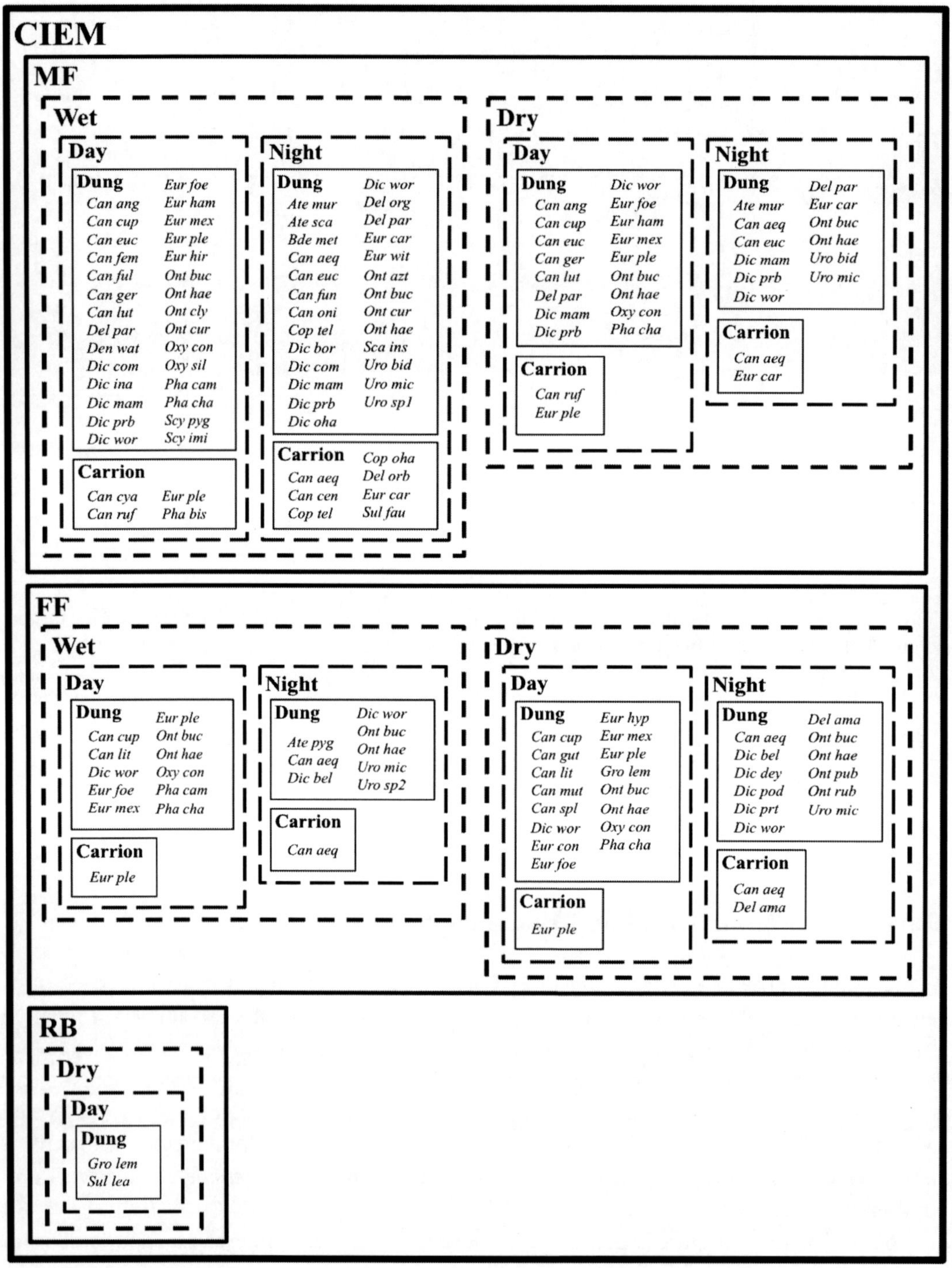

Figure 9. Coexistence scheme of the principal aspects of the niche segregation structure of the dung beetles assemblage at the CIEM - Tinigua National Park, Meta Department, Colombia.

Finally, to more thoroughly characterize assemblage composition, exhaustive sampling should take place over a full year and other kinds of traps should be used (Newton & Peck 1975, Moron & Terrón 1984, Noriega 2011). Furthermore, more specialized baits should be utilized; these baits could include invertebrate feces, such as from earthworms (Matthews 1965, Howden & Young 1981), dead spirobolid millipedes (Bernon 1981, Cano 1998, Brühl & Krell 2003), excrement from vertebrates found in the area (Martinez 1952, Young 1981a, Janzen 1983, Gill 1991, Estrada et al. 1993), rotten fungi (Navarrete & Galindo 1997), and other rare resources (Young 1980, Howden & Young 1981, Monteith & Storey 1981, Gill 1991, Villalobos et al. 1998, Pfrommer & Krell 2004).

In addition, it may be important to sample certain specific microhabitats, such as bromeliads (Pereira et al. 1960, Huijbregts 1984, Cook 1998), caves (Zunino & Halffter 1988), mammal burrows (Zunino & Halffter 2007), ant nests (particularly made by the genera *Acromymex*, *Atta*, *Pheidole*, and *Solenopsis*), and termite nests (Vaz-de-Mello et al. 1998, Navarrete 2001). Lastly, the close (e.g., phoretic) relationships that some beetle species may have with mammals such as sloths should not be forgotten (Ratcliffe 1980, Young 1981b).

ACKNOWLEDGMENTS

To the Japanese-Colombian cooperative agreement between the University of Los Andes and the University of Myiagi. To the National Park Unit for granting the permits to carry out research in Tinigua National Park. To Carlos Arturo Mejia, Ivan Jimenez, Daniel Cadena, Alejandra Vega, Libia Carolina, Juan Cristobal Calle and Ramiro Montealegre for their help in the field samplings. To Fernando Vaz-de-Mello, David Edmonds, Bruce Gill, Francois Genier and Alejandro Lopera for their help in identifying the specimens. To Luis M. Carrascal and Daniel Monroy for their important and valuable contributions and their help with the statistical analyses in this paper. To Edison Torrado-Leon from Naturavision (www.naturavision.com) for his significant support and collaboration in taking the photos of the species. Ricardo Botero Trujillo made comments and suggestions that improved the manuscript. To Pablo Stevenson for providing valuable information and references. To David Morris and Jessica Pearce-Duvet for kindly checking the English version of the manuscript.

REFERENCES

Almeida, S. & J. Louzada. 2009. Estrutura da comunidade de Scarabaeinae (Scarabaeidae: Coleoptera) em fito sionomias do cerrado e sua importância para a conservação. *Neotropical Entomology,* 38(1): 32-43.

Amat, G.G., A.T. Lopera & S. Amezquita. 1997. Patrones de distribución de escarabajos coprófagos en relicto del bosque altoandino Cordillera Oriental de Colombia. *Caldasia,* 19 (1-2): 191-204.

Amezquita, M.S.J., A. Forsyth, A. Lopera & A. Camacho. 1999. Comparación de la composición y riqueza de especies de escarabajos coprófagos (Coleoptera: Scarabaeidae) en remanentes de bosque de la Orinoquía Colombiana. *Acta Zoológica Mexicana* (n.s.), 76: 113-126.

Andresen, E. 2001. Effects of dung presence, dung amount and secondary dispersal by dung beetles on the fate of Micropholis guyanensis (Sapotaceae) seeds in Central Amazonia. *Journal of Tropical Ecology,* 17: 61-78.

Arango, L. & J.M. Montes. 2009. Caracterización entomológica parcial de la cuenca del río la Miel en el departamento de Caldas (Colombia). *Boletín Científico Centro de Museos Museo de Historia Natural, 13*(2): 249-268.

Atkinson, W.D. & B. Shorrocks. 1981. Competition on a divided and ephemeral resource: A simulation model. *Journal of Animal Ecology,* 50: 461-471.

Barragan, F., C.E. Moreno, F. Escobar, G. Halffter & D. Navarrete. 2011. Negative impacts of human land use on dung beetle functional diversity. *Plos One,* 6(3): 1-8.

Bernon, G. 1981. Species abundance and diversity of the Coleoptera component of a South African cow dung community, and associated insect predators. Ph.D. diss., Univ. of Bowling Green, Ohio. 183 p.

Boonrotpong, S., S. Sotthibandhu & C. Pholpunthin. 2004. Species composition of dung beetles in the primary and secondary forests at Ton Nga Chang Wildlife Sanctuary. *ScienceAsia,* 30: 59-65.

Bornemissza, G.F. 1976. The Australian dung beetle project 1965-1975. *Australian Meat Research Committee Review,* 30: 1-32.

Brühl, C. & K.T. Krell. 2003. Finding a rare resource: bornean Scarabaeoidea (Coleoptera) attracted by defensive secretions of diplopoda. *The Coleopterist Bulletin,* 57: 51-55.

Bustos-Gómez, F. & A. Lopera. 2003. Preferencias por cebo de los escarabajos coprofagos (Coleoptera: Scarabaeidae: Scarabaeinae) de un remanente de bosque seco tropical al norte del Tolima (Colombia). 59-65 pp. En: Escarabeidos de Latinoamerica: estado del conocimiento. Onore, G., P. Reyes-Castillo & M. Zunino (Eds.). m3m, Monografias Tercer Milenio, SEA, Zaragoza.

Cambefort, Y. 1991. Dung beetles in tropical savannas. 156-178 pp. In: I. Hanski and Y. Cambefort (Eds.), Dung Beetle Ecology. Princeton University Press, Princeton, NJ.

Camero, E. 2010. Los escarabajos del género Eurysternus Dalman, 1824 (Coleoptera: Scarabaeidae) de Colombia. *Boletín de la Sociedad Entomológica Aragonesa (S.E.A.),* 46: 147-179.

Cano, E.B. 1998. Deltochilum valgum acropyge Bates (Coleoptera: Scarabaeidae: Scarabaeinae): habits and distribution. *The Coleopterist Bulletin,* 52(2): 174-178.

Castellanos, M.C., F. Escobar & P.R. Stevenson. 1999. Dung Beetles (Scarabaeidae: Scarabaeinae) attracted to Woolly Monkey (Lagothrix lagothricha Humboldt) dung at Tinigua National Park, Colombia. *The Coleopterists Bulletin,* 53(2): 155-159.

Connor, E.F. & D. Simberloff. 1983. Interspecific competition and species co-occurrence patterns on islands: null models and the evaluation of evidence. *Oikos*, 41: 455-465.

Cook, J. 1998. A revision of the Neotropical genus of Bdelyrus Harold (Coleoptera: Scarabaeidae). *The Canadian Entomologist*, 130: 631-689.

Cultid, C.A., C.A. Medina, B.G. Martínez, A.F. Escobar, L.M. Constantino & N.J. Betancur. 2012. Escarabajos coprófagos (Scarabaeinae) del eje cafetero: Guía para el estudio etológico. Ed. WCS, CENICAFE & Federación Nacional de Cafeteros de Colombia. Villa Maria, Caldas, Colombia. 197 p.

Cupello, M. & F.Z. Vaz-de-Mello. 2013. Taxonomic revision of the South American dung beetle genus Gromphas Brullé, 1837 (Coleoptera: Scarabaeidae: Sacrabaeinae: Phanaeini: Gromphadina). *Zootaxa,* 3722(4): 439-482.

Damborsky, M.P., M.E. Bar, M.C. Alvarez & E.B. Oscherov. 2008. Comunidad de escarabajos copronecrófagos (Coleoptera: Scarabaeidae) en dos bosques del Chaco Oriental Húmedo, Argentina. *Revista de la Sociedad Entomológica Argentina,* 67(1-2): 145-153.

Davis, A.L.V. & C.H. Scholtz. 2001. Historical vs ecological factors influencing global patterns of scarabaeine dung beetle diversity. *Diversity and Distributions,* 7: 161-174.

Davis, A.L.V., C.H. Scholtz, U. Kryger, C.M. Deschodt & W.P. Strümpher. 2010. Dung beetle assemblage structure in Tswalu Kalahari Reserve: responses to a mosaic of landscape types, vegetation communities, and dung types. *Environmental Entomology,* 39(3): 811-820.

Delgado-Gómez, P., A. Lopera & J.O. Rangel-Ch. 2012. Variación espacial del ensamblaje de escarabajos coprófagos (Scarabaeidae: Scarabaeinae) en remanentes de bosque seco en Chimichagua (Cesar, Colombia). 833-849 pp. En: J.O. Rangel-Ch. (Ed.), Colombia Diversidad Biótica XII: La región Caribe de Colombia. Instituto de Ciencias Naturales, Bogotá.

Doube, B. 1990. A functional classification for analysis of the structure of dung beetle assemblages. *Ecological Entomology,* 15: 371-383.

Edmonds, W.D. 1994. Revision of Phanaeus Macleay, a new world genus of Scarabaeine dung beetles (Coleoptera: Scarabaeidae, Scarabaeinae). Contributions in Science, Natural History Museum of Los Angeles Country, 443: 1-105.

Edmonds, W.D. 2000. Revision of the Neotropical dung beetles genus Sulcophanaeus (Coleoptera: Scarabaeidae: Scarabaeinae). *Folia Heyrovskyana, suppl.* 6: 1-60.

Edmonds, W. & J. Zidek. 2004. Research of the Neotropical dung beetle genus Oxysternon (Scarabaeidae: Scarabaeinae: Phanaeini). *Folia Heyrovskyana,* 11: 1-58.

Edmonds, W & J. Zidek. 2010. A taxonomic review of the Neotropical genus Coprophanaeus Olsoufieff, 1924 (Coleoptera: Scarabaeidae, Scarabaeinae). *Insecta Mundi,* 0129: 1-111.

Escobar, F. 2000. Diversidad de coleópteros coprófagos (Scarabaeidae: Scarabaeinae) en un mosaico de hábitats en la reserva Natural Nukak, Guaviare, Colombia. *Acta Zoológica Mexicana* (n.s.), 79: 103-121.

Escobar, F. 2004. Diversity and composition of dung beetle (Scarabaeinae) assemblages in a heterogeneous Andean landscape. *Tropical Zoology,* 17: 123-136.

Escobar, F. & C.A. Medina. 1996. Coleópteros coprófagos (Scarabaeidae) de Colombia: estado actual de su conocimiento. 93-116 pp. In: G. Amat, G. Andrade & F. Fernández (Eds.). Insectos de Colombia. Academia Colombiana de Ciencias Exactas Físicas y Naturales y Pontificia Universidad Javeriana, Bogotá.

Escobar, F., J.M. Lobo & G. Halffter. 2005. Altitudinal variation of dung beetle (Scarabaeidae: Scarabaeinae) assemblages in the Colombian Andes. *Global Ecology and Biogeography,* 14: 327-337.

Esparza-Leon, A.C. & G.D. Amat-García. 2007. Composición y riqueza de escarabajos coprófagos (Coleoptera: Scarabaeidae: Scarabaeinae) en un gradiente altitudinal de selva húmeda tropical del Parque Nacional Natural Catatumbo-Barí (Norte de Santander), Colombia. *Actualidades Biologicas*, 29(87): 181-192.

Estrada, A. & E.R. Coates. 1991. Howler monkey (Alouatta palliata), dung beetles and seed dispersal: Ecological interactions in the tropical rain forest of Los Tuxtlas. *Journal Tropical Ecology,* 7: 475-490.

Estrada, A., G. Halffter, R. Coates-Estrada & D.A. Merrit JR. 1993. Dung beetles attracted mammalian hervivore (Allouatta palliata) and omnivores (*Nasua narica*) dung in the tropical rain forest of los Tuxtlas. *Journal of Tropical Ecology,* 9: 45-54.

Finn J.A. & T. Gittings. 2003. A review of competition in north temperate dung beetle communities. *Ecological Entomology,* 28: 1-13.

Fuentes, P.V. & E. Camero. 2006. Estudio de la fauna de escarabajos copröfagos (Coleoptera: Scarabaeidae) en un bosque húmedo tropical de Colombia. *Entomotropica,* 21(3): 133-143.

Gaston, K.J. & E. Hudson. 1994. Regional patterns of diversity and estimates of global insect species richness. *Biodiversity and Conservation,* 3: 493-500.

Genier, F. 1996. A revision of the neotropical genus Ontherus Erichson (Coleoptera: Scarabaeidae, Scarabaeinae). *Memoirs of The Entomological Society of Canada,* 170: 1-168.

Gill, B.D. 1991. Dung beetles in tropical American forests. 211-230 pp. In: I. Hanski & Y. Cambefort (Eds.), Dung Beetle Ecology. Princeton University Press, Princeton, NJ.

Giller, P.S. & B.M. Doube. 1989. Experimental analysis of inter- and intraespecific competition in dung beetle communities. *The Journal of Animal Ecology,* 58(1): 129-142.

Giller, P.S. & B.M. Doube. 1994. Spatial and temporal co-ocurrence of competitors in Southern African dung beetle communities. *The Journal of Animal Ecology,* 63(3): 629-643.

Giraldo, C., F. Escobar, J.D. Chará & Z. Calle. 2011. The adoption of silvopastoral systems promotes the recovery of ecological processes regulated by dung beetles in the Colombian Andes. *Insect Conservation and Diversity,* 4(2): 115-122.

González, F.A., F. Molano & C.A. Medina. 2009. Los subgéneros Calhyboma, Hybomidium y Telhyboma (Coleoptera: Scarabaeidae: Sacrabaeinae: Deltochilum) en Colombia. *Revista Colombiana de Entomología,* 35(2): 253-274.

Hairston, N., F.E. Smith & L.B. Slobodkin. 1960. Community structure population control and competition. *The American Naturalist,* 94(879): 421-425.

Halffter, G. & E.G. Matthews. 1966. The natural history of dung beetles of the subfamily Scarabaeinae (Coleoptera: Scarabaeidae). *Folia Entomológica Mexicana,* 12-14: 1-312.

Halffter, G. & E.G. Matthews. 1971. The natural history of dung beetles: A supplement on associated biota. *Revista Latinoamericana de Microbiología,* 13:147-163.

Halffter, G. & A. Martinez. 1977. Revision monografica de los Canthonina americanos, IV parte. Clave para géneros y subgeneros. *Folia Entomológica Mexicana,* 38: 29-107.

Halffter, G. & M.E. Favila. 1993. The Scarabaeidae (Insecta: Coleoptera) an animal group for analysing, inventoryng and monitoring biodiversity in tropical rainforest and modified landscapes. *Biology International,* 27: 1-21.

Halffter, G. & V. Halffter. 1989. Behavioral evolution of the nonrolling roller beetles. *Acta Zoológica Mexicana (n.s.),* 32: 1-53.

Halffter, G. & V. Halffter. 2009. Why and where coprophagous beetles (Coleoptera: Scarabaeinae) eat seeds, fruits or vegetable detritus. *Boletin Sociedad Entomológica Aragonesa,* 45: 1-22.

Hammer, Ø., D.A.T. Harper & P.D. Ryan 2001. PAST: Paleontological Statistics Software Package for Education and Data Analysis. *Palaeontologia Electronica,* 4(1): 1-9. URL: http://folk.uio.no/ohammer/past.

Hammond, P.M. 1994. Practical approaches to the estimation of the extent of biodiversity in speciose groups. *Philosophical Transactions of the Royal Society of London B,* 345: 119-136.

Hanski, I. 1980a. Spatial patterns and movements in coprophagous beetles. *Oikos,* 34: 293-310.

Hanski, I. 1980b. Patterns of beetle succession in droppings. *Annales Zoologici Fennici,* 17: 17-25.

Hanski, I. 1981. Coexistence of competitors in patchy environment with and without predation. *Oikos,* 37: 306-312.

Hanski, I. 1983. Distributional ecology and abundance of dung and carrion-feeding beetles (Scarabaeidae) in tropical rain forests in Sarawak, Borneo. *Acta Zoologica Fennica,* 167: 1-45.

Hanski, I. & Y. Cambefort. 1991. Competition in dung beetles. 283-304 pp. In: Hanski, I. & Y. Cambefort (Eds), Dung Beetle Ecology. New Yersey, Princeton University Press.

Hardin, G. 1960. The Competitive Exclusion Principle. *Science,* 131: 1292-1297.

Hastings, A. 1980. Disturbance, coexistence, history, and competition for space. *Theoretical Population Biology,* 18(3): 363-373.

Hernández, J.C., A.G. Hurtado, R.Q. Quijano & T.B. Walschburger. 1992. Unidades biogeográficas de Colombia. En: G. Halffter (Ed.). La Diversidad Biológica de Iberoamerica. 41-43 pp. Acta Zoológica Mexicana, vol. Especial, México.

Hernández, M.I. 2002. The night and day of dung beetles (Coleoptera: Scarabaeinae) in the Serra do Japi Brazil: elytra colour related to daily activity. *Revista Brasileira de Entomologia,* 46(4): 597-600.

Hernández, M.I., L.R. Monteiro & M.E. Favila. 2011. The role of body size and shape in understanding competitive interactions within a community of Neotropical dun beetles. *Journal of Insect Science,* 11(13): 1-14.

Hill, C.J. 1996. Habitat specificity and food preferences of an assemblage of tropical Australian dung beetles. *Journal of Tropical Ecology,* 12(4): 449-460.

Hirabuki, Y. 1990. Vegetation and landform structure in the study area of La Macarena. A physionomic investigation. *Field Studies of New World Monkeys La Macarena Colombia,* 3: 35-48.

Hirschberger, P. 1998. Spatial distribution, resource utilization and intraspecific competition in the dung betle Aphodius ater. *Oecologia,* 116: 136-142.

Holter, P. 1982. Resource utilization and local coexistence in a guild of scarabaeid dung beetles (Aphodius spp). *Oikos,* 39: 213-227.

Hood, G.M. 2010. PopTools version 3.2.3. URL: http://www.poptools.org.

Horgan, F.G. 2005. Aggregated distribution of resources creates competition refuges fro rainforest dung beetles. *Ecography,* 28: 603-618.

Horgan, F.G. 2006. Aggregation and coexistence of dung beetles in montane rain forest and deforested sites in central Peru. *Journal of Tropical Ecology,* 22: 359-370.

Horgan, F.G. & R.C. Fuentes. 2005. Asymmetrical competition between Neotropical dung beetles and its consequences for assemblage structure. *Ecological Entomology,* 30: 182-193.

Howden, H.F. & V.G. Nealis. 1975. Effects of clearing in a tropical rain forest on the composition of the coprophagous scarab beetle fauna (Coleoptera). *Biotropica,* 7(2): 77-83.

Howden, H.F. & O.P. Young. 1981. Panamenian Scarabaeinae: taxonomy, distribution and habits (Coleoptera: Scarabaeidae). *Contributions of the American Entomological Institute,* 18(1): 1-204.

Huijbregts, J. 1984. Bdelyrus beijskesi, a new scarab (Coleoptera: Scarabaeidae) from Suriname with Bromeliaceae. *Zoologische Mededelingen,* 59: 61-67.

Hutton, S.A. & P.S. Giller. 2004. Intra – and interspecific aggregation of north temperate dung beetles on standardized and natural dung pads: the influence of spatial scale. *Ecological Entomology,* 29: 594-605.

Janzen, D. 1983. Insects at carrion and dung. 640-42I pp. In: D.H. Janzen (Ed.). *Costa Rican* Natural History. University of Chicago Press, Chicago.

Jessop, L. 1985. An identification guide to Eurysternine dung beetles (Coleoptera, Scarabaeidae). *Journal of Natural History,* 19: 1087-1111.

Kimura, K., A. Nishimura, K. Isawa & C.A. Mejia. 1994. Annual changes of rainfall and temperature in the tropical seasonal forest at la Macarena field station Colombia. *Field Studies of New World Monkeys La Macarena Colombia*, 9: 1-3.

Klein, B.C. 1989. Effects of forest fragmentation on dung and carrion beetle communities in central Amazonia. *Ecology,* 70: 1715-1725.

Kohlmann, B. 1984. Biosistematica de las especies Norteamericanas del género *Ateuchus* (Coleoptera: Scarabaeidae: Scarabaeinae). *Folia Entomológica Mexicana*, 60: 3-81.

Kohlmann, B. & A. Solis. 1997. El género Dichotomius (Coleoptera: Scarabaeidae) en Costa Rica. Giornale *Italiano di Entomologia,* 8: 343-382.

Kohlmann, B. & A. Solis. 2001. El género Onthophagus (Coleoptera: Scarabaeidae) en Costa Rica. *Giornale Italiano di Entomologia,* 49(9): 159-261.

Krell-Westerwalbesloh, S., F.T. Krell & K.E. Linsenmair. 2004. Diel separation of Afrotropical dung beetle guilds-avoiding competition and neglecting resources (Coleoptera: Scarabaeoidea). *Journal of Natural History,* 38: 2225-2249.

Krikken, J. 2008. Two new species from Kenya in the physogastric termitophilous genus Termitoderus Mateu, 1966 (Coleoptera Scarabaeidae Aphodiinae). *Tropical Zoology*, 21: 153-162.

Larsen, T.H., A. Lopera & A. Forsyth. 2006. Extreme trophic and habitat specialization by Peruvian dung beetles (Coleoptera: Scarabaeidae: Scarabaeinae). *The Coleopterist Bulletin,* 60(4): 315-324.

Laverde, L.J., M.C. Castellanos & P.R. Stevenson. 2002. Dispersión secundaria de semillas por escarabajos coprófagos (Scarabaeidae) a partir de heces de churucos (Lagothrix lagothricha) en el Parque Nacional Tinigua, Colombia. *Universitas Scientiarum*, 7(1): 17-29.

Levins, R. & D. Culver. 1971. Regional coexistence of species and competition between rare species. *Proceedimngs of the National Academy of Sciences of the United States of America*, 68(6): 1246-1248.

Louzada, J. & F.S. Lopes. 1997. A comunidade de Scarabaeidae copro-necrofagos (Coleoptera) de um fragmento de mata Atlântica. *Revista Brasileña de Entomologia*, 41(1): 117-121.

Lumaret J.P., N. Kadiri & M. Bertrand. 1992. Changes in resources: consequences for the dynamics of dung beetle communities. *The Journal of Applied Ecology,* 29(2): 349- 356.

Magurran, A.E. 1988. Ecological diversity and its measurement. Princeton University, Princeton, New Jersery, EE.UU. 192 p.

Martínez, A. 1952. Scarabaeidae nuevos o poco conocidos III. *Mision de Estudios de Patología Regional Argentina,* 81-82: 53-118.

Martínez, A. 1967. Notas para una monografía del género Trichillum Harold, 1868 (Col. Scarabaeinae-Coprini). *Revista de la Sociedad Mexicana de Historia Natural,* 28: 119-147.

Martínez, N., H. García, A. Pulido, D. Ospino & J.C. Narváez. 2009. Fauna de escarabajos coprófagos (Coleoptera: Scarabaeinae) en un gradiente altitudinal en la vertiente Noroccidental de la Sierra Nevada de Santa Marta, Colombia. *Neotropical Entomology,* 38(6): 708-715.

Maruyama, M. 2012. Termitotrox cupido sp. n. (Coleoptera, Scarabaeidae), a new termitophilous scarab species from the Indo-Chinese subregión, associated with Hypotermes termites. *ZooKeys,* 254: 89-97.

Matthews, E.G. 1965. The taxonomy, geographical distribution, and feeding habits of the Canthonines of Puerto Rico (Coleoptera: Scarabaeidae). *Transactions of the American Entomological Society,* 91: 431-465.

Medina, C.A. & A. Lopera. 2000. Clave ilustrada para la identificación de los géneros de escarabajos coprófagos (Coleoptera: Scarabaeinae) de Colombia. *Caldasia,* 22(2): 299-315.

Medina, C.A., A. Lopera, A. Vitolo & B.D. Gill. 2001. Escarabajos coprófagos (Coleoptera: Scarabaeidae) de Colombia. *Biota Colombiana,* 2(2): 131-144.

Medina, C.A., F. Escobar & G.H. Kattan. 2002. Diversity and habitat use of dung beetles in a restored Andean landscape. *Biotropica,* 34(1): 181-187.

Monteith, G.B. & R.I. Storey. 1981. The biology of *Cephalodesmius*, a genus of dung beetles which synthesizes "dung" from plant material (Coleoptera: Scarabaeidae: Scarabaeinae). *Memoirs of the Queensland Museum,* 20: 253-277.

Montes de Oca, E. & G. Halffter. 1995. Daily and seasonal activities of a guild of the coprophagous, burrowing beetle (Coleoptera Scarabaeidae Scarabaeinae) in tropical grassland. *Tropical Zoology,* 8: 159-180.

Morelli, E., P. Gonzalez-Vainer & A. Baz. 2002. Coprophagous beetles (Coleoptera: Scarabaeoidea) in Uruguayan prairies: abundance, diversity and seasonal occurrence. *Studies on Neotropical Fauna and Environment,* 37(1): 53-57.

Moron, M.A. & R.T. Terron. 1984. Distribución altitudinal y estacional de los insectos necrofilos en la Sierra Norte de Hidalgo, México. *Acta Zoologica Mexicana (n.s.),* 3: 1-47.

Murillo, D.A., K. Quiros & A. Rodríguez. 2010. Estudio preliminar de la composición de escarabajos copronecrófilos (Scarabaeidae), en la estacion ambiental Tutunendo Quibdo, Choco, Colombia. *Investigacion, Biodiversidad y Desarrollo,* 29(1): 102-109.

Myers, N., R.A. Mittermeier, C.G. Mittermeier, G.A.B. Fonseca & J. Kent. 2000. Biodiversity hotspots for conservation priorities. *Nature,* 403: 853-858.

Navarrete, J.L. 2001. Beetles associated with Atta and Acromyrmex ants (Hymenoptera: Formicidae: Attini). *Transactions of the American Entomological Society,* 127: 381-429.

Navarrete, J.L. & N.E. Galindo. 1997. Escarabajos asociados a basidiomycetes en San Jose de los Laureles, Morelos, Mexico (Coleoptera: Scarabaeidae). *Folia Entomológica Mexicana,* 99: 1-16.

Neita, J.C., L.C. Pardo, D. Quinto & N. Cuesta. 2003. Los escarabajos copronecrófilos (Coleoptera, Scarabaeidae) en la parcela permanente de investigación en Biodiversidad (PPIB) en Salero, Unión Panamericana, Chocó. p. 79-90. En: F. García, Y. Ramos, J. Palacios, J. Arroyo, A. Mena & M. González (Eds.). Salero Diversidad biológica de un bosque pluvial tropical (bp-T). Universidad Tecnológica del Chocó, Chocó.

Neita, J.C. & F. Escobar. 2011. The potential value of agroforestry to dung beetle diversity in the wet tropical forests of the Pacific lowlands of Colombia. *Agroforest Systems,* 85(1): 121-131.

Newton, A. & S.B. Peck. 1975. Baited pitfall traps for beetles. *The Coleopterist Bulletin,* 29(1): 45-46.

Nichols, E., S. Spector, J. Louzada, T. Larsen, S. Amezquita, M.E. Favila & The Scarabaeinae Research Network. 2008. Ecological functions and ecosystem services provided by Scarabaeinae dung beetles. *Biological Conservation,* 141: 1461-1474.

Noriega, J.A. 2002a. First report of the presence of the genus *Digitonthophagus* (Coleoptera: Scarabaeidae) in Colombia. *Caldasia,* 24(1): 213-215.

Noriega, J.A. 2002b. Aportes a la biología del escarabajo sudamericano Sulcophanaeus leander (Waterhouse, 1891) (Coleoptera: Scarabaeidae). *Acta Zoológica Mexicana (n.s.),* 87: 67-82.

Noriega, J.A. 2004. Preliminary checklist of the scarab community (Coleoptera: Scarabaeidae) at CIEM, Tinigua National Park, Meta-Colombia. *Field Studies of Fauna and Flora La Macarena Colombia*, 14: 37-44.

Noriega, J.A. 2011. A new arboreal dung trap. *Scarabs,* 62: 1-7.

Noriega, J.A. 2012. Dung beetles (Coleoptera: Scarabaeinae) attracted to *Lagothrix lagotricha* (Humboldt) and Alouatta seniculus (Linnaeus) (Primates: Atelidae) dung in a colombian amazon forest. *Psyche,* 437589: 1-6.

Noriega, J.A., & J.C. Calle. 2008. Consumption of Gustavia hexapetala (Aublet) Smith (Lecythidales: Lecythidaceae) by the dung beetle Eurysternus plebejus Harold (Coleoptera: Scarabaeidae). *The Coleopterist Bulletin,* 62(4): 455-460.

Noriega, J.A. & G. Fagua. 2009. Monitoreo de escarabajos coprófagos (Coleoptera: Scarabaeidae) en la región neotropical. 165-188 pp. En: A. Acosta, G. Fagua & A. M. Zapata (Eds.). Técnicas de Campo en Ambientes Tropicales, manual para el monitoreo en ecosistemas acuáticos y artrópodos terrestres. Pontificia Universidad Javeriana, Bogotá.

Noriega, J.A. & A. Acosta. 2011. Population size and dispersal of Sulcophanaeus leander (Coleoptera: Scarabaeidae) on riverine beaches in the Amazonian region. *Journal of Tropical Ecology,* 27: 111-114.

Noriega, J.A., E. Realpe & G. Fagua. 2007. Diversidad de escarabajos coprófagos (Coleoptera: Scarabaeidae) en un bosque de galería con tres estadios de alteración. *Universitas Scientiarum edición especial I*, 12: 51-63.

Noriega, J.A., J.M. Palacio, J.D. Monroy-G. & E. Valencia. 2012. Estructura de un ensamblaje de escarabajos coprófagos (Coleoptera: Scarabaeinae) en tres sitios con diferente uso del suelo en Antioquia, Colombia. *Actualidades Biológicas,* 34(96): 43-54.

Noriega, J.A., E. Camero, J. Arias-Buriticá, L.C. Pardo-Locarno, J.M. Montes, A.A. Acevedo, A. Esparza, B. Murcia, H. Garcia & C. Solis. 2015. Grado de cobertura del muestreo de escarabajos coprófagos (Coleoptera: Scarabaeidae: Scarabaeinae) en Colombia. *Revista de Biología Tropical,* 63(1): en prep.

Otavo, S., A. Parrado-Rosselli & J.A. Noriega. 2013. Superfamilia Scarabaeoidea (Insecta: Coleoptera) como elemento bioindicador de perturbación antropogénica en un parque nacional amazónico. *Revista de Biología Tropical,* 61(2): 735-752.

Pardo, L.C., J.E. Arroyo & F. Quiñonez. 2004. Observaciones de los escarabajos copronecrófagos y sapromelífagos de San Luis Robles, Nariño. *Boletín Científico del Centro de Museos de la Universidad de Caldas,* 8: 113-139.

Peck, S.B. & A. Forsyth. 1982. Composition, structure, and competitive behaviour in a guild of Ecuadorian rain forest dung beetles (Coleoptera; Scarabaeidae). *Canadian Journal of Zoology,* 60: 1624-1634.

Pereira, F.S. 1954. O gênero *Scatonomus* Er. (Coleoptera, Scarabaeidae). *Revista Brasileira de Entomologia,* 1: 53-78.

Pereira, F.S., M.A. Vulcano & A. Martinez. 1960. O gênero Bdelyrus Harold 1869. Actas Trab. 1er Congreso Sudamericano de Zoologia, *La Plata,* 3: 155-164.

Peres-Neto, P., P. Legendre, S. Dray & D. Borcard. 2006. Variation partitioning of species data matrices: estimation and comparison of fractions. *Ecology,* 87: 2614-2625.

Pfrommer, A. & F.T Krell. 2004. Who steals the egg? Coprophanaeus telamon (Erichson) buries decomposing eggs western Amazonian rain forest (Coleoptera: Scarabaeidae). *The Coleopterist Bulletin,* 58: 21-27.

Pulido, L.A., R.A. Riveros, F. Gast & P. Von Hildebrand. 2003. Escarabajos coprófagos (Coleoptera: Scarabaeidae: Scarabaeinae) del Parque Nacional Natural "Serranía de Chiribiquete", Caquetá, Colombia (Parte I). *m3m-Monografías Tercer Milenio (SEA),* 3: 51-58.

Ratcliffe, B.C. 1980. New species of coprini (Coleoptera: Scarabaeidae: Scarabaeinae) taken from the pelage of three toed sloths (Bradypus tridactylus L.) (Edentata: Bradypodidae) in central Amazonia with a brief commentary on scarab-sloth relationships. *The Coleopterists Bulletin,* 34(4): 337-350.

Rougon, D. & C. Rougon. 1991. Dung beetles of the Sahel region. Cap. 13, pags. 230-241. In: I. Hanski and Y. Cambefort (Eds.), Dung Beetle Ecology. Princeton University Press, Princeton, NJ.

Santos-Heredia, C., E. Andresen & D.A. Zarate. 2010. Secondary seed dispersal by dung beetles in a Colombian rain forest: effects of dung type and defecation pattern on seed fate. *Journal of Tropical Ecology,* 26: 355-364.

Schoener, T.W. 1983. Field experiments on interspecific competition. *American Naturalist,* 122: 240.

Solis, A. & B. Kohlmann. 2002. El género Canthon (Coleoptera: Scarabaeidae) en Costa Rica. *Giornale Italiano di Entomologia,* 10: 1-68.

Solis, A. & B. Kohlmann. 2004. El género Canthidium (Coleoptera: Scarabaeidae) en Costa Rica. *Giornale Italiano di Entomologia,* 52(11): 1-73.

Solis, A. & B. Kohlmann. 2013. El género *Uroxys* (Coleoptera: Scarabaeidae) en Costa Rica. *Giornale Italiano di Entomologia,* 13(58): 289-340.

Solis, C., J.A. Noriega & G. Herrera. 2011. Escarabajos coprófagos (Coleoptera: Scarabaeinae) en tres bosques secos del departamento del Atlántico-Colombia. *Boletín del Museo de Entomología de la Universidad del Valle,* 12(1): 33-41.

Stevenson, P.R. 2002. Frugivory and seed dispersal by Woolly Monkeys at Tinigua National Park, Colombia. PhD. Dissertation. Interdepartamental Program in Anthropological Sciences. State University of New York at Stony Brook. Stony Brook. 344 p.

Stevenson, P.R. & M.J. Quiñones. 1993. Vertical stratification of four New World primates, at Tinigua National Park, Colombia. *Field Studies of New World Monkeys La Macarena Colombia*, 8: 11-18.

Stevenson, P.R., M. Suescun & M.J. Quiñones. 2004. Characterization of forest types at the CIEM, Tinigua Park, Colombia. *Field Studies of New World Monkeys La Macarena Colombia*, 14: 1-20.

Tilman, D. 1980. Resources: a graphical-mechanistic approach to competition and predation. *The American Naturalist,* 116: 362-393.

Vårdal, H. & M. Forshage. 2010. A new genus and species and a revised phylogeny of Stereomerini (Coleoptera, Scarabaeidae, Aphodiinae), with notes on assumedly termitophilic aphodiines. *ZooKeys,* 34: 55-76.

Vaz-de-Mello, F.Z. 2008. Synopsis of the new subtribe Scatimina (Coleoptera: Scarabaeidae: Scarabaeinae: Ateuchini), with descriptions of twelve new genera and review of Genieridium, new genus. *Zootoxa,* 1955: 1-75.

Vaz-de-Mello, F.Z., J.N.C. Louzada & J.H. Schoereder. 1998. New data and comments on Scarabaeidae (Coleoptera: Scarabaeoidea) associated with Attini (Hymenoptera: Formicidae). *The Coleopterist Bulletin,* 52(3): 209-216.

Vaz-de-Mello, F.Z., W.D. Edmonds, F.C. Ocampo & P. Schoolmeesters. 2011. A multilingual key to the genera and subgenera of the subfamily Scarabaeinae of the New World (Coleoptera: Scarabaeidae). *Zootaxa,* 2854: 1-73.

Vernes, K., L.C. Pope, C.J. Hill & F. Bärlocher. 2005. Seasonality, dung specificity and competition in dung beetle assemblages in the Australian wet tropics, north-eastern Australia. *Journal of Tropical Ecology,* 21: 1-8.

Villalobos, F.J., A. Diaz & M.E. Favila. 1998. Two species of Canthon Hoffmannsegg (Coleoptera: Scarabaeidae) feed on dead and live invertebrates. *The Coleopterists Bulletin, 52*(2): 101-104.

Vulcano, M.A. & F.S. Pereira. 1973. Duas novas espécies do género Scatonomus Er., 1835 (Col. Scarabaeidae). *Boletim de Zoologia e Biologia Marinha N.S.*, 30: 535-551.

Waterhouse, D.F. 1974. The biological control of dung. *Scientific American*, 230: 100-109.

Young, O.P. 1980. Bone burial by a neotropical beetle (Coleoptera: Scarabaeidae). *The Coleopterists Bulletin,* 34(2): 253-255.

Young, O.P. 1981a. The attraction of neotropical Scarabaeinae (Coleoptera: Scarabaeidae) to reptile and amphibian fecal material. *The Coleopterists Bulletin,* 35(3): 345-348.

Young, O.P. 1981b. The utilization of sloth dung in a neotropical forest. *The Coleopterists Bulletin*, 35(4): 427-430.

Zunino, M. & G. Halffter. 1988. Nueva especie de Onthophagus (Coleoptera, Scarabaeidae) asociada a cuevas. *Folia Entomológica Mexicana,* 75: 17-32.

Zunino, M. & G. Halffter. 2007. The association of Onthophagus Latreille, 1802 beetles (Coleoptera: Scarabaeinae) with vertebrate burrows and caves. *Elytron,* 21: 17-55.

In: Beetles: Biodiversity, Ecology and Role in the Environment ISBN: 978-1-63463-380-2
Editor: Camilla Stack

Chapter 8

BIODIVERSITY IN GEOGRAPHICALLY REMOTE NATURAL POPULATIONS OF *ADALIA* LADYBIRDS (COLEOPTERA: COCCINELLIDAE)

Elena V. Shaikevich* and Ilia A. Zakharov
N. I. Vavilov Institute of General Genetics, Moscow, Russia

ABSTRACT

The species of the genus *Adalia* are highly variable and demonstrate a clearly notable polymorphism in color and pattern of elytra and pronotum. This makes a problem to identify taxonomic status of some forms. Black-colored individuals (melanic morphs) are always present in populations of the most widely distributed Eurasian species, *A. bipunctata*. We studied frequency distribution of the melanics in population from France, Germany, Italy, Bulgaria, the Baltic States, Ukraine, and Russia in comparison with the data reported earlier for the population of the E Europe.

Several centers of melanism with high frequency of black individuals were identified. The highest percentage of the melanics (about 80%) was found in Saint Petersburg and in Rome, confirming the assumption of Y. Lusis that the melanic morphs are better adapted to the conditions of sea climate and polluted atmosphere in big cities.

We studied intraspecies and interspecies DNA polymorphism: the sequence of the mitochondrial gene *COI* and the nuclear sequence ITS2 were analyzed. It was shown that the forms often considered as distinct species, *A. revelierei, A. fasciatopunctata* and *A. turanica*, are in fact the forms of geographically polymorphic populations of the species *A. bipunctata*. As indicated by DNA markers, *A. tetraspilota, A.decempunctata, A.frigida*, showing sympatric distribution with *A. bipunctata* should be considered as distinct species, *A. frigida* being the closest species to *A. bipunctata*. In geographically remote populations of *A. bipunctata* symbiotic male killing bacteria are found: *Spiroplasma* and *Rickettsia*. These bacteria infect the individuals of specific mitotypes (H1 and related mitotypes in case of *Spiroplasma*, H10 in case of *Rickettsia*).

* Corresponding author: Elena V. Shaikevich, E-mail: elenashaikevich@mail.ru

INTRODUCTION

The interest to the ladybird beetles as the objects of ecological-genetic research aroused long ago and remains high at present. The species of the *Adalia* ladybirds have vast distribution areas, allowing the researches to study geographical trends in gene pool variations and to evaluate the factors affecting genetic structure of the populations. Studies of the melanic morphs present examples of a rapid genetic response within populations to climate change; these examples should be evaluated in detail in order to correctly estimate possible contribution of adaptation to the local conditions (Brakefield, de Jong, 2011). While morphological variation of *Adalia* ladybirds was studied long ago (Dobzhansky, 1924), molecular polymorphism was not yet studied systematically.

The main aim of our study was comparative molecular-genetic analysis of the population genetic structure of *A. bipunctata* and its related species across the vast area covering the European part of Russia, Baikal area and some countries to the south from Russia. Two regions of the ladybirds genome were investigated: ITS2 ribosomal region of nuclear DNA and the mitochondrial gene *COI*. Infection of the ladybirds with cytoplasmic symbionts of three genera: *Rickettia*, *Wolbachia* and *Spiroplasma* was also studied. Molecular study of *Adalia* populations from very geographically distant habitats was performed for the first time. The obtained results allowed us to estimate the infection rate of symbiotic bacteria in populations, to analyze ladybird DNA polymorphism and geographical variability, and to clarify the evolutionary relationships between species of the genus *Adalia* and taxonomic relationships between the forms within *A. bipunctata*.

MATERIALS AND METHODS

The ladybirds *A. bipunctata* (Coleoptera: Coccinellidae) were collected at imago and pupae stages in W Europe (Table 1), and also collected in Russia: in St. Petersburg, June 2009; Arkhangelsk, August 2010; Kem, August, 2010; in Ulan-Ude, Baikal Area, September 2010 and also in Uzbekistan, Tashkent, June 2011; *A. decempunctata* was collected in Sweden, Stockholm, Kungens Kurva area, July 2001; *A. frigida* was collected in Arkhangelsk, August 2005; *A. tetraspilota* was collected in Uzbekistan, Tashkent, June 2011.

The morphs of *A. bipunctata* were identified following Ya. Lussis (1928, 1973).

The DNA was isolated from adult beetles using the DIAtom™ DNA Prep Kit (Isogen, Moscow, Russia). Polymerase chain reaction (PCR) was run in thermocyclers GeneAmp^R PCR System 2700 (Applied Biosystems, Foster City, CA, USA) with amplification kits GenePak™ PCR Core (Isogene, Moscow, Russia), 1 mkg of extracted DNA and 5 pmol of each primer.

The mtDNA gene *COI* was amplified using primers LCO and HCO (Folmer et al., 1994) and C1-j-1951 and C1-N-2618 (Schulenburg et al., 2002). The ITS2 region was amplified using primers complementary to the 5,8S and 28S rRNA (Porter &Collins 1991).

The amplicons were revealed by electrophoresis in 1% agarose gel (Sigma, St. Louis, Mo, USA). The amplified DNA fragments were isolated from the gel using JETQUICK Gel Extraction Spin Kit (Genomed, Löhne, Germany) for subsequent sequencing. The sequencing was run from both primers on an ABI 310 automated sequencer using the ABI PRISM

BigDye Terminator Cycle Sequencing kit (Applied Biosystems, Foster City, CA, USA). The newly identified sequences were submitted to GenBank, accession numbers of the gene *COI* HM150667-HM150700; JQ757048-JQ757053 and of the ITS2 JX459794-JX459830.

Table 1. Composition of the *A. bipunctata* populations in W Europe

Locality	date	n	typica	anulata	6-pustulata	4-maculata	melanics others	melanics, %
France								
Marseille	06.1996	260	106	1	57	79	17	58.8±3.05
Avignon	06.1996	22	18	1		3		13.6±7.31
Lille	07.1996	61	47	1	5	8		21.3±5.24
Paris	07.1996	94	62	1	5	26		33.0±4.85
Tours	07.1996	210	128	2	37	43		39.8±3.33
Italy								
Bari	06.2002	28	17	1	8	2		35.7±9.05
Salerno	06.2002	116	103		11	2		11.2±2.93
Roma-Ostia	06.2002	163	25		112	26		84.7±2.82
Bulgaria								
Slynchev briag	06.2000	185	149		13	23		19.5±2.91
Nessebar	06.2000	177	151		7	19		16.7±2.51
Sofia	06.2000	130	101		6	23		22.3±3.65
Germany								
Munchen	06.2002	87	79	2	1	5		6.9±2.72
Greifswald	05.1981	160	142	3	2	11	2	9.4±2.31
Sweden								
Stockholm	07.1998	189	180	3	5	1		3.2±1.28

The obtained sequences were analyzed using software Chromas (http://www.technelysium.com.au) and aligned with MEGA5 (Tamura et. al. 2011). The evolutionary distances for *COI* were computed using five methods: 1) the p-distance method; 2) the Jukes-Cantor method commonly used to evaluate evolution of protein molecules; 3) the Tamura-Nei method commonly used to estimate the number of nucleotide substitutions in the control region of mitochondrial DNA; 4) the Kimura 2-parameter method, which is a simple way of estimating the evolutionary rate of base substitutions by comparing nucleotide sequences; and 5) the Tajima-Nei method in MEGA5 developed for estimating the evolutionary distance between nucleotide sequences (Tamura et. al. 2011). The values of divergence for the sequences of the *COI* gene obtained using all the above methods were practically identical (not shown). Tables present the data obtained using the p-distance method.

Phylogenetic analyses were conducted in MEGA5 (Tamura et. al. 2011). The bootstrap consensus trees inferred using the Neighbour-Joining method were generated to represent the evolutionary history of *Adalia* based on *COI* and ITS2. The percentage of replicate trees in which the associated taxa clustered together in the bootstrap test (1000 replicates) is shown next to the branches. The trees were drawn to scale, with branch lengths in the same units as those of the evolutionary distances used to infer the phylogenetic tree. The evolutionary distances were computed using the Maximum Composite Likelihood method and are expressed in terms of the number of base substitutions per site. Codon positions included were 1st+2nd+3rd. All positions containing gaps and missing data were eliminated from the dataset (Complete deletion option). Nucleotide polymorphism from aligned DNA sequences of both *COI* and ITS2 data was analyzed using the DNA Sequence Polymorphism (DnaSP) software (Librado & Rozas, 2009).

Infection of the beetles by symbiotic bacterium *Spiroplasma* was detected using PCR with the primers MGSO and FP specific to the small subunit of the bacterial 16S RNA (Kuppeveld et al., 1992). *Rickettsia* were identified with the help of the primers specific to the 17-kDa *Rickettsia* antigen, Rr17.61p and Rr17.492n (Williams et al., 1992). Identification by bacteria of the genus *Wolbachia* in total ladybird beetle DNA was performed by means of PCR with the primers complementary to the wsp gene for the *Wolbachia* surface proteins, wsp 81F and wsp 691R (Braig et al., 1998).

Taxonomy of the ladybird beetles *Adalia*

The ladybird beetles of the genus *Adalia* (Coleoptera, Coccinellidae) demonstrate notable intra-populational polymorphism and intraspecies geographic variations for two common morphological traits: color and pattern of pronotum (dorsal prothorax) and elytra (forewings). The two-spot ladybird *Adalia bipunctata* is widely distributed in the Palaearctic region, and it is assumed that the distribution area of *A. bipunctata* covers Europe, Siberia, and Central Asia. In the Palearctic region some taxonomists identify ten species of the genus *Adalia*. Three related species of the genus *Adalia*: *A. decempunctata*, *A. tetraspilota*, *A. frigida* are close to *A. bipunctata*.

Adalia bipunctata

The two-spotted ladybird *Adalia bipunctata* inhabits the Palearctic zone of both the Old World and the New World, often showing high abundance in towns and settlements. It is considered as a synanthropic species. In the Old World the habitat covers all Europe, Siberia, and Middle Asia. At the periphery of the distribution area, especially to the south and to the south-east, the typical form is substituted by the forms of different color (Figure 1). Basing on morphological differences, Lusis (1973) assumed that *A. bipunctata* is a polytypic species, consisting of four geographic populations or subspecies: *A. b. bipunctata* L. (Europe and Asia), *A. b. fasciatopunctata* Fald. (Mongolia, Tuva, Transbaikalia and the adjacent part of Siberia), *A. b. turanica* Lus (Central Asia) and *A. b. revelierei* Muls. (Middle East). The taxonomic status of these forms is still a matter of study and debate.

A typical *A. bipunctata* has two black spots, one on each of the two red-colored elytra. Black morphs of *A. bipunctata* also exist and are called melanics, they have 2-4-6 red spots on black foreground color. Other forms of *A. bipunctata* show multiple spots forming specific patterns on elytra (Figure 1). Basing on morphological traits and knowing that a specific population has a certain specific ecologo-geographic habitat and, evidently, its own history of development, many researches considered the forms of distinct morphology as distinct species or subspecies: *A. bipunctata* Linneaus, 1758 (Eurasia), *A. revelierei* Mulsant, 1866 (Western Asia, Asia minor, South Caucasus), *A. turanica* Lusis, 1947 (Central Asia), *A. fasciatopunctata* Mader, 1931 (Mongolia, Tyva, Transbaikalia, and adjacent part of Siberia). Ya. Lussis (1973) basing on the analysis of multiple crossing of individuals from various geographic sites has proved that *A. revelierei* and *A. turanica* early identified as distinct species produce fertile hybrids with *A. bipunctata* and considered them as subspecies of *A. bipunctata*.

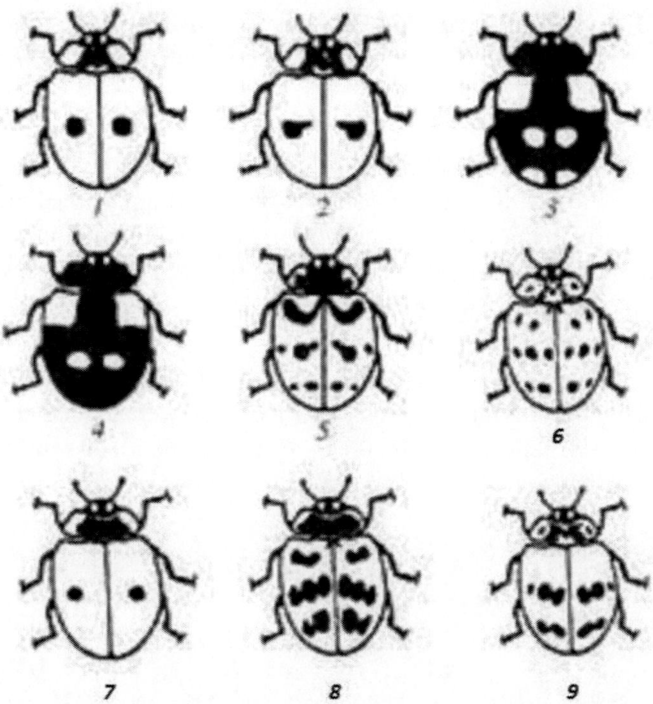

Figure 1. Ladybird beetles of the genus *Adalia*. 1 – 6 - *A. bipunctata*: 1 – f. typica, 2 – f. anulata, 3 – f. 6-pustulata, 4 –f. 4-maculata, 5 – f.fasciatopunctata, 6 – f.revelierei; 7, 8 - *A.tetraspilota*; 9 – *A.frigida*.

Other Species of the Genus *Adalia*

Close to *A. bipunctata* are three species of the genus *Adalia*: *A. decempunctata* Linneaus, 1758, *A. tetraspilota* Hope, and *A. frigida* Schneider, 1792. The species *A. decempunctata* and *A. tetraspilota* produce no hybrids with *A. bipunctata* and with each other, this confirms their status of distinct species (Lussis, 1973). The habitat of *A. decempunctata* covers the areas of moderate climate in Europe, where *A. decempunctata* coexists with *A. bipunctata*. Mating between them is possible, but the eggs do not develop. *A. tetraspilota* (Figure 1) is an Indian species found in the Southern part of countries formed within the former Soviet Central Asia. A circumpolar species *A. frigida* is found in near-polar areas of North Eurasia, from Scandinavia to Yakutia; its morphology is very similar to morphology of the forms included in the *A. bipunctata* complex (Lussis, 1976). Taxonomical status of *A. frigida* still remains unclear. Different authors consider *A. frigida* as a subspecies or a form of *A. bipunctata* (Kuznetsov, 2006, Korsun, 1997) or as a distinct species *A. frigida* (Lussis, 1976; Bielawski, 1984).

Mixed populations and mating of *A. frigida* with typical *A. bipunctata* were observed in Transbaikal area, however, notable ecological and behavioral differences were also reported. A.V. Butko has revealed different preferable wintering sites and feeding for the two species. The author suggested the existence of partial spatial (biotope), temporal and behavioral isolation of *A. frigida* individuals within the transbaikalian populations of *A. bipunctata* (Butko, 2005).

Morphological Polymorphism of *A. bipunctata* in Populations of the Western Europe

Composition *A. bipunctata* population on the territory of the former USSR was reported by Lussis (1961) and Zakharov (1990, 2003), Any population of *A. bipunctata* always contains individuals with various color patterns on pronotum and elytra. The populations from Europe and Siberia include the beetles with red elytra and two black spots (forms typica and anulata) and individuals with black elytra and 4 or 6 red spots (forms 4-maculata and 6-pustulata). At South periphery of the distribution area some individuals have red or pink elytra and some black spots (Figure 1).

Here we report the results of the original research for the populations from the W Europe (Table 1, Figure 2). The ladybird beetles *A. bipunctata* were collected by I. Zakharov in several countries in 1981 and in 1996-2002, i.e. before mass dissemination in Europe of *Harmonia axyridis* , a species which competes with *A. bipunctata* and may affect populations of the latter.

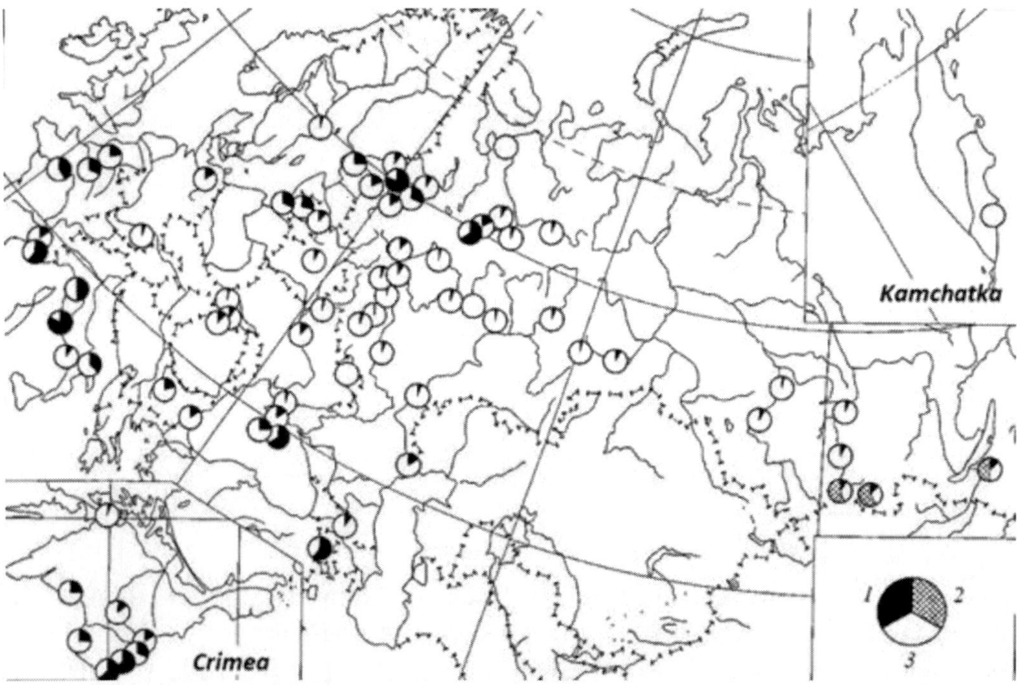

Figure 2. Composition of *A. bipunctata* populations in North Eurasia. Size of the black (1) and dashed (2) sectors reflects the relative number of melanics and of f. *fasciatopunctata* individuals. Sector (3) shows the relative number of *A. bipunctata typica* with red elytra and two black points.

Among the populations studied by us, higher percentage of the black individuals (melanics) was found in Marseille and in Rome. Accumulation of melanics in these populations supports suggestion of Ya. Lusis (1961) that the melanic morphs are advantageous compared with the red morphs in big cities with polluted atmosphere and sea climate. This hypothesis also explain high percentage of the melanics in Saint Petersburg, in Glasgow, and in Edinburgh (Creed, 1971; Zakharov, 1990, 2003).

Polymorphism of mtDNA in Populations of *Adalia bipunctata*

The sequences of the gene *COI* were studied in individual *A. bipunctata* from populations of St. Petersburg (34 specimen), Kem (32), Arkhangelsk (14) and Tashkent (10), *A. b. bipunctata* and *A. b. fasciatopunctata* from Buryatia (20), *A. b. bipunctata* and *A. b. revelierei* from Erevan (20) and also *A.b. turanica* from Kirgizia (1 individual).

Basing on composition of the mt-haplotypes, the population of *A. bipunctata* from Saint Petersburg appeared highly polymorphic, 6 mt haplotypes out of 10 described earlier by Schulenburg and coworkers (2002) were found, haplotypes №5, 6, 7, 8 were not found. The numbers of the mt-haplotypes correspond to the numbers suggested earlier (Schulenburg et al., 2002). Beside these, we found new haplotypes numbered 11, 12, 13, 14, 15, 16 and 17 (Figure 3).

The specimen from Kem had three mt-haplotypes: H2 (1 individual ladybirds), H7 (15). and H10. The population from Arkhangelsk contained four mt-haplotypes: H1 (8 specimen), H7 (5), H9 (4), and H10 (3 specimen) (Shaikevich et al., 2012).

The population of Buryatia included the forms *A. bipunctata* typica and *A.bipunctata* f. *fasciatopunctata.* Four specimen of the studied population had mt-haplotype H10, which was found in two specimen of *A.bipunctata* f. *fasciatopunctata* and in two specimen of *A. bipunctata* typica, other studied specimen, 8 *A. bipunctata* typica and 8 *A. bipunctata* f. *fasciatopunctata* had mt-haplotype H1 (Table 2).

The available individual of *A. b. turanica* from Kirgizia had mt-haplotype H1. The specimen of *A. bipunctata*, collected in Tashkent had only two haplotypes: 14 specimen had H1, and one individual had mt-haplotype H18, which differs from H1 by one nucleotide substitution.

In the *Adalia* population from Erevan we studied the sequences of the gene *COI* for 8 specimen with phenotype of f. *revelierei* and 12 specimen of phenotype *A. bipunctata* typica: identical mt haplotypes were found for both forms: 20 beetles had 5 already known mt-haplotypes and two new mt-haplotypes which differ from the known ones by one additional nucleotide substitution. Following the available numbering we defined them as H19 and H20. The mt-haplotype H1 was the most frequent type for specimen with phenotype *A. b.* f. *revelierei* (Table 2), the same mt-haplotype prevailed in other populations of the *A. bipunctata* complex. One individual had mitotype H3. And one had a new mt-haplotype defines by us as H20. Among the typical *A. bipunctata* from Erevan, 8 specimen had the mt-haplotype H1, one had H4 and one had H19. Two specimen of *A. bipunctata* typica from the Erevan population had mt-haplotype H9.

Our study has demonstrated that all the analyzed populations show mtDNA polymorphism. We found 14 mt-haplotypes in populations from Russia. Except the population from Kem, the mt-haplotype H1 was dominant (Table 2). The population from Tashkent had H1 and a new mt-haplotype 18, similar to the mt-haplotype 1 (Figure 3). Two more new mt-haplotypes, H19 and H20 were found in the population from Erevan. In total, for the 126 specimen of the species *A. bipunctata* we found 16 distinct haplotypes. The main part of the sample, 52 specimen of *A. bipunctata* had the mt-haplotype H1, 35 specimen had the haplotype H10, 19 had the haplotype H7, 7 specimen had H9, 4 specimen had H3, 2 specimen were found for each H2, H4 and H11, and one individual was found to have each of the haplotypes H12-H20.

```
[                             111111 1111111111 1222222222 2222222333 3333333333 3333333444 4444444444 4445555555 55555]
[                  12333455 5666022445 5566678889 9023344456 6677889000 0111233335 5667789011 2334455578 9990011122 33346]
[                3907349812 7036209140 3725670395 8181836752 4536284036 9289403691 7162879814 9584556976 2581703625 47894]
#A.bipunctataH1   ATCAATTCAC TATCTACATA CCATTAGTGC ATATTCTTTG AAGATAATAT TTTAATGTTA TTCTTTATTT ATGTCCATTA TAATAAATAA TATTA
#A.bipunctataH2   .......... .......... .......... .......... .......... .......... .......... .......... .......... .....
#A.bipunctataH3   .......... .......... .......... .......... .......... .......... .......... .......... .......... .....
#A.bipunctataH4   .......... .......... .......... .......... .......... .......... .......... .........G .......... .....
#A.bipunctataH7   .......... .......... .......... .......... .......... .......... .......... .......... .......... .....
#A.bipunctataH9   .AT...CT.. C.....AG.. .....GA.A. ...C....C. ..A......C ..C....... .......... ..ACT..... ....TGG... ..C..
#A.bipunctataH10  .AT.GCAT.T .......G.. .T.GCGA.AT G..C.T.C.. ..A..G.CC. .....C...G .C.C...... .AT.T..... .T....GC.. ..CC.
#A.bipunctataH11  .......... .......... .......... .......... .......... .......... .......... .......... .......... .....
#A.bipunctataH12  .......... .......... .......... .......... .......... .......... .......... .......... .......... .....
#A.bipunctataH13  .......... .......... .......... .......... .......... .......... .......... .....T.... .......... .....
#A.bipunctataH14  .......... .......... .......... .......... .......... .......... .......... .......... A......... .....
#A.bipunctataH15  .......... .......... .......... .......... .......... .......... .......... .........G .......... .....
#A.bipunctataH16  .......... .......... .......... .......... .......... .......... .......... .......... .......... .....
#A.bipunctataH17  .......... .......... .......... .......... .......... .......... .......... .......... .......... .....
#A.bipunctataH18  .......... .......... .......... .......... .......... .......... .......... .......... .......... .....
#A.bipunctataH19  .......... .......... .......... .......... .......... .......... .......... .......... .......... .....
#A.bipunctataH20  .......... .......... .......... .......... .......... .......... .......... .......... .......... .....
#A.frigida1       .A....CTG. C.....AG.. .....GA.A. ........C. ..AG.....C ..C..A.... .......C.. T..CT..... ....T.G... ..C.G
#A.frigida2       .A....CTG. C.....AG.. .....GA.A. ........C. ..AG.....C ..C..A.... .......C.. T.ACT..... ....T.G... ..C.G
#A.tetraspilota1  GA...CAT.. ...T.G.G.. TT.A..A... .C.CCTA... ..AG.G.C.C ...C..A... ...C...... ..TC....C. .T.A...... .....
#A.tetraspilota2  GA...CAT.. ...T.G.G.. TT.A..A.A. .C.CCTA... ..AG.G.C.C ...C..A... ...C...... ..TC....C. .T.A...... .....
#A.decempunctata1 TC.G..AT.T .TC.C...AT TTT...TA.T ..G..T...A TGAGATTCC. CA.TC.AAGT CCTACCT.CA ..T.T.CAA. .TT.T...GG CG.CT
#A.decempunctata2 TC.G..AT.T .TC.C...AT TTT...TA.T ..G..T...A TGAGATTCC. CA.TC.AAGT CCTACCT.CA ..T.T.CAA. .TT.T...GG CG.CT

[                 5555555556 6666666666 6777777777 7777777777 7888888888 8888888889 9999999999 999999]
[                 6778889990 1123356677 9011222234 5566677889 9001122334 5556889990 0112334445 556667]
[                 7692581476 2870673958 0214036985 0602517395 8170925146 2587581273 6581065681 243690]
#A.bipunctataH1   ACTTAAGTAC CTTTAAAATT AGTTACATTT TTCAATATAT TACCATCTGT AAACAAACTT ATTTACTTAT CATGGC
#A.bipunctataH2   .......... .......... .......... .......... .......... .......... ....G..... ......
#A.bipunctataH3   .......... .......... ........C. .......... .......... .......... .......... ......
#A.bipunctataH4   .......... .......... .......... .......... .......... .......... .......... ......
#A.bipunctataH7   .......... .......... .A........ .......... .......... .......... .......... ......
#A.bipunctataH9   .....GA... T......... .A..G.G... .CT....... ....G...AC .TT......C .......... ..CAA.
#A.bipunctataH10  GTC....C.. ....G..... .A...T.... C.T..C.CG. CGTT....A. GGT......C ..CCGT.... ....T.
#A.bipunctataH11  ...G...... .......... .......... .......... .......... .......... .......... ......
#A.bipunctataH12  ........G. .......... ........C. .......... .......... .......... .......... ......
#A.bipunctataH13  ...G...... .......... .A......C. .......... .......... .......... .......... ......
#A.bipunctataH14  .......... .......... .......... .......... .......... .......... .......... ......
#A.bipunctataH15  ...G...... .......... .......... .......... .......... .......... .......... .G....
#A.bipunctataH16  .......... .......... ........C. .......... .......... .......... .......... ....T.
#A.bipunctataH17  .......... .......... .......... .......... .......... .......... .......... .G....
#A.bipunctataH18  .......... .......... .......... .......... ...A...... .......... .......... ......
#A.bipunctataH19  .......... .......... .......... .......... .......... .......... .....T.... ......
#A.bipunctataH20  .......... .......... .......... .......... ......T... .......... .......... ......
#A.frigida1       .T........ T...G..T.. .A.C...... ..T....... ....G...AC ..T...C..C .......... ...AA.
#A.frigida2       .T........ T...G..... .AC...G... ..T....... ....G...A. ..T...C..A .......... ...AA.
#A.tetraspilota1  ....G..... T.CC.G.... .T...T.... ..T..A.C.. .G.TG..CA. ..TT.T.T.C ..C..T...C TG.A..
#A.tetraspilota2  ....G..... T.CC.G.... .T...T.... ..T..A.C.. .G.TG..CA. ..TT.T.T.C ..C..T...C TG.A..
#A.decempunctata1 .....GA..T TC.A..GTCC CA..GT.GCC CA.TCAG..A A.TA.CTCA. ...TT..TC. GC...TCCC. .T..CT
#A.decempunctata2 .....GA..T TC.A..GT.C CA..GT.GCC CA.TCAG..A A.TA.CTCA. ...TT..TC. GC...TCCC. .T..CT
```

Figure 3. Alignment of variable nucleotide sequences of the central region of the gene *COI* from *Adalia ssp.* H1-H20 are polymorphic mt-haplotypes of *Adalia bipunctata*.

Variability of the mtDNA gene *COI* within and between populations of the *A. bipunctata* complex, excluding the mt-haplotypes H9 and H10, is below 0.4%, essential contribution to this variation is made by the populations from St. Petersburg and Arkhangelsk (Table 3). The mean number of nucleotide substitutions per site for the 29 studied specimen of various forms of the *A. bipunctata* complex (mt haplotypes H1, H3, H4, H19, and H20) is 0.001, while the difference between mt-haplotypes H9 and H10 and other haplotypes is at the level of interspecies variation (Table 3).

Table 2. Composition of the *Adalia bipunctata* populations and infection of the beetles with symbiotic bacteria

Population	*COI type*	N	N of infected with	
			Spiroplasma	*Rickettsia*
S.Petersburg	1	18	11	
	2	1	1	
	3	3	1	
	4	1		
	9	1		
	10	2		
	11	2	1	
	12	1		
	13	1		
	14	1		
	15	1		
	16	1		
	17	1	1	
Total		34	15	
Arkhangelsk	1	8		
	7	5		
	9	4		
	10	3		
Total		20		
Kem	2	1		
	7	15		
	10	16		16
Total		32		16
Moscow*	1	11	4	
	2	1		
	3	4		
	5	1		
	6	1		
	8	1		
	9	1		
	10	2		2
Total		22	4	2
Buryatia *A. bipunctata*	1	8		
	10	2		
A.b. fasciatopunctata	1	8		
	10	2		1
Total		20		1

Population		COI type	N	N of infected with	
				Spiroplasma	*Rickettsia*
Tashkent		1	14		
		18	1		
Total			15		
Erevan	*A. bipunctata*	1	8		
		4	1		
		9	2		
		19	1		
	A.b.revelierei	1	6		
		3	1		
		20	1		
Total			20		
Kirgizia	*A.b.turanica*	1	1		
Germany*		1	7	3	
		2	4	1	
		3	1		
		4	1		
		7	1		
		9	3		2
		10	2		2
Total			19	4	4
In all populations			173	24	23

Notes: * Data from Schulenburg et al., 2002.

Comparison of mtDNA for Different Species of the Genus *Adalia*

Structure and variability of the mtDNA gene *COI* were studied in three species close to *A. bipunctata*. The mt-haplotypes were analyzed for two specimen of *A. decempunctata* (Stockholm, Sweden) and *A. tetraspilota* (Tashkent, Uzbekistan). We also studied an individual of *A. frigida* (№1) and its progeny from Arkhangelsk, and one more individual from Yakutia (№2) (Figure 3). All found haplotypes were compared. For the 23 haplotypes revealed in four species, 171 variable nucleotide sites were found, 144 being informative basing on the principle of parsimony. Five nucleotide substitutions, three of them are at the first codon position, result in amino acid substitution (Figure 3). Most of the studied sequences of the gene *COI* differ within the species by less than 0.4 %. The minimal evolutionary divergence (number of nucleotide substitutions per site), 4.3 % was found between *A. bipunctata* and *A. frigida*. The highest difference is found between *A. decempunctata* and other species, about 12.5 % (Table 4).

We compared the peculiar DNA sequences of *A. bipunctata*, mt-haplotypes H9 and H10, with the gene *COI* sequences from other species. The haplotype H9, reported for the *A. bipunctata* populations from North regions of Europe (Schulenburg et al., 2002; Jiggins and Tinsley, 2005), was found by us not only in Saint Petersburg and Arkhangelsk, but also in Erevan. This mt-haplotype differs from the group of the common haplotypes of *A. bipunctata* by 4.3 %, from *A. frigida* by 2.3 %, from *A. tetraspilota* by 6.7 %, and from *A. decempunctata* by 13.6%. The sequences of the type H10 differ from the group of the common *A. bipunctata* haplotypes by almost 7 %, showed same difference from *A. frigida* and *A. tetraspilota*, and differed from *A. decempunctata* by 13.5 %.

Table 3. Evolutionary divergence values (number of nucleotide differences per site) between *COI* haplotypes H1 to H20

	H1	H2	H3	H4	H5	H6	H7	H8	**H9**	**H10**	H11	H12	H13	H14	H15	H16	H17	H19	H20
H1																			
H2	0,002																		
H3	0,002	0,004																	
H4	0,002	0,004	0,004																
H5	0,002	0,004	0,004	0,004															
H6	0,004	0,006	0,002	0,006	0,006														
H7	0,002	0,004	0,004	0,004	0,004	0,006													
H8	0,006	0,008	0,004	0,008	0,008	0,006	0,008												
H9	**0,045**	**0,047**	**0,047**	**0,047**	**0,047**	**0,049**	**0,043**	**0,047**											
H10	**0,064**	**0,062**	**0,066**	**0,066**	**0,062**	**0,068**	**0,062**	**0,066**	**0,075**										
H11	0,002	0,004	0,004	0,004	0,004	0,006	0,004	0,008	**0,047**	**0,066**									
H12	0,004	0,006	0,002	0,006	0,006	0,004	0,006	0,006	**0,049**	**0,068**	0,006								
H13	0,008	0,009	0,006	0,009	0,009	0,008	0,006	0,009	**0,049**	**0,068**	0,006	0,008							
H14	0,002	0,004	0,004	0,004	0,004	0,006	0,004	0,008	**0,047**	**0,066**	0,004	0,006	0,009						
H15	0,006	0,008	0,008	0,004	0,008	0,009	0,008	0,011	**0,051**	**0,070**	0,004	0,009	0,009	0,008					
H16	0,002	0,004	0,000	0,004	0,004	0,002	0,004	0,004	**0,047**	**0,066**	0,004	0,002	0,006	0,004	0,008				
H17	0,002	0,004	0,004	0,004	0,004	0,006	0,004	0,008	**0,047**	**0,066**	0,004	0,006	0,009	0,004	0,004	0,004			
H18	0,002	0,004	0,004	0,004	0,004	0,006	0,004	0,008	**0,047**	**0,062**	0,004	0,006	0,009	0,004	0,008	0,004	0,004		
H19	0.002	0.004	0.004	0.004	0.004	0.006	0.004	0.008	**0.047**	**0.064**	0.004	0.009	0.006	0.004	0.008	0.004	0.004	0.004	
H20	0.002	0.004	0.004	0.004	0.004	0.006	0.004	0.008	**0.047**	**0.066**	0.004	0.009	0.006	0.004	0.008	0.004	0.004	0.004	0.004

Differences between haplotypes H9 and H10 and other *A. bipunctata* haplotypes are shown in bold. Haplotypes H5, H6 and H8 were absent from the populations examined and were taken for comparison from the study of Schulenburg et al. (2002). The data are based on pairwise comparison of 20 sequences of the COI gene, 532 bp in size, using the maximum likelihood method in the program MEGA 5 (Tamura et al., 2011).

Table 4. The level of evolutionary divergence between the sequences of the gene *COI* in the studied subspecies and species of the genus *Adalia*

	*A.bipunctata**	*A.b.fasciatopunctata**	*A.b.turanica*	*A.b.revelierei*	*A.frigida*	*A.tetraspilota*	*A.decempunctata*	H9**
*A.bipunctata**								
*A.b.fasciatopunctata**	0.001							
A.b.turanica	0.001	0.001						
*A.b.revelierei**	0,001	0,001	0,001					
A.frigida	0.043	0.043	0.043	0.043				
A.tetraspilota	0.062	0.062	0.062	0.062	0.066			
A.decempunctata	0.120	0.120	0.120	0.120	0.138	0.129		
H9**	0.043	0.043	0.043	0.043	0.023	0.067	0.136	
H10***	0.067	0.067	0.067	0.067	0.070	0.069	0.135	0.067

Levels of evolutionary divergence are presented in the values indicating the average number of nucleotide substitutions per site between the pairwise compared sequences.

* *A. bipunctata* (including all mtDNA haplotypes except H9, H10).

** *A. bipunctata* mt-haplotype H9.

*** *A. bipunctata* mt-haplotype H10.

Our analysis of DNA polymorphism of the gene *COI* has revealed that, 64 variable sites are different between H10 and *A.frigida*, 66 – 68 nucleotide sites are different from *A. bipunctata*, 64 from *A. tetraspilota* and 116 sites are different from *A. decempunctata*. The mt-haplotype H10 appears genetically similarly distant from *A. bipunctata*, *A. frigida*, and *A. tetraspilota*, the distance from *A. decempunctata* being the highest. The mt-haplotype H9 is genetically similar to *A. frigida*, it is closer than the H10 to the other mt-haplotypes of *A. bipunctata*. The genetic distances for the H9 with *A. tetraspilota* and *A. decempunctata* are practically identical to those for the H10 (Table 4, Figure 4).

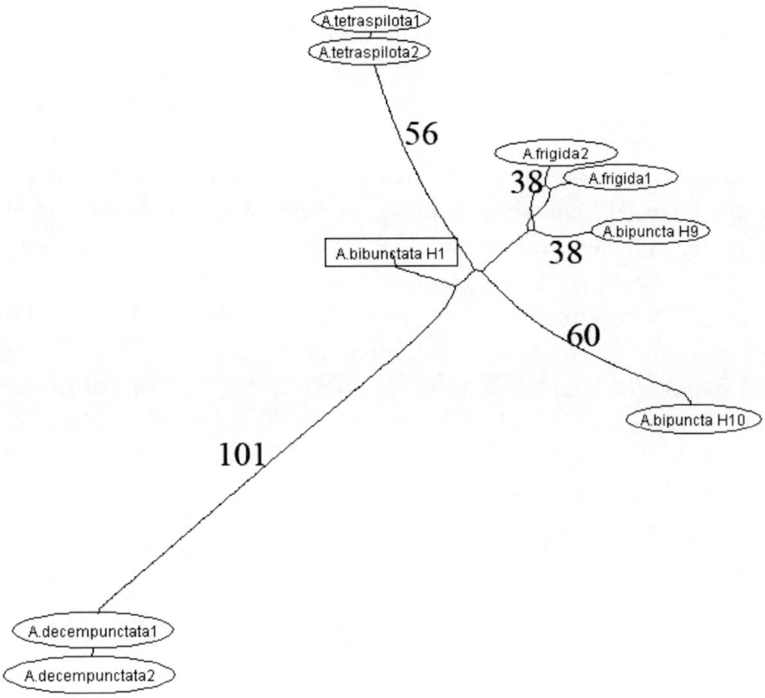

Figure 4. Median net of mt-haplotypes of *Adalia*. Numbers on the lines joining the haplotypes show the number of nucleotide substitutions.

Relation of mtDNA Variability and Infection with Symbiotic Male-Killing Bacteria

Aiming to reveal a relationship in the patterns of geographic distribution of the *Adalia* mt-haplotypes and symbiotic bacteria, we studied infection of the beetles by *Spiroplasma, Rickettsia* and *Wolbachia*. The latter were not found in the studied material. The obtained results are shown in Table 2 in comparison with the data for the populations from Moscow and Germany published by Schulenburg and coworkers (2002).

In Saint Petersburg population in 2009, 21 out of 84 sampled beetles were infected with *Spiroplasma* (the frequency of infection correspond to the data obtained by us in 1998 - 2000 (Zakharov et al., 1998, 2000). The males were not separated from the females. If the females form about 50% of the sample, than about 50% of the females should be infected (the infected males cannot be detected, since *Spiroplasma* is a male-killing agent). Most of the infected beetles had mt-haplotype H1 (18 specimen); the mt-haplotypes 2, 3, 17, 11, each different from the H1 by one nucleotide substitution were also found for the infected individuals. At the same time, the specimen uninfected by *Spiroplasma* had 10 mt-haplotypes, 9 of them differ from H1 by 1-4, 24 and 34 nucleotides (Table 2).

Known that the Saint Petersburg population is close to the northern border of the distribution area of the species and considering the bottleneck effect at formation of the Saint Petersburg population of *A. bipunctata*, low level of polymorphism could be expected. However, this population appeared highly polymorphic (13 mt-haplotypes). It should be noted that among the beetles infected with *Spiroplasma* is considerably lower level of mtDNA variability.

Mean nucleotide variation was 0.002 for the sample of infected specimen and 0.021 for the sample of uninfected specimen (Table 5). High level of polymorphism in sample of not infected by *Spiroplasma A. bipunctata* is manly related with the two mt-haplotypes notably different from the other, H9 and H10. However, even if we exclude H9 and H10, variability in the sample of the infected specimen (S+ = 0,002) remains two times lower than in the sample of uninfected specimen (S- = 0,004).

In Arkhangelsk, 20 specimen had four mt-haplotypes, H1, H7, H9 and H10, no infection was detected. In Kem, 32 specimen had three mt-haplotypes H2, H7 and H10, all 16 specimen with haplotype H10 were found to be infected by *Rickettsia*. In Buryatia, 20 specimen had mt-haplotypes H1 and H10, one individual was infected with *Rickettsia* and had mt-haplotype H10. In Tashkent, 15 specimen had mt-haplotypes H1 and H18, no infection was found. In Erevan, six mt-haplotypes were found H1, H3, H4, H9, H19 and H20, the symbionts were not detected (Table 2).

Polymorphism of Nuclear DNA in Ladybird Beetles of the Genus *Adalia*

For our studies of nuclear DNA variation the sequence of the rRNA second internal transcribed spacer (ITS2) was used as a marker.

The size of the amplified fragment of the ITS2 region was 900 bp. Analyzed were 8 specimen of *A. bipunctata* typica, one f. *turanica* , and 2 f. *fasciatopunctata* with the typical haplotype H1, 5 specimen with the haplotype H9, 7 f. typica and 2 f. *fasciatopunctata* with the haplotype H10, and also 3 *A. frigida*, 8 *A. tetraspilota* , and 1 *A. decempunctata*. Contrary to mtDNA, the ITS2 appeared absolutely conserved within a species. All studied *A. bipunctata*, independently of the haplotype, morphological traits and geographical distribution had practically identical ITS2 sequences.

Mean frequency of nucleotide substitutions per site was 0.001 for 25 specimen of *A. bipunctata*. Monomorphic ITS2 were also found for the other studied species (Figure 5). The evolutionary divergence between *A. bipunctata* and *A. frigida* are 1,8 %, reaching 41.2 % for *A. bipunctata* and *A. decempunctata* and 57.8 % for *A. bipunctata* and *A. tetraspilota*. Based on ITS2 analysis, *A. decempunctata* differs by 41.6% from *A. frigida*; *A. tetraspilota* differs from *A. frigida* by 58,5% and from *A. decempunctata* by 57.7%.

Table 5. Difference in *COI* gene sequences between *A. b. bipunctata* infected and not infected with *Spiroplasma*.
The number of base substitutions per site from averaging over all sequence pairs between infected and uninfected groups is shown

[	1	2	3	4	5	6	7	8	9	10	11	12	13	14	15	16	17	18	19]
[1]		0.002	0.046	0.066	0.000	0.000	0.066	0.000	0.004	0.000	0.004	0.000	0.009	0.004	0.006	0.002	0.002	0.004	0.000
[2]			0.048	0.068	0.002	0.002	0.068	0.002	0.006	0.002	0.006	0.002	0.011	0.006	0.004	0.004	0.004	0.006	0.002
[3]				0.077	0.046	0.046	0.077	0.046	0.050	0.046	0.050	0.046	0.052	0.050	0.052	0.048	0.048	0.048	0.046
[4]					0.066	0.066	0.000	0.066	0.070	0.066	0.070	0.066	0.072	0.070	0.072	0.068	0.068	0.066	0.066
[5]						0.000	0.066	0.000	0.004	0.000	0.004	0.000	0.009	0.004	0.006	0.002	0.002	0.004	0.000
[6]							0.066	0.000	0.004	0.000	0.004	0.000	0.009	0.004	0.006	0.002	0.002	0.004	0.000
[7]								0.066	0.070	0.066	0.070	0.066	0.072	0.070	0.072	0.068	0.068	0.066	0.066
[8]									0.004	0.000	0.004	0.000	0.009	0.004	0.006	0.002	0.002	0.004	0.000
[9]										0.004	0.007	0.004	0.009	0.007	0.009	0.002	0.002	0.004	0.004
[10]											0.004	0.000	0.009	0.004	0.006	0.002	0.002	0.004	0.000
[11]												0.004	0.006	0.004	0.006	0.006	0.006	0.007	0.004
[12]													0.009	0.004	0.006	0.002	0.002	0.004	0.000
[13]														0.009	0.011	0.007	0.007	0.009	0.009
[14]															0.009	0.006	0.006	0.007	0.004
[15]																0.007	0.007	0.009	0.006
[16]																	0.000	0.002	0.002
[17]																		0.002	0.002
[18]																			0.004
[19]																			

Average over all sequence pairs within uninfected with *Spiroplasma* - **0.021**

[	1	2	3	4	5	6	7	8	9	10	11	12	13	14	15	16]
[1]																
[2]	0.004															
[3]	0.002	0.002														
[4]	0.002	0.002	0.000													
[5]	0.002	0.002	0.000	0.000												
[6]	0.006	0.006	0.004	0.004	0.004											
[7]	0.002	0.002	0.000	0.000	0.000	0.004										
[8]	0.002	0.002	0.000	0.000	0.000	0.004	0.000									
[9]	0.002	0.002	0.000	0.000	0.000	0.004	0.000	0.000								
[10]	0.002	0.002	0.000	0.000	0.000	0.004	0.000	0.000	0.000							
[11]	0.006	0.006	0.004	0.004	0.004	0.004	0.004	0.004	0.004	0.004						
[12]	0.002	0.002	0.000	0.000	0.000	0.004	0.000	0.000	0.000	0.000	0.004					
[13]	0.002	0.002	0.000	0.000	0.000	0.004	0.000	0.000	0.000	0.000	0.004	0.000				
[14]	0.002	0.002	0.000	0.000	0.000	0.004	0.000	0.000	0.000	0.000	0.004	0.000	0.000			
[15]	0.002	0.002	0.000	0.000	0.000	0.004	0.000	0.000	0.000	0.000	0.004	0.000	0.000	0.000		
[16]	0.004	0.004	0.002	0.002	0.002	0.002	0.002	0.002	0.002	0.002	0.002	0.002	0.002	0.002	0.002	

Average over all sequence pairs within infected with *Spiroplasma* - **0.002.**

The respective intergroup genetic differentiation, Gst, (Nei 1973) of mtDNA for the four studied *Adalia* species (Gst: 0.25945) is about 4 times smaller than that of the ITS2 rRNA (Gst: 0.82393), though the haplotype diversity, Hd of mtDNA (Hd 0.98814) in higher than of ITS2 (Hd 0.59459). Mutation rate in the intergene spacer is higher, however the absence of recombination allows preservation of rare mutational events in the mitochondrial genome. Consequently, for the different species of the genus *Adalia* the haplotype diversity is higher for mtDNA, while genetic differentiation is higher for rRNA.

Our studies of the nuclear marker confirmed taxonomical independence of *A. bipunctata, A. decempunctata, A. tetraspilota* and *A. frigida* , while the status of *A. b. fasciatopunctata, A. b. turanica* and *A. b. revelierei* as distinct species was not confirmed.

CONCLUSION

It is known that the studied forms of *A. bipunctata* with multiple spots on red or pink elytra found in polymorphic populations of the subspecies *A. b. revelierei, A. b. turanica, A. b. fasciatopunctata,* are phenotypes of the specific alleles of the gene *S*. Other alleles of this gene determine 'European' color and patterns of elytra: red with two spots or black with red spots, *A. bipunctata typica* (also called *A b. bipunctata* (Lussis, 1973). A notable fact is that the forms of *A. bipunctata* with multiple spots were never found in continental Europe. They are common only for areas with dry continental climate and for the areas with arid steppe and semi-desert vegetation: Armenia, Kirgizia, Uzbekistan, Mongolia, Tyva, Buryatia, Transbaikalia. It may be suggested that the alleles corresponding to this pattern either have a strong pleiotropic physiological effect by themselves, or are tightly linked (form stable haplotypes) with the genes favoring survival under conditions of continental and arid climate. In more humid areas of Europe and Siberia these genotype variants are non-adaptive and are not maintained in the populations.

Our DNA analysis of *A. b. turanica, A. b. revelierei* and *A. b. fasciatopunctata* revealed no difference in haplotype diversity of the mtDNA gene *COI.* For *A. b. bipunctata, A. b. fasciatopunctata, A. b. revelierei* and *A. b. turanica* 20 variable mt-haplotypes were found for the gene *COI,* none of them being unique for a certain form. Mt-haplotype H1 is the most common haplotype in all samples (Table 2).

Two mitotypes, H9 and H10, are notably different from the others. The haplotype H9, found in populations of *A. bipunctata* from N Europe (Schulenburg et al., 2002; Jiggins, Tinsley, 2005), from St. Petersburg and Arkhanhelsk (Zakharov, Shaikevich, 2011; Zakharov, Shaikevich, 2013) was detected by us also in the beetles from Erevan. The mt-haplotype H10 was detected by us in populations of *A. b. typica (bipunctata*) in North part of the European Russia in Sain-Petersburg, Arkhangelsk, Kem (Shaikevich et al., 2012). It was also detected for *A. b. fasciatopunctata* in the E part of Russia in Transbaikalia, Ulan-Ude, Buryatia (Figure 6).

Contrary to the mtDNA, the ITS2 region was found to be absolutely conserved. All studied individuals of *A. bipunctata*, independently of their mitochondrial haplotype, had practically identical ITS2 sequences (Figure 5). Therefore, the DNA analysis has not confirmed the status of the distinct species for morphologically different forms *A. b. bipunctata, A. b. fasciatopunctata, A. b. revelierei* and *A. b. turanica*. It may be concluded that *A. b. fasciatopunctata, A. b. revelierei* and *A. b. turanica,* sometimes described as distinct species, in fact are geographic forms of the polytypical species *Adalia bipunctata*.

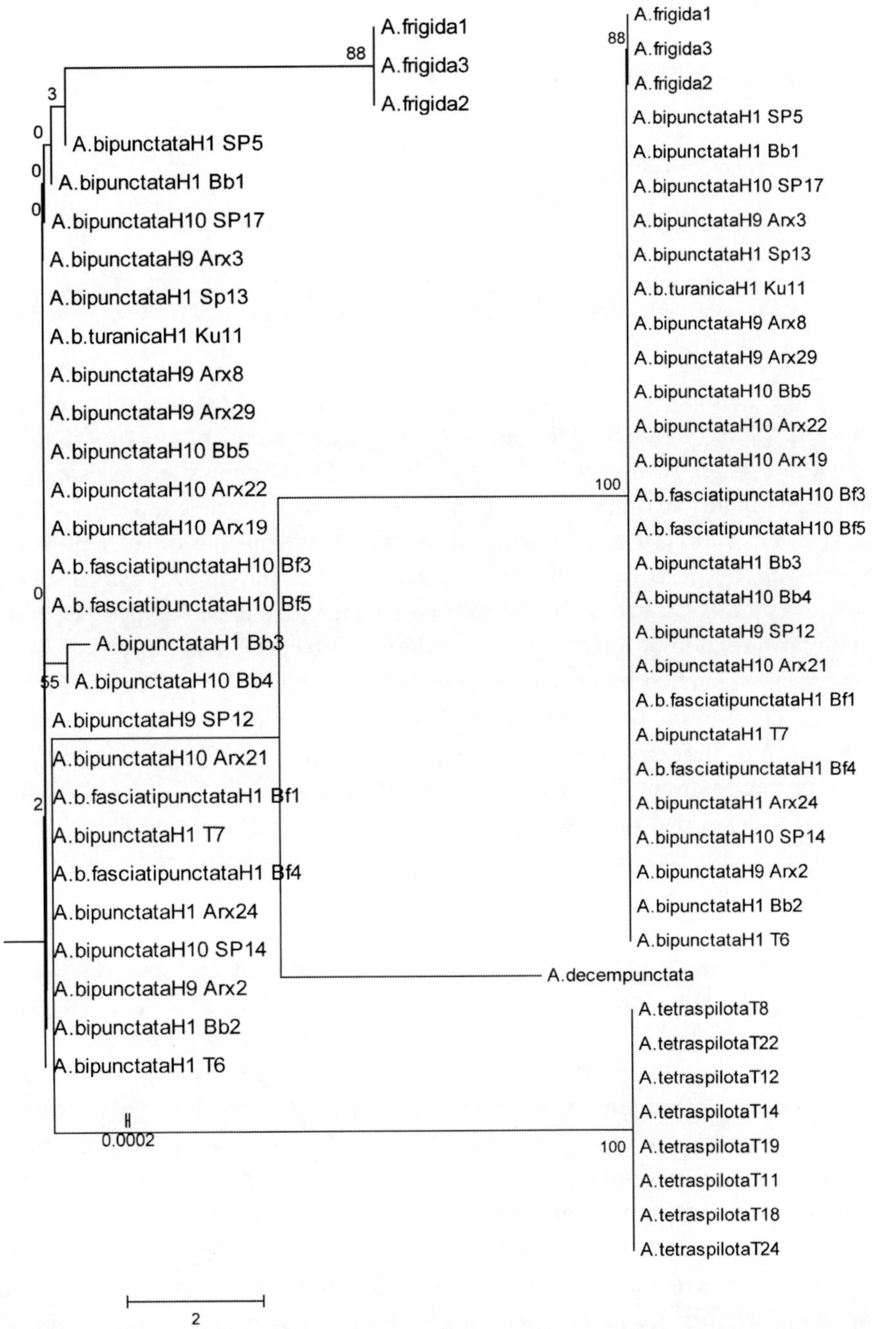

Figure 5. Phylogenetic tree of the studied species of *Adalia*, generated basing on the **ITS2** rRNA analysis using the Neighbor-Joining method**.** The numbers are the bootstrap coefficients calculated for 1000 replication. The additional dendrogram shows the subclade of *A. bipunctata* and *A. frigida* at higher scale. SP, Saint Petersburg; Arx, Arkhangelsk; Bb, Buryatia f. typical (bipunctata); Bf, Buryatia f. fasciatopunctata; Ki, Kirgizia; T, Tashkent.

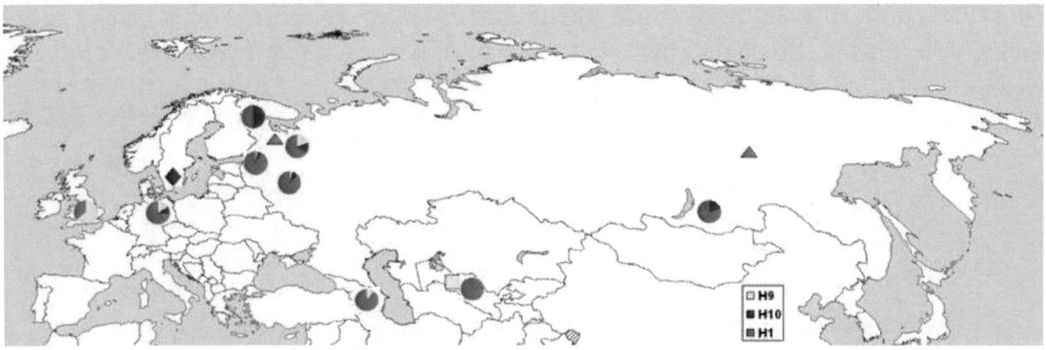

Figure 6. Distribution map of the *Adalia* species and of mt-haplotypes of *Adalia bipunctata*. H1 **includes all related haplotypes except H9 and H10.** *A. decempunctata* is shown as green diamonds, *A.tetraspilota* as yellow squares, and *A. frigida* - as red triangles.

Among the studied ladybird species of the genus *Adalia* the minimal evolutionary divergence, 4.3 % in the gene *COI* and 1.8 % in ITS2 sequences, was revealed between two closely related species *A. bipunctata* and *A. frigida*. The species *A. decempunctata* and *A. tetraspilota* are equally evolutionary distant from the other species and from each other. The mt-haplotypes H9 and H10 considerably differ not only from the other haplotypes common for *A. bipunctata,* but also from DNA of the gene *COI* of the other ladybird species close to *A. bipunctata* (Figure 4). Probably, the source species of the haplotypes H9 and H10 were involved in introgression of mtDNA, but at present these species no more exist in nature.

The population studied by us had considerably variable levels of bacterial infection, from up to 50% infection of the population from St Petersburg with bacterium *Spiroplasma* and up to 50% infection of the population from Kem with bacterium *Rickettsia*, to the total absence of bacteria (or their extreme rareness) in the population from Arkhangelsk. It should be noted that all three populations are under similar climatic conditions. Absence of infection in Southern populations, Tashkent and Erevan, may be explained by elimination of bacteria at high temperature. Effect of the temperature on survival of the symbionts was reported for other representatives of Coccinelidae (Elnagdy et al., 2013).

A. bipunctata demonstrates correlation of the mt-haplotype H10 and *Rickettsia* infection, and earlier reports show that *Rickettsia* infection affects ladybirds with mt-haplotype H9 (Schulenburg et al., 2002, Jiggins, Hurst, 2005). Our studies of the population from Kem, and also of the population from Buryatia, confirmed the «linkage» of the mt-haplotype 10 with infection by *Rickettsia*. Bacteria of the genus *Spiroplasma* infect the beetles of mt-haplotypes 1,2,3,11,17. The mt-haplotypes 2,3,11, 17 differ from the mt-haplotype 1 by one nucleotide substitution, therefore probably the most common haplotype H1 compared with the other haplotypes is the ancestral one. Consequently, it can be suggested that infection by bacteria of the genus *Spiroplasma* occurred before differentiation of the mt-haplotypes 1,2,3,11,17.

Our results confirm the statement about relation of infection with *Spiroplasma* and *Rickettsia* with specific mt-haplotypes of the host. These results indicate the absence or rareness of horizontal transfer of the symbionts between *A. bipunctata* ladybirds and point to the ancient time of the first contact of bacretia, at least *Rickettsia*, with ladybirds of the species *A. bipunctata*. As shown by our studies of the populations from Stockholm, Sweden, about 23% of the beetles of the related sympatric species *Adalia decempunctata* are also infected by *Rickettsia* (Zakharov, Shaikevich, 2001). Probably, the infection occurred before the divergence of the two species. Preservation of bacteria carrying for millions of years suggests biological advantage of the male-killing effect which is demonstrated in *Adalia* infected by all the three known species of symbiotic bacteria.

ACKNOWLEDGMENTS

This work was supported by the program "Live Nature: Current State and Developmental Problems" of the Presidium of the Russian Academy of Sciences, the Russian Foundation for Basic Research (project no. 11_04_01776). We thank Dr. Ekaterina Gupalo for her helpful assistance and comments during the preparation of the chapter.

REFERENCES

Bielawski, R. (1984). Coccinellidae (Coleoptera) of Mongolia. *Ann. Zool. Warszawa*, V. 38(14), P. 281–460.

Brakefield, P. M. & de Jong, P. W. (2011). A steep cline in ladybird melanism has decayed over 25 years: a genetic response to climate change. *Heredity*, V. 107, P. 574–578.

Braig, H. R.; Zhou, W.; Dobson, S. L. & O'Neil, S. L. (1998). Cloning and characterization of a gene encoding the major surface protein of the bacterial endosymbiont Wolbachia. *J. Bacteriology*, V. 180, P. 2373-2378.

Butko, E. V. (2005). Environmental and intraspecific variation of lady beetles (Coleoptera, Coccinellidae) in East Transbaikalia. Doctor of Philosophy theses. Ulan-Ude, Russia. (in russian).

Creed, E. R. (1971). Melanism in the two-spot ladybird, Adalia bipunctata, in Great Britain. In: *Ecological Genetics and Evolution*. ed. Creed E.R. Oxford: Blackwell. P. 134-151.

Dobzhansky, T. (1924). Über geographische und individuelle Variabilität von *Adalia bipunctata* und *A. decempunctata*. Russk. Entomol. *Obozrenie,* V. 18, P. 201–211.

Elnagdy, S.; Messing, S. & Majerus, M. E. N. (2013). Two Strains of Male-Killing Wolbachia in a Ladybird, Coccinella undecimpunctata, from a Hot Climate. PLoS ONE, V. 8(1): e54218. doi:10.1371/journal.pone.0054218.

Folmer, O.; Black, M.; Hoeh, W.; Lutz, R. & Vrijenhoek, R. (1994). DNA primers for amplification of mitochondrial cytochrome c oxidase subunit I from diverse metazoan invertebrates. *Mol.Mar. Biol. Biotechnology,*. V. 3 (5), P. 294–299.

Jiggins, F. M. & Tinsley, M. C. (2005). An Ancient Mitochondrial Polymorphism in Adalia bipunctata Linked to a Sex-Ratio-Distorting Bacterium. *Genetics,* V. 171, P. 1115–1124.

Kuznetsov, V. N (2006). The fauna and distribution of the lady beetles (Coleoptera: Coccinellidae) in the Russian Far East. *Proceedings of the Russian Entomological Society*, St. Petersburg, V. 77, P. 192–199.

Korsun, O. V. (2000). Phenotypes in natural populations of *Adalia bipunctata* L. (Coleoptera: Coccinellidae) in East Transbaikalia. *Contemporary Problems of Ecology* (Sibirskiy ekologicheskiy zhurnal), V. 6. P. 723 726.

Librado, P. & Rozas, J. (2009). DnaSP v5: A software for comprehensive analysis of DNA polymorphism data. *Bioinformatics*, V. 25, P. 1451-1452.

Lus, Y. Y. (1928). O nasledovanii okraski i risunka u bozh'ih korovok *Adalia bipunctata L.* i *Adalia decempunctata* L. [On the inheritance of color and pattern in lady beetles *Adalia bipunctata* L. and *Adalia decempunctata* L.]. *Izvestie Byuro Genetiki*, Leningrad, V. 6, P. 89-163 (in Russian, English abstr.).

Lusis, Y. Y. (1961). On the biological meaning of colour polymorphism of ladybeetle *Adalia bipunctata*. *Latvijas Entomologs*, V. 4, P. 3-29.

Lusis (Lus), Y. Y. (1973). Taksonomicheskie otnoshenija i geograficheskoe rasprostranenie form zhukov roda *Adalia* Mulsant [Taxonomic relations and geographic distribution of ladybirds of the genus *Adalia* Mulsant]. *Proc. Latvian State University*, V.184 (1), P. 5-128 (in Russian, English abstr.).

Lusis (Lus), Y. Y. (1976). O sistematicheskom polozhenii *Adalia frigida* Schneider (Coleoptera, Coccinellidae) [On taxonomy of *Adalia frigida* Schneider (Coleoptera, Coccinellidae)]. Genetic and selectional studies in LatSSR. *Proc of the Conf. Riga*, Zinatne, P. 3-6 (in Russian).

Porter, C. H. & Collins, F. H. (1991). Species-diagnostic differences in a ribosomal DNA internal transcribed spacer from the sibling species Anopheles freeborni and Anopheles hermsi (Diptera: Culicidae). *Am. J. Trop. Med. Hyg*, V. 45, P. 271-279.

Shaikevich, E.V.; Ivshina, E.V. & Zakharov I.A. (2012) Polymorphism of mtDNA and distribution of cytoplasmic symbionts in populations of the two-spot ladybird beetle *Adalia bipunctata*. *Russian Journal of Genetics*, V.48 (5), P. 566–570.

Schulenburg, J. H.; Hurst, G. D.; Tetzlaff, D.; Booth, G. E.; Zakharov, I.A. & Majerus, M. E. (2002). History of infection with different male-killing bacteria in the two-spot ladybird beetle *Adalia bipunctata* revealed through mitochondrial DNA sequence analysis. *Genetics,* V. 160, P. 1075-1086.

Tamura, K.; Peterson, D.; Peterson, N.; Stecher, G.; Nei, M. & Kumar, S. (2011). MEGA5: molecular evolutionary genetics analysis using maximum likelihood, evolutionary distance, and maximum parsimony methods. *Molecular Biology and Evolution*, V. 28, P. 2731–2739.

Van Kuppeveld, F. J. M.; van der Logt, H. T. M.; Angulo, A. F.; Zoest, M. J.; Quint, W. G. V.; Niesters, H. D. G.; Galama, J. M. D. & Melchers, W. J. G. (1992). Genus- and species-specific identification of mycoplasmas by 16S rRNA amplification. *Appl. Environ. Microbiolology,* V. 58, P. 2606-2615.

Williams, S. G.; Sacci, J.B.; Schriefer, M.E.; Andersen, E.M.; Fujioka, K. K.; Sorvillo, F. J.; Barr, A. R. & Azad, A. F. (1992). Typhus and typhuslike rickettsiae associated with opossums and their fleas in Los Angeles County, California. *J. Clin. Microbiology,* V. 30, P. 1758-1762.

Zakharov, I. A. (1990). Interaction of anthropogenic and natural factors in the development of urban melanism in west European populations of Adalia bipunctata L. *Genetika,* V. 26, P.1932-1941.

Zakharov, I. A. (2003). Industrial melanism and its dynamics in populations of the two-spot ladybird Adalia bipunctata L. *Advances in current Biology,* V. 123(1), P.3-15.

Zakharov, I. A.; Shaikevich, E. V. & Goriacheva, I. I. (1998). Bacteria of the *Spiroplasma* genus infect two-spot ladybugs (*Adalia bipunctata* L.) in Russia. *Dokl Akad Nauk*, V. 362(4), P. 570-573.

Zakharov, I.A.; Goriacheva, I.I.; Shaikevich, E.V. & Dorzhu, Ch. M. (2000). Distribution of cytoplasmically inherited bacteria *Spiroplasma* causing female bias in the Eurasian populations of *Adalia bipunctata*. *Genetika,* V. 36(2), P.191-4.

Zakharov, I. A. & Shaikevich, E. V. (2001). The Stockholm populations of Adalia bipunctata (L) (Coleoptera: Coccinellidae)--a case of extreme female-biased population sex ratio. *Hereditas,* V. 134(3), P. 263-266.

Zakharov, I. A. & Shaikevich, E. V. (2011). An mtDNA polymorphism in the St. Petersburg population of *Adalia bipunctata* and its correlation with infection by the symbiotic bacterium *Spiroplasma*. *Ekologicheskaya Genetika,* V. 9 (1), P. 27–31.

Zakharov, I. & Shaikevich E. (2013). Comparative study of mtDNA in species of the genus *Adalia* (Coleoptera: Cocinellidae) and the origin of ancient mitochondrial haplotypes in gene pool of *Adalia bipunctata*. *Eur. J. Entomology*, V. 110 (3), P. 427-433.

In: Beetles: Biodiversity, Ecology and Role in the Environment ISBN: 978-1-63463-380-2
Editor: Camilla Stack

Chapter 9

LADYBIRD DIVERSITY ON CROPS AND ECOLOGY OF *COCICNELLE ALGERICA* KOVAR IN TUNISIA

Ben Halima Kamel Monia * ***and Mdellel Lassad***

13AGRO4: Développement de la lutte biologique et intégrée au niveau de la parcelle en agriculture biologique, Institut Supérieur Agronomique, Chott Mariem 4042, Université de Sousse, Tunisie

ABSTRACT

A regular prospection was carried out from 1993 to 2014 to identify the ladybeetles predators of fruits trees and vegetables crops aphids in Tunisia. The study allows detecting five Coccinellids with a frequency of 6% were the most abundant predators: *Scymnus aptezi* Mulsant, *Oenopia doublieri* Mulsant, *Hippodamia variegata* Goeze, *Coccinella undecimpunctata* L. *currently Hippodamia (Semiadalia) undecimnotata* Schneider, *Hippodamia tredecimpunctata* L., *Psyllobora 22-punctata* L. and *Coccinella algerica* Kovar. This latest was the most abundant coccinellids and was observed as larva and adult respectively on population of *Aphis gossypii* Glover, *Aphis fabae* Scopoli, *Aphis punicae* Shinji, *Capitophorus elaeagni* Del Guercio, *Myzus persicae* Sulzer, *Macrosiphum euphorbiae Thomas, Macosiphum rosae L., Hyalopterus pruni Geoffrey and Pterochloroides persicae* Cholodkovsky *on several host plant: eggplant, pepper,* pomegranate, tomato, lettuce, melon, peach, almond, plum...

The association's plants, aphids and preys of *C. algerica* which has an important action on the control of aphids were analyzed. *C. algerica* is found on several host plants belonging to 22 botanical families who may be vegetable crops, fruit trees or spontaneous plants. *C. algerica* consumed alternative or essential food present on hosts plants. These plants supported eggs, larva and nymph of *C. algerica*. This predator consumes twenty species of aphids. The consumption varies depending on the state of physiological activity: It feeds on *Aphis*, *Rhopalosiphum* and *Uroleucom* genera at quiescence. *C. algerica* shows a preference for *C. elaeagni* in autumn and *H. pruni* in spring. Furthermore, this ladybird on diapauses was collected in the summer nears the root of *Stipa tenacissima* (Gramineae).

* E-mail: kamonia_tn@yahoo.fr.

Also, it's interesting to mention that *C. algerica* was attacked by several parasites like *Dinocampus coccinellae* Schrank (Braconidae, Hymenoptera), *Homolotylus flaminius* Dalm (Encyrtidae, Hymenoptera) and *Podapolipus* Rovelli et Grassi (Acarina).

Keywords: Ladybird species, crops, *Coccinella algerica*, aphids prey, host plant, ecology, natural enemies, Tunisia

INTRODUCTION

The Coccinellidae, whose members are commonly known as ladybirds, ladybugs or lady beetles, is a cosmopolitan family with approximately 6,000 described species to date (Vandenberg 2002), including1400 South American species (González 2010). These ladybirds can be fungivorous, phytophagous, or entomophagous, the latter ones mainly developing on a variety of sternorrynchan prey (Majerus, 1994; Hodek and Honek, 1996). As for coccinellids entomophagous, they are of interest and importance in agriculture in agriculture and forestry, since adult and larvae of most species are predators of herbivorous such as aphids, adelgids, mealy bugs and scales insects (Obrycki and Kring, 1998). Most aphidophagous coccinellids belong to the subfamilies Coccinellinae and Scymninae but there are also aphidophagous species in other subfamilies. Coccinellids are probably the best studied aphid predators due to their considerable visibility, economic importance in a variety of crops, ability to feed on a wide range of prey, to be very voracious and to have a rapid numeric response (Hodek and Honek, 1996; Obrycki and Kring, 1998; Dixon, 2000). They are found in many habitats, including those as diverse as cities, fields, gardens, sea coasts and mountains (Majerus and Kearns, 1989; Hodek and Honek, 1996). Both larvae and adults feed on the same type of prey species and occur in identical habitats (Majerus, 1994; Hodek and Honek, 1996; Dixon, 2000). The lifetime fecundity varies greatly between species and may range from slightly more than 100 to more than 1500 eggs per female (Kawauchi, 1991). Eggs are usually laid in clusters (average: 11-30 eggs, depending on the species), but some species on the subfamily of Scymninae may deposit eggs singly (Hodek and Honek, 1996). Developmental time, adult longevity, eggs hatching and oviposition period per female varies greatly between species and is influenced by temperature, the amount of food consumed and prey species (Obrycki and Orr, 1990; Majerus, 1994; Hodek and Honek, 1996). For example, at 25°C, the percentage of hatching eggs of *Adonia Variagata* Goeze and *Cocinella undicempuntata* L. were respectively 81.8 and 74.5%, the adult longevity were 71.8 and 70 days and the fecundity was 370.5 and 276.3. Concerning coccinellids multiplication, Hodek and Honek (1996) proved that most species are univoltine or bivoltine in temperate regions. While, the multivoltinism behavior is more common in tropical regions. Most species in temperate regions hibernate as adults at select sites, where extremely high ladybird numbers may be found (van Emden and Harrington, 2007). This enables them to respond very quickly and precisely to spatial variation in aphid availability in the following spring. Often, coccinellids hibernate in packed masses under and between rocks above the snow line on mountains near valleys with agriculture (Hagen, 1962). Paoletti (1999) revealed that coccinellids before migratory of lights are positively phototoxic. After landing, they change their behavior and walk on the soil searching for shelter, darkness and thigmotactic conditions. Depending on the geological structure and botanic richness on the larding area, they aggregate among rocks, under piles of stones or at the base of plants.

They reach these sites by orienting their flight towards the silhouette of upstanding objects. At such low temperatures, the beetles can maintain their metabolism at a very low level, and pathogens that have infected them or might do so be able to multiply. In the present chapter, we cited the prominent ladybird species found on crop in Tunisia. Then, we presented the aphid's preys and host plants of *C. algerica* and finally we noted their natural enemy in a coastal area of Tunisia.

Ladybird Species Found on Crop in Tunisia

The follow-up of the capture of ladybird in the different regions and on diver's crops in Tunisia from 1993 to 2014 demonstrates the presence of 8 species of coccinellids which belonging to the subfamilies Coccinellinae and Scymninae (See table 1). The Coccinellinae was represented by respectively *Oenopia doublieri* Mulsant (1846) (Figure 4),, *Hippodamia variegata* Goeze (1777) (Figure 2), *Coccinella undecimpunctata* L. (1758) (Figure 6) currently *Hippodamia (Semiadalia) undecimnotata* Schneider (1792), *Hippodamia tredecimpunctata* L. (1758) (Figure 5), *Psyllobora 22-punctata* L. (1899) and *Coccinella algerica* Kovar (1977) (Figure 3) (Ben Halima Kamel and Ben Hamouda, 1993; Ben Halima et al., 1994, 2013; Guesmi et al., 2011; Mdellel and Ben Halima Kamel, 2012; Rebhi et Ben Halima Kamel, 2009). Nevertheless, the Scymninae was figured by *Scymnus aptezi* Mulsant (1846) (Figure 1) (Ben Halima Kamel et al., 2013). Among these coccinellids species, *C. algerica* was the most abundant and it was observed on various crops (Ben Halima Kamel and Ben Hamouda, 1993; Ben Halima Kamel et al., 1994, 2013; Guesmi et al., 2011; Mdellel and Ben Halima, 2012; Rebhi and Ben Halima Kamel, 2009)

Table 1. Information about Ladybird species found on crop in Tunisia

Coccinellids	Family	Subfamily	locality	Aphid host	Host crops
Scymnus aptezi Mulsant	Coccinellidae	Scymninae	Jammel, Akouda Chott Mariem	*Hyalopterus pruni* Geoffroy	Almond, peach, plum
Hippodamia variegata Goeze	Coccinellidae	Coccinellinae	Jammel, Akouda Chott Mariem, Utique	*H. pruni, Brachycaudus cardui*	Almond, Peach, artichoke
Coccinella algerica Kovar	Coccinellidae	Coccinellinae	Jammel, Akouda Chott Mariem, Utique, Mornag, Sidi Thabet,	*H. pruni, Brachycaudus cardui, Aphis citricola, Aphis gossypii, Toxoptera aurantii, Aphis craccivora, Brachycaudus helichrysi, Capitophorus elaeagni, Myzus persiace*	Almond, peach, plum, pepper, citrus, artichoke, tomato, cucumber,
Oenopia doublieri Mulsant	Coccinellidae	Coccinellinae	Jammel, Chott Mariem	*H. pruni*	Almond, peach, plum
Hippodamia tredecimpunctata L.	Coccinellidae	Coccinellinae	Jammel, Chott Mariem, Utique	*H. pruni, B. cardui*	Almond, peach, plum, artichoke
Coccinella undecimpunctata L.	Coccinellidae	Coccinellinae	Utique	*B. cardui*	Artichoke

Figure 1. *Scymnus aptezi* Mulsant.

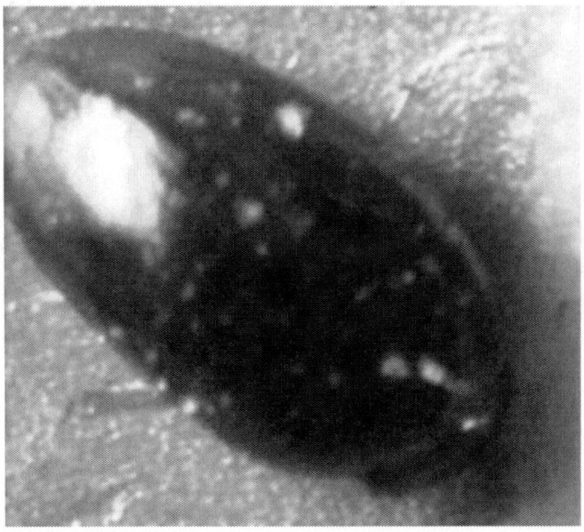

Figure 2. *Hippodamia variegata* Goeze.

Figure 3. *Coccinella algerica* Kovar.

Figure 4. *Oenopia doublieri* Mulsant.

Figure 5. *Hippodamia tredecimpunctata* L.

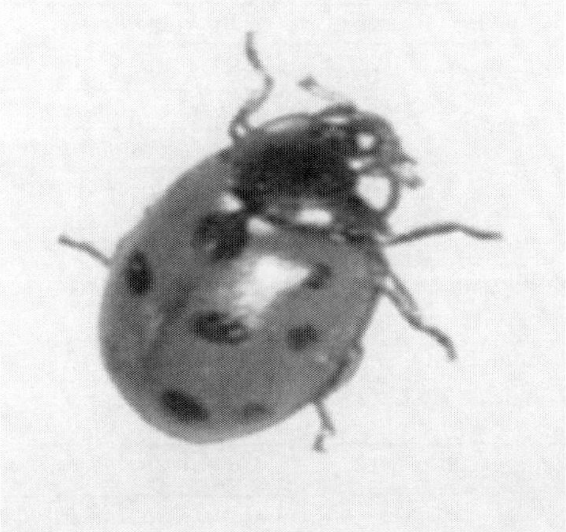

Figure 6. *Coccinella undecimpunctata* L.

APHID'S PREYS AND HOST PLANTS OF *C. ALGERICA*

C. algerica was observed on diversity aphid's species on different cultivated or spontaneous hosts (Ben Halima Kamel and Ben Hamouda, 1993, Ben Halima Kamel et al., 1994, 2013; Guesmi et al., 2009; Mdellel and Ben Halima Kamel, 2012, Rebhi and Ben Halima, 2009). Indeed, table 2 summarizes the principal botanic family of host plants and aphid's subfamily, tribes and species predated by *C. algerica.* Also table 3 demonstrated the spatio-temporal repartition of this coccinellid on different host plant and consumed aphid's preys. Furthermore, this ladybird on diapauses was collected in the summer nears the root of *Stipa tenacissima* (Gramineae) (Rebhi and Ben Halima Kamel, 2009).

Table 2. Aphids preys and hosts plants of *C. algerica* *: cultivated plant, **: spontaneous plant

Sub Family	Tribu	Aphids	Hosts plants	Botanic Family
Aphidinae	Aphidini	*Rhopalosiphum maidis* (Fitch)	*Sorghum halepense***	Poaceae
		Rhopalosiphum insertum (Walker)	*Malva parviflora***	Malvaceae
			*Cyperus rotundus***	Cyperaceae
		Aphis craccivora Koch	*Amaranthus angustiflora***	Amaranthaceae
			*Chenopodium murale***	Chenopodiaceae
		Aphis fabae Scopoli	*Urtica urens***	Urticaceae
			*Chenopodium murale***	Chenopodiaceae
			*Pisum sativum**	Fabaceae
			*Capsicum annuum**	Solanaceae
			*Solanum nigrum***	Solanaceae
		Aphis sp	*Medicago soleinoli***	Fabaceae
			*Chenopodium album***	Chenopodiaceae
			*Chenopodium murale***	Chenopodiaceae
			*Rumex bucephalophorus***	Amarantaceae
			*Urtica urens***	Urticaceae
		Aphis punicae Passerini	*Punica granatum**	Punicaceae
		Aphis gossypii Glover	*Cucurbita pepo**	Cucurbitaceae
			*Capsicum annuum**	Solanaceae
			*Chenopodium murale***	Chenopodiaceae
			*Amaranthus angustiflora***	Amarantaceae
			*Chenopodium album***	Chénopodiaceae
	Macrosiphini	*Hyalopterus pruni* Geoffroy	*Prunus persicae**	Rosaceae
		Uroleucon compositae (Theobald)	*Centauria nicaeensis***	Composeae
			*Sonchus tenerrimus***	Composeae
		Uroleucon ambrosiae (Thomas)	*Centauria nicaeensis***	Composeae
			*Centauria dimorpha***	Composeae
			*Sonchus tenerrimus***	Composeae
		Hyperomyzus lactucae (L.)	*Sonchus tenerrimus* **	Composeae

Sub Family	Tribu	Aphids	Hosts plants	Botanic Family
		Dysaphis crataegi (Kaltenbach)	*Daucus carotta**	Umbelliferae
		Macosiphum euphorbiae (Thomas)	*Solanum tuberosum**	Solanaceae
		Brachycaudus cardui (L.)	*Cynara Scolymus**	Composeae
			*Carduus pycnocephalus***	Composeae
		Capitophorus elaeagni (del Guercio)	*Cynara Scolymus**	Composeae
		Acyrthosiphon pisum (Harris)	*Vicia fabae**	Fabaceae
			*Pisum sativum**	Fabaceae
		Dysaphis sp	*Amaranthus angustiflora***	Amarantaceae
			*Chenopodium murale***	Chenopodiaceae
Calaptidinae	Panaphidini	*Therioaphis trifolii* (Monell)	*Melilotus indica***	Fabaceae
			*Cyperus rotundus***	Cyperaceae
Chaito phorinae	Siphini	*Sipha maydis* Passerini	*Lolium rigidum***	Gramineae
Lachininae	Lachinini	*Pterochloroides persicae* (Cholodkovsky)	*Prunus persica**	Rosaceae
			*Prunus amygdalus**	Rosaceae

Table 3. Spatio-temporal repartition of *C. algerica* on different host plant and consumed aphids preys

Aphids	Hosts plants	Month											
		S	*O*	*N*	*D*	*J*	*F*	*M*	*A*	*Ma*	*J*	*Ju*	*A*
R. maidis	*Sorghum halepense*	---	---	---									
R. insertum	*Malva parviflora*				---	---	---	---					
	Cyperus rotundus	---	---	---									
A. craccivora	*Amaranthus angustiflora*						---	---					
	Chenopodium murale				---	---	---	---	---	---			
A. fabae	*Urtica urens*				---	---	---	---					
	Chenopodium murale				---	---	---	---	---	---			
	Pisum sativum								---	---	---		
	Capsicum annuum				---								
	Solanum nigrum						---			---	---		
	Emex spinosa		---	---	---	---	---	---	---	---			
A. sp	*Medicago soleinolii*										---		
	Chenopodium alba	---											
	Chenopodium murale					---	---	---	---	---	---		
	Rumex bucephalophorus									---			
	Urtica urens					---	---	---	---	---	---		

Table 3. (Continued)

Aphids	Hosts plants	Month											
		S	*O*	*N*	*D*	*J*	*F*	*M*	*A*	*Ma*	*J*	*Ju*	*A*
A. punicae	*Punica granatum*									---	---	---	
A. gossypii	*Cucurbita pepo*									---	---	---	
	Capsicum annuum				---	---							
	Chenopodium murale				---	---	---	---	---	---			
	Amaranthus angustiflora									---			
	Chenopodium alba										---		
H. pruni	*Prunus persicae*									---	---		
U. compositae	*Sonchus tenerrimus*			---	---	---	---			---	---		
U. ambrosiae	*Centauria nicaeensis*										---	---	
	Centauria dimorpha										---	---	
	Sonchus tenerrimus		---	---	---	---	---		---	---			
H. lactucae	*Sonchus tenerrimus*		---	---	---	---	---		---	---			
D. crataegi	*Dauccus carotta*									---	---		
M. euphorbiae	*Solanum tuberosum*						---	---		---	---		
B. cardui	*Cynara Scolymus*							---	---	---	---		
	Carduus pycnocephalus									---	---		
C. eleaegni	*Cynara Scolymus*		---	---	---								
A. pisum	*Vicia fabae*							---	---	---			
	Pisum sativum							---	---	---	---		
D. sp	*Amaranthus angustiflora*										---		
	Chenopodium murale				---	---	---	---	---	---			
T. trifolii	*Melilotus indica*						---	---	---				
	Cyperus rotundus	---	---	---									
S. maydis	*Lolium rigidum*									---	---	---	
P. persicae	*Prunus persica*				---	---	---				---	---	
	Prunus amygdalus										---	---	

S: September, O: October, N: November, D: December, J: January, F: February, M: March, A: April, Ma: May, J: June, Ju: July, A: August.

Natural Enemy of *Coccinella Algerica* Kovar in a Coastal Area of Tunisia

The occurrence of *C. algerica* in coastal area revealed that this predator was affected by parasites and diseases at the natural conditions (Ben Halima Kamel, 2010). Indeed, authors demonstrated that three parasites were present, *Dinocampus coccinellae* (Braconidae, Hymenoptera) (Figures 7 and 8), *Homalotylus flaminius* Dalm (Encyrtidae, Hymenoptera) (Figure 9 and 10) and *Podapolipus* (Acarina) (Figure 11 and 12). Therefore, the *Beauveria* causes the diseases of the predator (Figure 13 and 14).

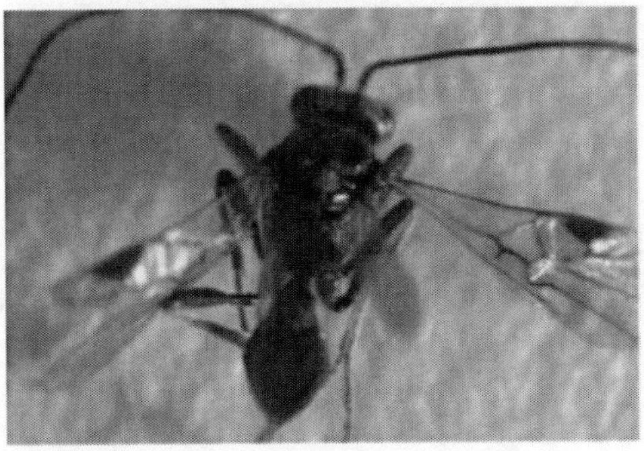

Figure 7. *Dinocampus coccinellae*.

Figure 8. Cocoon of de *D. coccinellae* detached from *C. algerica.*

Figure 9. *Homalotylus flaminius.*

Figure 10. Fourth instars of *C. algerica* with 3 exit hole of *H. flaminius.*

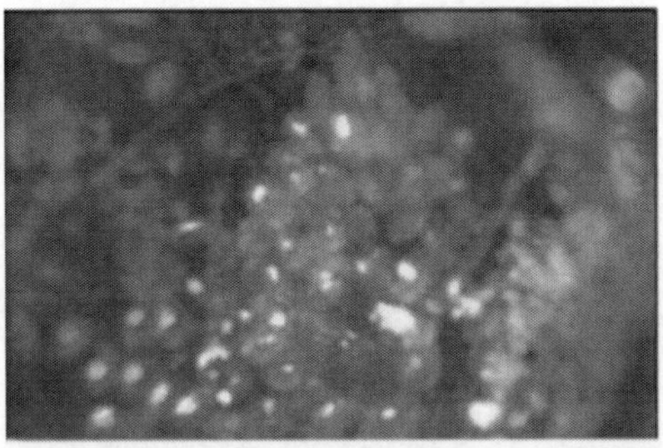

Figure 11. *Podapolipus* under elytra of *C. algerica.*

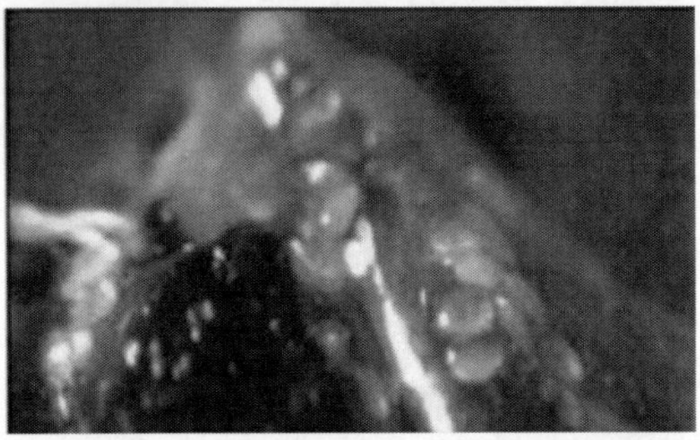

Figure 12. *Podapolipus* attached to the underside abdomen of *C. algerica.*

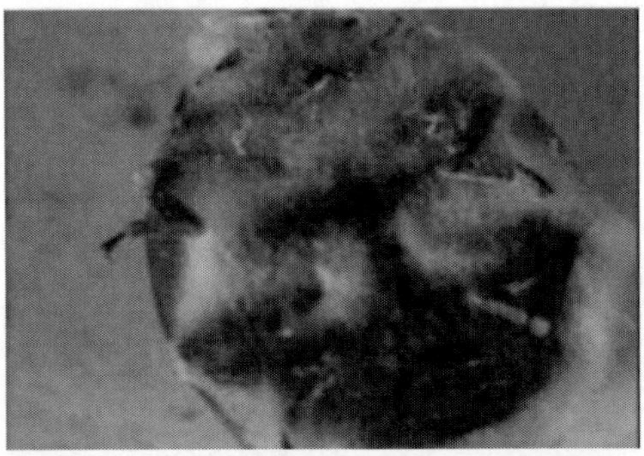

Figure 13. *C. algerica* achieved by *Beauveria* where mycelial hyphae surround the appendices.

Figure 14. *Beauveria* at the head and the of prothorax *C. algerica.*

REFERENCES

Ben Halima-Kamel M. (2005). Biological control of *Aphis gossypii* on pepper plant using *Coccinella algerica.* Comm. *Appl. Biol. Sci. Ghent University* 70/4, p. 737-743.

Ben Halima-Kamel M. (2006). Utilisation des différents stades biologiques de *Coccinella algerica Kovàr* dans la lutte contre *Aphis gossypii Glover* en serre de piment. CIFE VI, 2 à 6 juillet. Rabat, Maroc.

Ben Halima Kamel M. (2010). Utilisation des différents stades biologiques de *Coccinella algerica Kovàr* dans la lutte contre *Aphis gossypii Glover* en serre de piment. O. Himmi (Ed.). Actes de la CIFE VI, Travaux de l'Institut Scientifique, Série Zoologie, Rabat, 2010, N° 47, 109-112.

Ben Halima-Kamel M, Ben Hamouda M.H. (1993). Les pucerons des cultures protégées et leurs ennemis naturels. *Tropicultura* 11 (2): 50–53.

Ben Halima Kamel M., Rabasse J. M and Ben Hamouda M.H. (1993). Les pucerons des agrumes et leurs ennemis en Tunisie. *Tropicultura,* 12, 4, 145-147.

Ben Halima Kamel M., Rebhi R. and Ommezine A. (2011). Habitats et proies de Coccinella algerica Kovar dans différentes régions côtières de la Tunisie. *Entomologie faunistique – Faunistic Entomology, 63* (1), 35-45.

Dixon A.F.G. (2000). *Insect Predator-Prey Dynamics: Ladybird Beetles & Biological Control.* Cambridge University Press, Cambridge, 257 pp.

Guesmi J., Ben Halima Kamel M. and Al Mohandes Dridi B. (2011). Diversity of Natural Enemy Associated to Aphid's Artirhttkc in Different Areas of Production in Tunisia. Proc. *VIIth IS on Artichoke* and their Wild Relatives. Ed. Bazinet, Acta Hortic. 942. HIS.

González G. (2010). Actualización de la bibliografía y nuevos registros en Coccinellidae de América del Sur (Insecta: Coleoptera). *Boletín de la Sociedad Entomológica Aragonesa* 47: 245-256.

Hagen K.S. (1962). Biology and ecology of predaceous Coccinellidae . *Ann. Rev. Entom.,* 7, 289-326.

Hodek I. and Honek A. (1996). *Ecology of Coccinellidae.* Boston: Kluwer Academic Publishers. 464 p.

Majerus M.E.N. (1994). *Ladybirds.* New Naturalist Series No. 81. Harper Collins, London, 320 pp.

Kawauchi S. (1991). *Selection for highly prolific females in three aphidophagous coccinellids.* In: Behaviour and Impact of Aphidophaga, (Eds.: L. Polgár, R.J. Chambers and A.F.G. Dixon), SPB Academic Publishing, The Hague, pp. 177-181.

Mdellel L. and Ben Halima Kamel M. (2012). Prey conception efficiency and fecundity of the ladybird beetle, *Coccinella algerica* Kovar (Coleoptera, Coccinellidae) feeding on the giant Brown bark aphid, *Pterochloroides persicae* (Cholokovsky) (Hemiptera: Lachninae). *African Entomology* 20 (2): 292-299.

Majerus M. and Kearns P. (1989). *Ladybirds Naturalists' Handbooks 10.* Slough: Richmond Publishing. 103 pp.

Majerus M.E.N. (1994). *Ladybirds.* New Naturalist Series No.81. Harper Collins, London, 320 pp.

Obrycki J.J. and Orr C.J. (1990). Suitability of three prey species of reartic populations of *Coccinella septumpunctata*, *Hippodamia variegata* and *Propylea quatuordecipunctata* (Coleoptera, Coccinellidae). *J. Econ. Entomol,* 83. 1292-1297.

Obrycki JJ and Kring T J (1998). Predaceous Coccinellidae in biological control. *Annal Review of Entomology.* 43, 295-321.

Paoletti M.G. (1999). *Biodiversity in Agroecosystems: Role for sustainability and Bioindication.* Elsevier 446 pp.

Vandenberg N.J. (2002). Coccinellidae Latreille 1807. InAmerican Beetles, Volume 2: *Polyphaga: Scarabaeoidea through Curculionoidea.* Edited by R.H. Arnett, Jr., M.C. Thomas, P.E. Skelley, and J.H. Frank. CRC Press, Boca Raton, USA. 371-389 p..

van Emden H.F. and Harrington R. (2007). *Aphids as Crop Pests.* Wallingford, UK: CAB International. 717 pp.

INDEX

C

D

E

F

G

H

I

J

K

L

M

Q

R

S

V

W

X

Z